# Drug Delivery Systems

**World Scientific Series:**
**From Biomaterials Towards Medical Devices**

**Series Editors:**

**Roger Guilard**
Université de Bourgogne
UFR Sciences et Techniques
Faculté des Sciences Mirande
ICMUB – UMR CNRS 6302
9 Avenue Alain Savary, BP 47870
21078 Dijon Cedex, France
Email: roger.guilard@u-bourgogne.fr

**Karl M Kadish**
Department of Chemistry
University of Houston
Houston, TX 77204-5003, USA
Email: kkadish@uh.edu

*Forthcoming*

Vol. 1 *Drug Delivery Systems*
edited by Pieter Stroeve (UC Davis) & Morteza Mahmoudi (Tehran University of Medical Sciences, Iran & Harvard Medical School, USA)

Vol. 2 *Dental Biomaterials*
edited by Edward Sacher (Ecole Polytechnique de Montréal, Canada) & Rodrigo França (University of Manitoba, Canada)

*Biomaterials and Bioengineered Device Research for Joint Prostheses*
edited by L'Hocine Yahia (Ecole Polytechnique de Montreal, Canada) & Véronique Migonney (Université Paris 13, France)

*Biomaterials and Bioengineered Device Research for Ligament Reconstruction*
edited by Véronique Migonney ((Université Paris 13, France) & L'Hocine Yahia (Ecole Polytechnique de Montréal, Canada)

*Biomaterials and Bioengineered Device Research for Cardiovascular Applications*
edited by Laurence Bordenave (Université de Bordeaux, France)

World Scientific Series in

**From Biomaterials Towards Medical Devices – Vol. 1**

# Drug Delivery Systems


*editors*

**Pieter Stroeve**

*University of California Davis, USA*

**Morteza Mahmoudi**

*Tehran University of Medical Sciences, Iran*
*Harvard Medical School, USA*

*Series Editors*

**Roger Guilard**

*Université de Bourgogne, France*

**Karl M. Kadish**

*University of Houston, USA*



World Scientific

NEW JERSEY • LONDON • SINGAPORE • BEIJING • SHANGHAI • HONG KONG • TAIPEI • CHENNAI • TOKYO

*Published by*

World Scientific Publishing Co. Pte. Ltd.
5 Toh Tuck Link, Singapore 596224
*USA office:* 27 Warren Street, Suite 401-402, Hackensack, NJ 07601
*UK office:* 57 Shelton Street, Covent Garden, London WC2H 9HE

**Library of Congress Cataloging-in-Publication Data**
Names: Stroeve, Pieter, 1945– editor. | Mahmoudi, Morteza, 1979– editor.
Title: Drug delivery systems / edited by Pieter Stroeve & Morteza Mahmoudi.
Other titles: Drug delivery systems (Stroeve) | World Scientific series: from biomaterials towards medical devices ; vol. 1.
Description: New Jersey : World Scientific, 2017. | Series: World Scientific series: from biomaterials towards medical devices ; vol. 1 | Includes bibliographical references and index.
Identifiers: LCCN 2017026738 | ISBN 9789813201040 (hardcover : alk. paper)
Subjects: | MESH: Drug Delivery Systems | Nanoparticles--therapeutic use
Classification: LCC RS199.5 | NLM QV 785 | DDC 615/.6--dc23
LC record available at https://lccn.loc.gov/2017026738

**British Library Cataloguing-in-Publication Data**
A catalogue record for this book is available from the British Library.



Desk Editors: Suraj Kumar/Joy Quek

Typeset by Stallion Press
Email: enquiries@stallionpress.com

Printed in Singapore

# About the Editors

**Morteza Mahmoudi** is a biomedical investigator with a multidisciplinary background in nanoscience, bioengineering, and cardiac cellular biology. His specific research interest is controlling nano–bio interfaces to develop new nano-based therapies for prevention/treatment of life-threatening conditions such as cardiomyopathy, cancer, and neurodegenerative diseases. He has a productive track record of extensive multidisciplinary collaboration, so far yielding over 160 publications in high-profile journals with more than 10,000 citations and an h-index of 48. His group has developed several approaches/assays to probe fundamental cell/molecular–nanoparticle interactions, including cell toxicity, cellular labelling, protein corona, and personalized protein corona. In addition, he introduced novel nanotechnology approaches for imaging tests, anti-biofilm bacterial applications, and probing fundamental cell/molecular–nanoparticle interactions.

**Pieter Stroeve** is a distinguished emeritus professor of chemical engineering at the University of California Davis, where he has taught and conducted research for 36 years. Previously he was a professor at the Department of Chemical Engineering at the University of Buffalo in New York. He also has had two postdoctoral appointments: at the Weizmann Institute of Science, Israel, and at the University of Nijmegen, the Netherlands. He obtained his Sc.D. in chemical engineering at the Massachusetts Institute of Technology and his BS in chemical engineering at the University of California Berkeley. Professor Stroeve was a visiting professor at Queensland University in Australia, the Max-Planck Institute of Polymer Science, Mainz, Germany, Wageningen University and Research Center, the Netherlands, and the Polytechnic University of Valencia in Spain.

Professor Stroeve conducts research in biological separations, nanotechnology, bio-nanotechnology, colloid and surface science, biomass conversion, image analysis, electrochemistry, and renewable energy. He has published on mass transfer with non-equilibrium chemical reactions in heterogeneous media. His focus in nanotechnology is the synthesis of nanoporous membranes, nanotubes, nanowires, nanocables, nanoparticles, and nanostructured surfaces for biosensors and sensors, nuclear reactors, nanostructured solar cells, and for molecular separations of biological molecules. In colloid and surface science, he has worked on the nanostructure of polyion-surfactant complexes, polycation–polyanion complexes, self-assembled monolayers, supported lipid bilayers, and nanoparticle–polymer composites. In renewable energy, he has conducted research on solar energy and thermoelectrics. In biomass conversion, he has worked on using pulsed electrical field to make plants more permeable to catalysts and acid molecules, by creating nanopores in the plant cell membranes, and on acid hydrolysis to make chemicals and liquid fuels. He has collaborations with scientists in Spain, Germany, the Netherlands, Iran, Japan, Taiwan, and China. He has published 300 scientific articles. Professor Stroeve has been cited over 13,000 times, and his h-index is 54. He also has won several teaching awards and is a fellow of the American Institute of Chemical Engineers.

# Contents

# Chapter 1

# Drug Delivery Systems: Possibilities and Challenges

Ryan Spitler*, Saeid Zanganeh*, Tahereh Jafari†, Nasser Khakpash‡, Mohsen Erfanzadeh§, Jim Q. Ho¶, and Nastaran Sakhaie||

** Department of Radiology, Molecular Imaging Program at Stanford, Stanford University, CA, USA*

*† Institute of Materials Science, University of Connecticut, 97 North Eagleville Road, Storrs, CT, USA*

*‡ Department of Chemical and Biomolecular Engineering, University of Connecticut, 191 Auditorium Road, Storrs, CT, USA*

*§ Department of Biomedical Engineering, University of Connecticut, Storrs, CT, USA*

*¶ Middlebury College, Middlebury, VT, USA*

*|| Department of Nanotechnology, Faculty of Pharmacy, Tehran University of Medical Sciences, Tehran, Iran*

## Abbreviations

| | |
|---|---|
| CPP | cell penetrating peptides |
| CRT | controlled-release technology |
| DSPE | distearoylphosphatidylethanolamine |
| DDS | drug delivery systems |
| EPR | enhanced permeability and retention |
| EMEA | European Medicine Evaluation Agency |
| FDA | Food and Drug Administration |
| LDL | low density lipoprotein |
| LUV | large unilamellar liposomes |
| mPEG | methoxy polyethylene glycol |
| MNs | micro needles |
| NPs | nanoparticles |
| PEG-PE | poly(ethylene) glycol-phosphatidylethanolamine |
| PEG | polyethylene glycol |
| RES | reticuloendothelial system |
| SPIONs | superparamagnetic iron oxide |
| TDDS | targeted drug delivery system |
| mAb | monoclonal antibodies |
| TATp | transcription protein |
| TDDS | transdermal drug delivery systems |

## I. Introduction

Drug delivery systems (DDS) have the capability to improve the administration and efficacy of treatment for pharmaceutical compounds such as antibodies, peptides, vaccines, drugs, enzymes, and many others. DDS have the potential to improve the quality of life and longevity of millions of patients. The ability to create controlled release formulations of existing drugs has changed the economics of drug development. Thereby, DDS can improve performance and efficacy, as well as extend the patent life of an existing compound by creating a new product. New DDS are less costly and can be developed much faster in comparison to (approximately \$20–50 million and 3–4 years) new drugs (approximately \$500 million in 10 years).[1] The US market for DDS has demonstrated substantial growth from \$75 million in 2001 to as much as \$121 billion in 2010.[2] In 1990, Liposomal amphotericin B was the first Food and Drug Administration (FDA) approved DDS. There are currently more than 10 DDS which are commercially available for the treatment of a broad range of diseases from cancer, fungal infections, and muscular degeneration.[3] Figure 1 shows the timeline of FDA-approved DDS in the market.[4]

However, a significant limitation for any drug therapy is the inherent cytotoxicity. Upon administration, many compounds will circulate systemically and may cause substantial toxicity. The possible occurrence of these harmful effects, post treatment, can significantly limit the clinical application of many new compounds. In many cases controlled delivery by DDS mitigates many of these drawbacks by improving the following properties: bioavailability, preventing premature degradation, enhancing uptake, maintaining and controlling the drug release rate, and reducing adverse effects by using highly specific targeting approaches.

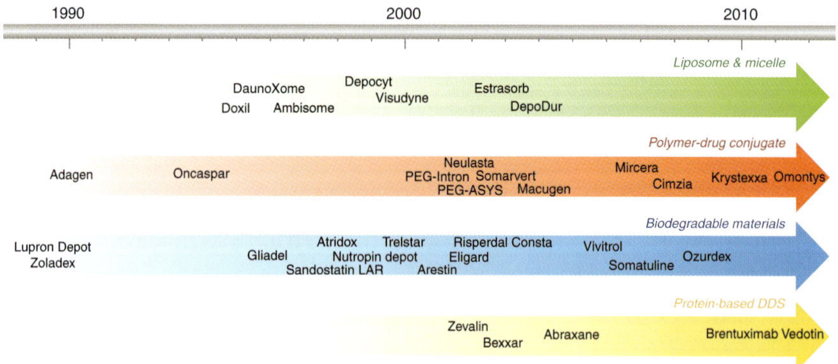


Figure 1. Timeline showing FDA approved DDS in the market. Taken from Ref. [4] with permission of Elsevier.

One example of a targeting approach is the use of site-specific pharmaceutical nanocarriers. The drug targeting of these nanocarriers has been extensively studied and they possess the following advantages: altered drug distribution dynamics, increased delivery to target sites, simplified drug administration methods, improved efficacy and reduced cost.[5] One example are liposomes as they are well-characterized nanocarriers and they will be discussed in detail in the following DDS sections.

Nanocarriers have been used in numerous animal and human investigations and have utilized alternate routes of administration for analgesic and anesthetic, using delivery through the skin, buccal, and nasal mucous membranes. In work to improve these delivery methods, other compounds have lead to an abundance of new devices, concepts and techniques that have collectively been termed controlled-release technology (CRT). Some examples of CRT include transdermal controlled release delivery systems, ml6 nasal and buccal aerosol sprays, drug-impregnated lozenges, encapsulated cells, oral soft gels, drugs administered through the skin and various programmable, and implanted drug-delivery devices. There are numerous motivating factors driving the development of these new devices, concepts and techniques including unmet clinical need, sustained efficacy of treatment and the demand for multiple forms of a given deliverable drug.[6] Some recent DDS platforms have implemented oral, injectable systems, transdermal,[7] ocular[8] and nasal routes[9] that are entering clinical trials and with some already on the market. In order to achieve successful clinical translation, DDS must be safe, perform the intended therapeutic function, offer a convenient administration method and must be scalable.

Of the many options for controlled release DDS, transdermal delivery has received much attention due to a number of distinct advantages, which will be discussed later in this chapter. One example of a popular drug delivery platform is the microemulsion. This drug delivery system is highly attractive for pharmaceutical formulators due to the relatively simple formulation process. Microemulsions can also be easily utilized for developing novel carriers for phytoactives and extracts.[10] In the following sections, microemulsion DDS will be discussed in more detail.

Nanotechnology has interfaced very closely in the past decade with drug delivery.[4,11,12] The size and shape of nanoparticles (NPs) are a critical component, since this affects their ability to navigate biological carriers in the body. The need for improved delivery systems motivated the development and application of nanofabrication technologies to be more effective DDS. For example, the size of a particular type of NPs determines their bio-distribution. Small particles in the range of 20 nm will be cleared from the circulation by the reticuloendothelial system (RES) within hours if administered intravenously, while larger particles become trapped in the liver and the spleen rapidly in minutes.[13]

All the DDS described will be very useful especially as new molecular targets are identified. Drug delivery will play a pivotal role in the next generation of therapeutics now and in the future. In this chapter, the focus is on DDS that are discussed often in the literature, including liposomal, transdermal, and microemulsion DDS. Nanotechnology for DDS also has been reviewed.

## II. Liposomal and Targeted Drug Delivery System

### A. Evolution of liposomal drug targeting

Liposomes can be described as being composed of phospholipids, which self-enclose to form spheres of lipid bilayers and contain aqueous cores (Figure 2). Phospholipids are amphiphilic in nature and result in polar shells in aqueous solutions due to the hydrophobic effect of hydrophobic acyl chains when subjected to an aqueous environment. This process is thermodynamically favorable and is further enhanced by hydrogen bonding, van der Waals forces and other electrostatic interactions.[14] Liposomes represent an attractive drug delivery system and have the ability to resemble cell membranes in both structure and composition. Additionally, liposomes are readily formed using nontoxic, nonimmunogenic, natural, and biodegradable amphiphilic molecules.[15]

Liposomes are known to incorporate both hydrophilic and lipophilic molecules. The solubility and *in vivo* fate of the incorporated

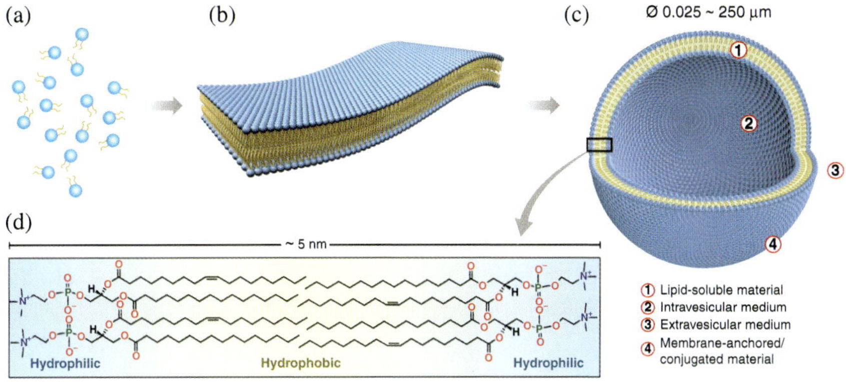


Figure 2. Schematic illustration of the fundamental self-assembly process from individual phospholipid molecules (a) to bilayer membrane leaflets, (b) followed by transformation into liposomes (c) Taken from Ref. [16] with permission of Annual Reviews.

molecules however are highly dependent on the particular liposomes used. Thus, for drug delivery purposes, liposomes have numerous advantageous properties, including biocompatibility, biodegradability, reduced toxicity, and influence the pharmacokinetic profile of the loaded drug. These advantageous properties allow for target drug delivery. While liposomes have been previously used for decades as pharmaceutical carriers, there is a need for improvement so that selective targeting can be achieved at the cell and molecular level.[17]

The first description of liposomes was by Bangham in the 1960s.[18] Soon after this discovery, a multiple enclosed phospholipid bilayer structure consisting of single bilayers, which were initially termed "bangosomes" followed by "liposomes" were described.[19] Work from pioneers such as Gregoriadis, established how liposomes could enclose drugs to be used as DDS.[20,21] Others demonstrated that liposomes have the ability to change the *in vivo* distribution of entrapped drugs.[22]

Many advances made during the late 1980s and early 1990s set the foundation to overcome many of the early challenges. Additionally, development efforts led to a comprehensive understanding of lipid polymorphisms, physiological mechanisms of *in vivo* liposome disposition and lipid-drug, and lipid-protein interactions. This progress allowed for improved liposome designs for increased stability both *in vitro* and *in vivo*, improved biodistribution and optimized stability of liposomes in circulation. By the mid-1990s, liposomes were regarded as drug carriers for pharmaceutical applications.[23]

At the same time, new methods were also being developed to enable the preparation of large unilamellar liposomes (LUV), which resulted in improvements in trapping efficiency and homogeneity. Further work lead to the ability to produce LUV by extrusion of multilamellar vesicles through 100 nm polycarbonate filters. Initially liposomes were originally formed under low pressure conditions, which was low throughput. However, eventually higher pressure systems were developed to enable large scale production. These developments also lead to the production of LUV with diameters of less than 50 nm, which previously wasn't possible until sonication or homogenization was used. Furthermore, the advancement of microfluidic mixing techniques allowed scalable production of LUV in the range of 20–50 nm.[24]

The first generation liposomes were rapidly cleared by the RES soon after administration. The first strategy implemented in order to generate long-circulating liposomes modified the characteristics of liposomes including composition and size. It was also discovered that small liposomes were less likely to be cleared by the RES than larger liposomes. Allen and Everest also reported that circulation times of small unilamellar vesicles happened to be higher than those of larger multilamellar vesicles.[25] Further optimization enabled the formulation of more stable, longer circulating liposomes with increased accumulation at target sites. One process that enhances accumulation is referred to as the enhanced permeability and retention (EPR) effect.[26] The EPR effect can be described as a phenomenon by which enhanced extravasation of macromolecules are first derived from tumor blood vessels, and then are retained more in tumor tissues, infarcts and inflamed regions more than in normal tissues. Furthermore, one study demonstrated that 100 nm liposomes were found to be optimal for tumor delivery.[27] However, if liposomes are formulated using saturated phospholipids with higher TM instead of unsaturated phospholipids, the circulation time and drug retention are better for larger vesicles as well.[28]

Additionally, surface modifications of liposomes have also been used to avoid RES uptake. Initial studies used ganglioside and sialic derivatives like GM1 (monosialoganglioside) to mimic the erythrocyte membrane surface to increase stability.[29] Later studies used hydrophilic polymers like PEG[17] to improve circulation and stability of the liposomes. Furthermore, the incorporation of polyethylene glycol–lipid conjugates (e.g. methoxy polyethylene glycol (mPEG)–distearoylphosphatidylethanolamine (DSPE)) within the bilayer membrane resulted in prolonged blood circulation half-life of the liposomes and promoted liposome accumulation where leaky vasculature resided by the EPR effect.[30]

It was discovered that by modifying liposomes such that a steric boundary was created, it prevented plasma protein binding and subsequent RES uptake.[31] These surface modifications lead to the development of so-called "stealth liposomes", which function more effectively by having the ability to mimic the biomembranes. For example, Doxil is a typical stealth liposome with a PEG surface coating that has demonstrated an improved circulation time and safety profile when used in DOX chemotherapy. Eventually the next generation of liposomes enabled direct molecular targeting facilitated by the attachment of site-specific ligands to the liposomal surface.[32] This strategy increases intracellular drug levels at target sites utilizing receptor-mediated endocytosis. The endocytosed material is then subjected to acidic lysosomal compartmentalization and hydrolysis by various enzymes resulting in reduced biological activity. This problem can be compounded for drugs that are particularly sensitive to this mechanism of degradation, such as nucleic acid and peptidic drugs.[33] For such drugs, methods enabling the release of the cargo into the cytosol are advantageous and help to mitigate this issue. Later development involved the formulation of a liposome, which combined long circulation properties and targetability. Doxil was used as an anti-cancer drug, and demonstrated improved *in vivo* activity of liposome-entrapped drugs in animal models. This drug resulted in significant survival times improvement,[34] and became a popular "model system" for validating the effects of a wide range of liposome characteristics on therapeutic efficacy.

Development continues for long-circulating liposomes, which is further driven by their potential to carry imaging agents and treating various infections.[33,36] However, evidence has shown that PEG liposomes can induce side reactions through activation of a complementary system.[37] In this section, the focus is on drug delivery by liposomes; however its applications are not discussed.

## B. Liposome classification

Liposome structures are typically categorized using their parameters including composition and applications (Figure 3), structural parameters (Figure 4)[38] and the method of preparation including thin-film hydration, sonication, French press, ethanol injection, detergent dialysis, reverse-phase evaporation and extrusion.[39] The conventional

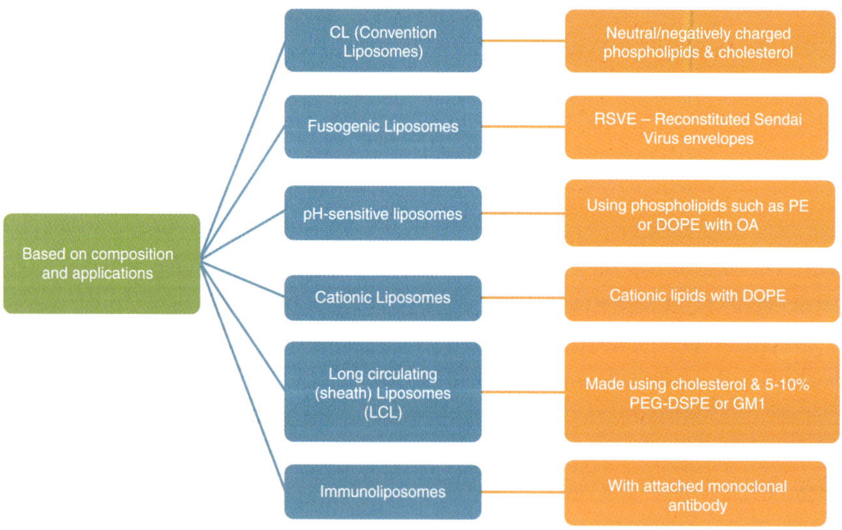


Figure 3. Classification of liposome structure based on composition and application. Taken from Ref. [38] with permission of Society of Pharmaceutical Technocrats.

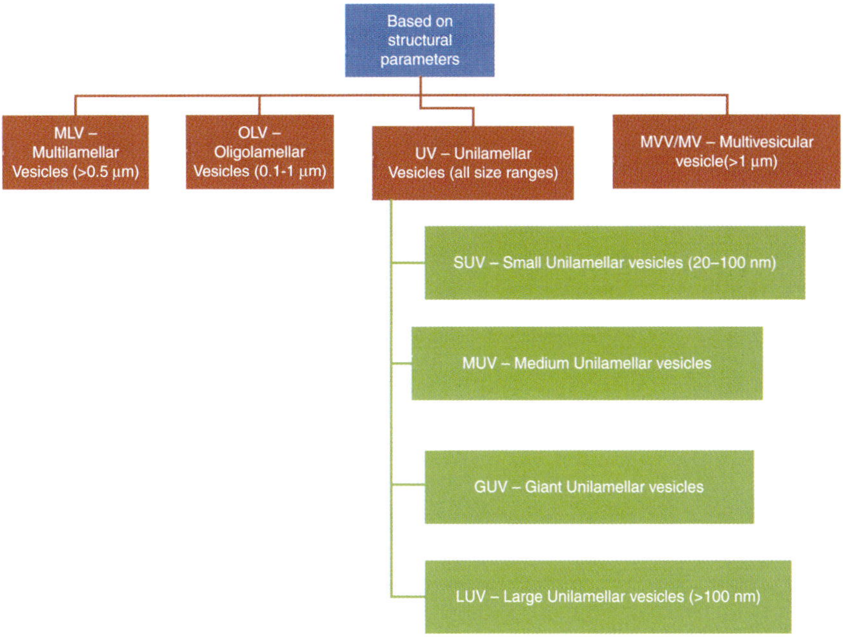


Figure 4. Classification of liposome structure based on structural parameters. Taken from Ref. [38] with permission of Society of Pharmaceutical Technocrats.

classification of liposomes is based on the number of lipid bilayers present in the colloidal structure, with unilamellar liposomes containing one lipid bilayer and multilamellar liposomes containing multiple lipid bilayers. It is known that due to their amphiphilic nature liposomes have the capacities to encapsulate both polar and nonpolar compounds for drug delivery.[40]

## C. Types of liposomal drug delivery platforms

The four main types of liposomal delivery systems are conventional liposomes, sterically stabilized liposomes, ligand-targeted liposomes and a combination of the above (Figure 5). Liposomes consist of a lipid bilayer containing cationic, anionic, or neutral (phospho) lipids and cholesterol, which enclose the aqueous volume (Figure 5).[41]

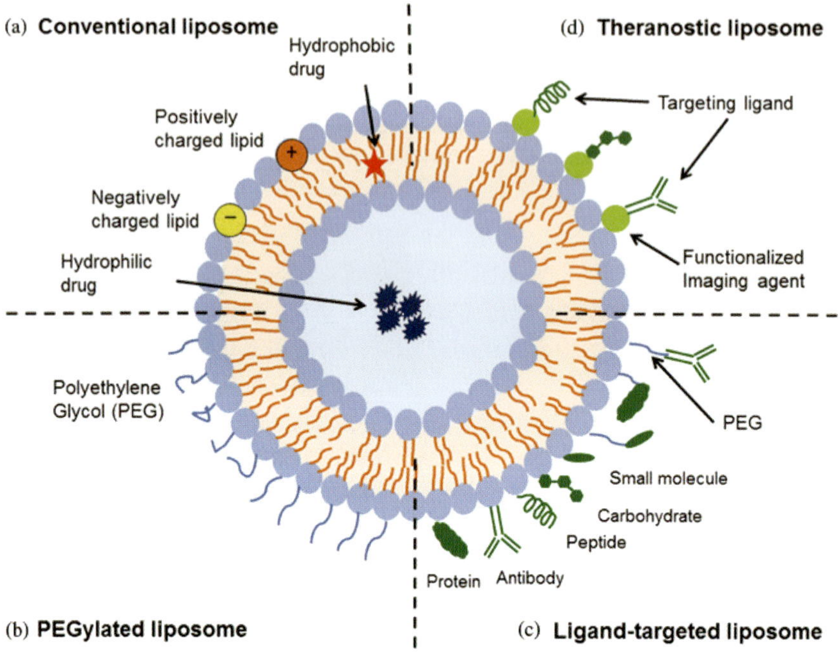


Figure 5. Schematic representation of the different types of liposomal DDS. Taken from Ref. [41] with permission of Frontier.

Sterically-stabilized liposomes were introduced in order to improve liposome stability and enhance their circulation times in the blood. It has been shown that the hydrophilic polymer, polyethylene glycol (PEG), is the optimal for producing sterically-stabilized liposomes (Figure 5). The steric barrier is able to improve the efficacy of the encapsulated agents by reducing *in vivo* opsonization with serum components, and thus rapid recognition and clearance by the RES. This approach has several distinct advantages including reducing the elimination of drugs by prolonging blood circulation, providing accumulation at pathological sites and minimizing side effects. Liposomes coated with PEG have the advantage of prolonged circulation times; however there can be a reduction in their ability to interact with the intended targets.[41]

There are several significant advantages of ligand-targeted liposomes. For example, they offer site-specific delivery of drugs to targeted cell types or organs *in vivo*, which selectively express or over-express specific ligands (e.g. receptors or cell adhesion molecules) at the disease site. There are several types of ligands available, which allow for better targeting specificity such as antibodies, peptides/proteins and carbohydrates (Figure 5). One of the most versatile ligands is monoclonal antibodies (mAb), since they can be coupled to liposome surfaces to create immunoliposomes. However, *in vivo* performance of immunoliposomes is limited, due to poor pharmacokinetics and immunogenicity. This has been a major challenge for liposomes in achieving their potential efficacy as site-specific drug carriers.

Several studies have focused on complex multi-functional liposomal formulations in order to create more efficient DDS. In particular, theranostic NPs are of great interest as an all in one therapeutic and diagnostic (Figure 5) tool. Typical theranostic delivery systems are characterized by including some type of NPs, an imaging component, targeting ligand and the therapeutic agent. Numerous studies have demonstrated effective diagnostic imaging as well as therapeutic delivery of encapsulated drugs *in vivo* using theranostic approaches.

## D. Passive targeting

Liposomes can use passive delivery as a method to reach diseased areas as well as utilizing altered structures and function. Many nanocarriers such as liposomes use passive targeting of large molecules to accumulate at tumor sites based on the EPR effect.[30] This passive delivery effect has been proven in many tumors models.[42,43] Tumor tissues are known to be highly vascularized, because of upregulated angiogenesis. However, the blood vessels are not formed in an orderly fashion, which makes them abnormally leaky. This haphazard formation of leaky vasculature resulting from enlarged endothelial cell gaps allows for convenient passage of liposomes and other nanocarriers in the moderate nanometer size range by extravasation into the tumor interstitium.[43] A vesicle size threshold of ~400 nm was reported for extravasation into tumors, however some work has demonstrated more effective extravasation using particles that are less than 200 nm.[43,44] Figure 6 depicts the schematic illustration of passive targeting via the EPR effect.[45]

Additionally, tumor tissues are known to have poor lymphatic drainage. These histological features make tumors susceptible to targeting for increased cargo delivery by long-circulating liposomes, which is further enhanced by the EPR effect.[46] Leaky tissue morphology has also been reported within inflamed tissues of inflammatory bowel disease and inflammatory rheumatoid arthritis.[47] In all cases, the EPR effect allows for the increased local tissue concentration of a particular drug to promote a therapeutic effect. Commercially, approaches for liposome-based formulations utilize the enhanced passive delivery to preferred sites of action for targeting. It can be the case that PEGylated liposomes become concentrated in the target area by the EPR effect, but remain unable to efficiently release the drug. It has also been demonstrated that the PEG coating can inhibit endosomal escape of the drugs post endocytosis.[48] One challenge is that each tumor type carries its own individual identity and is unique in the development of its blood supply. For example, one study compared the distribution of PEGylated liposomal DOX in different murine tumors (4T1, breast and 3LL, lung) *in vivo* and revealed that

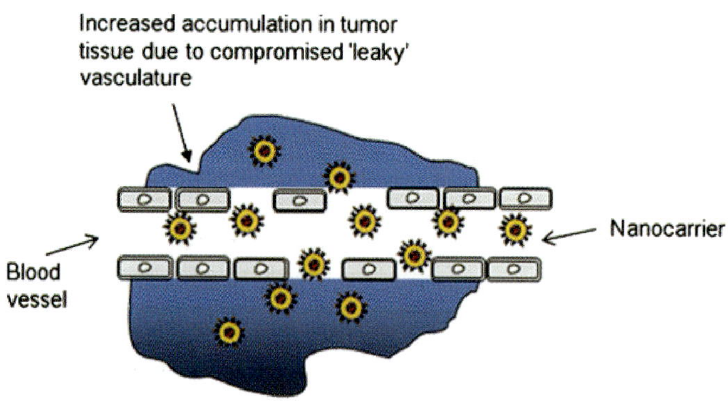


Figure 6. Schematic illustration of passive targeting via EPR effect. Taken from Ref. [45] with permission of Springer.

liposomes accumulated more efficiently in the 4T1 as compared to the 3LL cells, demonstrating increased vascular permeability in 4T1 and variance in vascular permeability between tumor types.[49] Thus, achieving an even distribution of the liposomes within a tumor can be extremely challenging. However, even after drugs are released from the liposomes, many agents will then need to enter the cells to exhibit a cytotoxic effect. When considering the complexity of tumor microenvironments, it is possible and expected that some drugs will be unable to produce the desired treatment outcome. Another compounding effect of these treatments is that they can also give rise to multidrug resistance by the tumor.[50,51] Other approaches involve the development of active drug targeting moieties and triggered/stimuli-sensitivity-based targeting techniques. Often times these techniques are used together to promote more efficient chemotherapy.

### 1. *Limitations of passive targeting*

While passive targeting is generally preferred for clinical therapy, it does have several limitations. For example, the porosity and pore size of tumor vessels vary with the type and status of tumors making treatment efficacy inconsistent.[50] Thus, it is possible that a passive targeting effect may not be reasonably achievable in all tumors types. In some cases drugs cannot readily diffuse throughout the tumor making homogeneous targeting within tumor cells extremely challenging. It is known that in most solid tumors, elevated interstitial fluid pressure[52] can cause inhibition of homogeneous distribution of nanocarriers within tumors.[3] Additionally, subsequent treatments can induce multiple-drug resistance.[53]

## E. Active targeting

Active targeting using modifications of the liposomal surface with a targeting ligand can improve efficiency and specificity, and can result in increased accumulation at the target site or facilitate intracellular delivery to target cells. Upon binding certain ligands, it has the ability to release the liposomal contents intracellularly via the receptor-mediated endocytosis pathway.[54] This effect can improve the overall efficacy of a treatment by reducing the diffusion of the drug from the tumor. In certain cases, liposomes targeted to internalizing receptors may be beneficial by improving internalization. Some of the factors that play a key role in mediating active targeting include: ligand or antibody attachment to the liposomes, the presence of a targeted receptor/antigen presentation on a diseased or normal tissue, internalizing properties of the target ligand and the long-circulation capabilities of a liposome bearing the targeting moiety.

Ligands can be attached to a variety of liposome formulations including directly to the surface of a non-PEGylated liposome, PEGylated liposome or to the free end of a PEGylated liposome. Particular attention should be taken when making surface-modifications of liposomes with ligands, especially antibodies or peptides, to ensure the conjugation does not alter ligand structure or restrict activity.[55] Targeting liposomes should be developed on a case-by-case basis to determine the optimum ratio of targeting ligand to liposome surface for optimum targeting. Several studies have already demonstrated how ligand attachment affects the efficiency of the targeted drug delivery system (TDDS).[56]

In general, it has been found that ligand or antibody presentation at the distal ends of the PEG is more efficient for TDDSs. Typically, targeting liposomes are prepared with the targeting ligand shielded during circulation, with the ligand becoming presentable once it reaches a targeted site. Therefore, a strong fundamental understanding of the targeted receptors or antigen presentation is essential for the development of a TDDS. Effective targeting ligands need to be able to distinguish the target cells from normal cells in order to avoid nonspecific binding and toxic or adverse effects. In order for this system to function optimally, the receptor or antigen should be exclusively expressed on the target site or the target site should have relatively high expression relative to normal cells. Moreover, in the ideal case, the ligand will efficiently induce receptor mediated endocytosis or fusion of the liposome with the cell membrane.

Similarly, the targeted antigen must be carefully selected based on selectivity or overexpression of the tumor tissue or on the angiogenic blood vessels feeding the tumor. Numerous targeting ligands have been investigated in order for further development of targeted liposomes including proteins (antibodies or antibody fragments), nucleic acids (aptamers) and other receptor ligands (peptides, carbohydrates, and vitamins). There are multiple important considerations for selecting a target antigen such as extent of overexpression, selective expression on the target site only,[56] internalization of the targeting-ligand formulation[57] and any shedding of the target antigen.[58] A ligand that is not internalized has little to no practical use, since it will remain in the interstitial space.[59] It is also ideal for a TDDS to promote a long circulation time, unless the targeted cells happen to be macrophages. The pharmacokinetics of targeted PEGylated liposomes are very similar to non-targeted PEGylated liposomes, because the liposomes are dependent on the blood circulation and the EPR effect for entry into the target site. Therefore, the targeting ligand is only effective if the liposomes are able to reach the target tissue. TDDS interacting with cell receptors should be internalized into the target cells, while the non-targeted liposomes will tend to accumulate in the tumor interstitium.[60]

The use of antibodies in order to modify the surface of liposomes comprise the majority of active targeting studies.[61] For example, specific antibody fragments (Fc, F(ab′)) and mAb attached to the surface of the liposome serve as TDDS. Additionally, highly upregulated HER2/neu growth factor receptors, which are specific to breast cancer cells have been targeted with anti-HER2 immunoliposomes. Over expression of the HER2 receptor in some cancers can be utilized for TDDS approaches. For example, patients with advanced HER2+ breast cancer were administered with HER2-targeted DOX-loaded immunoliposomes (MM-302) for a Phase I study. It was shown that MM-302 bound specifically, and entered the tumor cells, while demonstrating little to no effect on normal cells with low level of HER2 expression.[62] Several tumor types are known to overexpress surface-bound nucleosomes, which have been targeted effectively using the 2C5 mAb. This work involved binding of the mAb 2C5 to PEGylated DOX liposomes (Doxil), which resulted in retention of DOX from the liposomes. Also *in vitro* and *in vivo* studies have demonstrated enhanced binding to a broad range of cancer cell types as compared to the non-targeted liposomes.[63] Similarly, DOX-loaded PEGylated immunoliposomes targeted using the F(ab′) fragment of a tumor-specific human mAb-GAH were shown to be well tolerated by patients that have metastatic cancer in Phase I trials.[64]

Other receptors such as the transferrin receptor can also be a useful target for these approaches. The transferrin receptor is highly overexpressed in tumors and is necessary to enable their higher metabolic requirements for iron. This receptor can also be targeted using anti-transferrin receptor antibodies or transferrin itself. A phase I study consisting of 39 patients, which included different tumor types such as colorectal, pancreatic and neuroendocrine was performed with transferrin-conjugated oxaliplatin-loaded liposomes (MBP-426). This formulation was dependent on the EPR effect to initially reach the tumor site, followed by binding to the transferrin receptors. After the initial dose-limiting toxicity was recorded as reference, a modified dose was administered in patients with gastro-esophageal adenocarcinoma in combination with 5-fluorouracil/leucovorin in Phase II.[65] Other receptors such as the folate receptor has also been studied in several *in vitro/in vivo* applications, because of its overexpression by many different tumor types.[66] A published US patent (US 20130071321 A1) reports that folate-targeted liposomes were able to effectively deliver anti-inflammatory agents to inflamed tissues, due to upregulation of the folate receptor.[67] Combination treatments such as active targeting used in conjunction with cell penetrating peptides (CPP) have been of particular interest. CPP are characterized as protein translocation domains, which translocate through the cell membranes and are able to facilitate the transport of cargo. Typically, CPPs are short peptides consisting of less than 40 amino acids. These peptides can be transported across the cell membrane by either endocytosis pathway or direct entry through the cell membrane. One example that has been used is the 228 trans activator of transcription protein (TATp) of the HIV virus, octa-arginine (R8) and amphipathic model peptide (MAP), a well-studied CPP that has been shown to be effective for drug and gene delivery.[68] Other studies have utilized liposomes, which have been modified with TATp or R8 peptide on their surface. For example, TATp has been used to transfect the pEGFP plasmid (expressing green fluorescent protein) in liposome/DNA complexes in mouse spleen-derived antigen presenting cells, mouse NIH/3T3 fibroblasts and rat H9C2 cardiomyocytes without observing any significant cytotoxicity. Additionally, the TATp-modified liposomes were able to efficiently transfect the pEGFP in LLC tumors and U-87 MG intracranial tumors in mice.[69] Other approaches include the R8 peptide being conjugated to poly(ethylene) glycol-phosphatidylethanolamine (PEG-PE) and was attached to the DOX-loaded PEGylated liposome surface using the post-insertion technique. These liposomes were able to inhibit tumor growth in non-small cell lung tumor xenografts in mice in comparison to unmodified PEGylated liposomes.[70] However, it is essential when using liposome-encapsulated drugs that the liposome is internalized into target cells. For instance, it was demonstrated that HER2-targeted doxorubicin-loaded liposomes were able to significantly improve anti-tumor efficacy using internalizing ligand[59] as opposed to targeting to HER2/neu, which did not result in internalization.[71] Another important factor to consider is the degree of shedding of the target antigen. Several cancer cell types are known to shed their membrane antigens when under conditions of high tumor load. Therefore, under conditions where lots of shedding occurs, the liposomes could potentially be cleared faster due to a result of binding with the solubilized antigen.[58]

## F. Challenges in liposomal drug delivery

One of the major limitations preventing the successful market development of liposomes is the possibility of cytotoxic effects. It has been observed that in some cases liposomes can release the majority of their payload essentially immediately upon administration, which confirms not advantage for their use. Additional concerns include that some studies have demonstrated charged liposomes to be toxic.[72,73] Furthermore, depending on the procedure used for liposome production, there is a chance that some organic solvents (ethanol, ether) could be present in trace amounts in the final preparation. Liposomes also have manufacturing-related challenges such as batch to batch variations, low drug entrapment, effective sterilization methods, stability problems and most importantly, scalability problems. Complications can also arise when the production of multifunctional liposomes is required. These issues when compounded with the need for costly raw materials results in liposomes being an expensive segment within the medicinal market. However, considering advancements in technology over the past decade, the continued development of liposomes and nanomedicine, these approaches represent potentially very powerful and useful DDS.

## III. Transdermal Drug Delivery

The drug delivery market can be segmented into eight categories based on the route of administration. Some common routes of administration include: oral, pulmonary, transdermal, injectable, ocular, nasal, implantable, and transmucosal drug delivery. Oral delivery has the largest market share at greater than 50%, followed by transmucosal at 26.2% and transdermal delivery having the lowest market share at 12%. However, parenteral delivery is one of the most commonly used methods for drug administration. The parenteral[7] and oral[74] routes have some challenges and, some of which can be overcome by transdermal applications, which represent an alternative for controlled (therapeutic) release of drugs. Other transdermal drug delivery systems (TDDS), also known as "patches", are topically given medications designed for the delivery of a therapeutic quantity of drug through the skin.[75] Transdermal patches were originally developed in the 1970s and approved by the FDA in 1979 as a treatment for motion sickness. They came in the form of a three-day patch that is able to deliver scopolamine. Patches containing nitroglycerin were approved in 1981, and currently there exists numerous patches for the delivery of drugs such as clonidine, fentanyl, lidocaine, nicotine, nitroglycerin, oestradiol, oxybutynin, scopolamine, and testosterone. Additionally, there exist combination patches for contraception, as well as for hormone replacement. The patches generally last from one to seven days, depending on the drug used. In order to produce an effective transdermal drug delivery platform it must be formulated with three factors in mind: the drug that will be used, the biological factors of the skin and the vehicles to achieve delivery. Factors affecting transdermal delivery can be divided in two classes, biological factors and physicochemical factors. Biological factors to consider include[76]; skin condition, skin age, blood supply, regional of the skin, skin metabolism and differences among species. Some of the physicochemical factors affecting transdermal drug delivery include skin hydration, temperature, pH, diffusion coefficient, drug concentration, partition coefficient, molecular size of the drug and shape. Thus, some of the ideal molecular characteristics for transdermal drug delivery[77] are: adequate solubility is required in both a lipid and water environment for enhanced penetration of the drug (1mg/ml), an optimum partition coefficient is required for effective therapeutic action, a low melting point of the drug is ideal (<200°C) and the pH of the saturated solution should be in the range of 5–9.

TDDS has been described by Prausnitz and Langer in three generations.[7] In the first generation, the sustained delivery of low molecular weight molecules is achieved using transdermal patches,[78] however there are variations of these systems that include topical formulations, which, after solvent evaporation, release molecules in the stratum corneum by diffusion. In the second generation, methods are employed such as using chemicals, physical, or mechanical techniques, such as ultrasound and iontophoresis that temporarily and reversibly alter or damage the stratum corneum structure, thus allowing the delivery of higher molecular weight molecules, and enhancing permeability of the skin. In the third generation, a combination of various enhancers is implemented resulting in more invasive disruption of the stratum corneum. Typical examples of these disruptive techniques include electroporation, dermabrasion, and micro-needles.[79]

The vehicles used are typically embedded in topical formulations,[80] which encapsulate and protect drugs on their way to the target site. However, the efficacy of topical treatments is largely dependent on several factors including the vehicle used for the transport, the morphological, biophysical and physicochemical characteristics of the skin and the properties of the drug employed. Some widely used vehicles include vesicles, nanocarriers, and ME. Vesicles can be described as lamellar phase lipid bilayers, which surround a drug or other payload core and can be unilamellar or multilamellar.[81,82] One such example is liposomes, where phosphatidylcholine is the main shell component and tends to accumulate in the stratum corneum. However, the mechanisms involved for effective penetration and enhancement of liposomes through the skin are still largely unknown. One possible hypothesis speculates that liposomes fluidize the intercellular lipid channels, which improve penetration down into to the deeper layers of the skin.[83] Other types of encapsulating vesicles, depending on the exact composition, include transfersomes,[84] ethosomes[85] and niosomes[86] whose efficacy and penetration is dependent on deformability.[78] Some other approaches that can be used include NPs, typically polymeric or lipid structures in the form of nanospheres or nanocapsules.[87] The typical size of NPs allows them to easily accumulate in the pores of the skin, allowing entry using the follicular route.[88–90] Some of the major advantages of transdermal drug delivery are: improved bioavailability, more uniform plasma levels, greater efficacy resulting in less required treatments, reduced side effects and improved therapy performance over other deliverables. Other additional advantages of transdermal delivery are that this approach is non-invasiveness, has sustained release, can be self-administered, improves therapeutic outcomes[91,92] and has specificity for the target site. This approach can also be used for applications including local treatment delivery, e.g., muscular inflammations,[93] but also for major pathologies, such as cancer.[94]

Transdermal patches have demonstrated utility as novel applicators for existing therapeutics and for the increased drug stability. Patches have also been shown to reduce side effects; for example, oestradiol patches are used by over a million patients annually and, do not result in liver damage in contrast to oral formulations. Other types of patches that exist are therapeutic, cosmetic, aroma patches, weight loss patches, and non-medicated patches including thermal and cold patches, nutrient patches and skin care patches (a category that consists of patches that measure sunlight exposure).

## A. Passive transdermal drug delivery

Transdermal drug delivery can occur as either a passive or active process. Passive delivery can be described as delivery which requires no physical disruption of the stratum corneum. Active delivery is characterized by delivery requiring physical disruption of the stratum corneum or use of some external driving force. Therefore, in order for passive diffusion to occur through the skin, specific physicochemical criteria must be met including molecular weight, lipophilicity, and melting point.[95]

There are both occlusive and non-occlusive transdermal patch products available. These patch products have been predominantly made using testosterone and estradiol delivery systems. Typically non-occlusive applications using these formulations require large application areas (where patch is ~20–60 $cm^2$ and the formulations are being applied to areas between 100–200 $cm^2$). One of the most common non-occlusive products are testosterone patches, which result in increasing circulating testosterone levels in men. This product is predominantly in the United States with a market that has exceeded $2 billion in 2013. The testosterone product market can be segmented based on the route of administration such as injectables, oral, and implants. However, the vast majority of products are transdermal, non-occlusive gels, with the best-selling product being Androgel®, which had sales approaching $1.5 billion in 2013.[95]

## B. Active transdermal drug delivery

In order to improve the effectiveness of transdermal drug delivery, making it a more applicable option for broad-ranging therapeutics, various invasive physical penetration enhancement techniques (active transdermal delivery) have been investigated. Some common examples of active transdermal delivery methods include iontophoresis, sonophoresis, electroporation, thermal and laser-assisted ablation, and micro needles (MNs). A schematic summary of these methods is represented in Figure 7.[95] The goal of these techniques is to reduce the barrier effects and improve penetration through the stratum corneum, which will facilitate the transport of drug molecules across the skin. Over the last 20 years, the FDA has approved a number of technologies for enhancing transdermal drug delivery. Surprisingly, the majority of these products are actually delivery systems for local anesthesia, which are not "transdermals" as described at the beginning of this section.[95]

As an alternative for transdermal delivery, hypodermic injection and needle-free injectors have been designed as a painless method of delivery

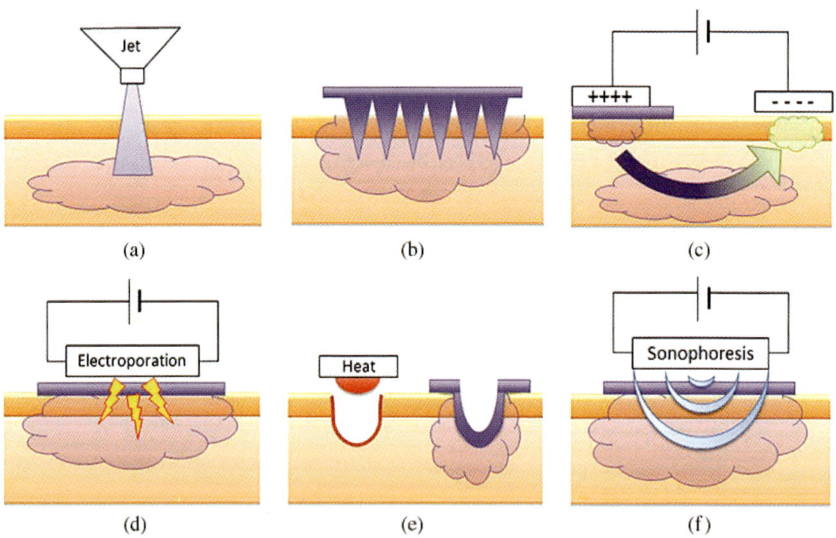


Figure 7. Schematic representation of enhanced transdermal drug delivery using (a) Jet injection, (b) Micro needles, (c) Iontophoresis, (d) Electroporation, (e) Thermal ablation and (f) Sonophoresis. Taken from Ref. [95] with permission of Taylor & Francis.

for liquids and powders into the skin. The way liquid injectors work is by puncturing the skin with a high-velocity jet in order to produce a stream containing the drug to be delivered with subsequent deposition into the skin. Another similar approach, using powder injectors in order to force powdered drugs into the skin using either compressed gas or an electrical power source, generates a high-velocity stream to penetrate the stratum corneum.[96] Most of the research contributing to this area is focused on the delivery of vaccines using PharmaJet, which has secured FDA approval for its influenza vaccine, Afluria®, in 2014.[97] However, this technology is not exclusive to vaccines and following an initial withdrawal, there was a relaunch of Zingo® in 2014, which is a powder injector used for delivery of lidocaine to the skin. The lidocaine was used for local anesthesia, that was developed in order for use in children (Marathon Pharmaceuticals 2015).[96] It should be noted, that neither of these applications is intended for frequent use, but rather "one-off" or periodic treatments.

Another interesting approach is the use of thermal energy technologies for enhanced delivery into the skin. Thermal ablation is characterized by heating the surface of the skin to a temperature that vaporizes the desired area of the stratum corneum in the time scale of micro to milliseconds, leading to enhanced permeability, without damaging the skin.[98]

## C. Transdermal delivery system design

Generally transdermal delivery systems include some drug to be dissolved or dispersed within an inert polymer matrix, which provides a support platform enabling drug release. There are two basic designs of the patch method that will determine the drug release properties and patch behavior including: (1) Matrix or Monolithic: The inert polymer matrix that binds and determines the controlled release of the drug. (2) Reservoir or Membrane: This polymer matrix does not control drug release, rather, it is a rate controlling membrane that sits between the drug matrix and the adhesive layer providing the actual rate-limiting barrier for drug release.[99]

## D. Transdermal drug delivery systems technologies

These technologies are classified in four basic approaches[76,100]:

### 1. *Polymer membrane partition-controlled TDD systems*

In this approach, the drug reservoir is positioned between a drug-impermeable backing laminate and polymeric membrane that controls the rate. The drug is then forced to permeate only through the rate controlling membrane. The solid drug is then homogeneously dispersed into the solid polymer matrix, and then suspended in an unleachable viscous liquid medium (e.g. silicone fluid) to form a suspension similar to paste, which is dissolved in a releasable solvent (e.g. alkyl alcohol) resulting in the clear drug solution.

### 2. *Polymer matrix diffusion-controlled TDD systems*

This method uses a drug reservoir which is formed when homogeneously dispersing the drug solids in a hydrophilic or lipophilic polymer matrix. This matrix is then molded into medicated polymer formed disks, which have a defined surface area and thickness. The drug reservoir containing the polymer disk can then by mounted on an occlusive baseplate specifically fabricated from a drug-impermeable plastic backing. As opposed to coating the adhesive polymer directly on the surface of the medicated disk, it is instead applied along the outside of the patch forming an adhesive rim, which surrounds the medicated disk.

### 3. *Drug reservoir gradient-controlled TDD systems*

Polymer matrix drug dispersion-type TDDS allow for the modification of drug loading levels to be adjusted in an incremental manner, which forms a gradient of drug reservoir along the diffusional path across the multilaminate adhesive layers.

### 4. *Micro-reservoir dissolution-controlled TDD systems*

This particular system is a hybrid between the reservoir and matrix dispersion-type drug delivery methods. This approach contains a drug reservoir formed by initially suspending the drug solids in some aqueous solution of water-miscible drug solubilizer (e.g. propylene glycol), followed by homogeneously dispersing the drug suspension using controlled aqueous solubility in a lipophilic polymer. Finally, the high shear mechanical force is able to form thousands of unleachable microscopic drug reservoirs.

## E. Transdermal drug delivery challenges

There are several challenges in TDDS, which are summarized as[101]:

- Disability of TDDS in delivering ionic drugs.
- Not able to achieve high drug levels in blood/plasma.
- TDDS is not suitable for drugs of large molecular size.
- Disability of TDDS in delivering drugs in a pulsatile fashion.
- TDDS cannot be developed if a drug or formulation causes irritation to skin.

# IV. Microemulsion Drug Delivery System

Hoar and Schulman introduced the concept of the microemulsion in the 1940s and were able to generate a clear single-phase solution by triturating a milky emulsion with hexanol.[102] They first prepared the initial microemulsion by dispersing oil into an aqueous surfactants solution and then adding alcohol as a co-surfactant, which lead to this transparent stable formulation. Microemulsions can be described as clear, transparent, thermodynamically stable dispersions of oil and water, which are stabilized by an interfacial film of surfactant that are frequently used in combination with a co-surfactant.[103] Some alternative names describing these systems include swollen micelle, transparent emulsion, solubilized oil and also micellar solution. Microemulsions can be characterized as having bulk phases of water and oil separated by a surfactant/co-surfactant rich interfacial zone. Some of the advantages of these systems over conventional emulsions are that they are thermodynamically stable liquid systems and can be spontaneously formed.[104] For this and other reasons, microemulsions have great potential as DDS.[105,106]

There are three types of microemulsions that are most likely to be formed and are dependent on the composition:

- O/W microemulsion where oil droplets are dispersed into the continuous aqueous phase.
- W/O microemulsion wherein water droplets are dispersed into the continuous oil phase.
- Bi-continuous microemulsion where micro-domains of oil and water are inter-dispersed within the system.

Drug delivery approaches based on w/o microemulsions have also been used where the dilution by an aqueous phase is unlikely to occur, such as after an intramuscular injection.[107]

Additionally, the use of o/w microemulsions in drug delivery is a simpler technique than with w/o microemulsions. This comes as the result of the droplet structure of o/w microemulsions being often retained after dilution by the biological aqueous phase, which permits both oral and parenteral delivery. However, there are particular advantages when conducting drug formulations using w/o microemulsion systems. For example, peptide drugs, are known to exhibit low levels of activity when delivered orally and can be highly susceptible to proteolysis when introduced into the gastrointestinal tract.[108]

One of the primary attractions of o/w microemulsion systems is that they have the ability to take hydrophobic drugs and incorporate them into the a polar oil phase, improving their solubility.[105,109] It should be noted that the majority of drugs are not soluble in hydrocarbon oils, rather these poorly water-soluble drugs favor solubilization of small to medium volumes of oils such as tributyrin or Miglyol 812. However, drug formulations in hydrocarbon oil-in-water microemulsions may not have much benefit in solubilization compared to micelles.[110]

There are three important considerations that should be observed when preparing microemulsions: (1) Surfactants should be carefully selected, to ensure an ultra-low interfacial tension ($<10^{-3}$ mN/m) is achieved at the oil/water interface, which is essential for the production of microemulsions. (2) A sufficient concentration of surfactant should be provided such that the necessary surfactant molecules are present in order to stabilize the microdroplets being produced for ultra-low interfacial tension. (3) In order to promote the formation of microemulsions the interface requires flexibility or fluidity.[111]

Pseudo-ternary phase diagrams consist of oil, water, and co-surfactant/surfactants mixtures and are composed at fixed co-surfactant/surfactant weight ratios. Figure 8 shows the different areas in the phase diagram.

Microemulsions have several advantages including thermodynamic stability, optical clarity, and simplicity of preparation.[112] Both water-soluble and oil-soluble substances can be solubilized at the same time if there exists microdomains of different polarity in the same single-phase solution. Amphiphilic drugs can also be incorporated into the microemulsion, which can increase the microemulsion region.

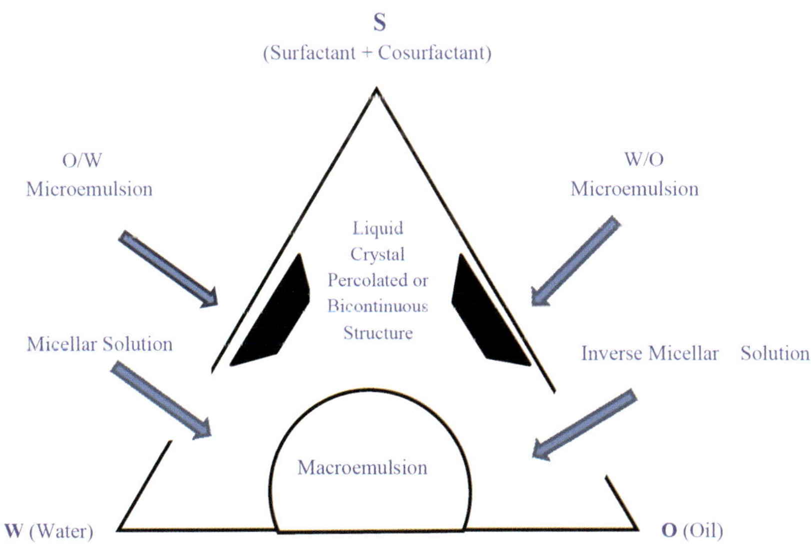


Figure 8. Hypothetical phase regions of microemulsion system. Taken from Ref. [111] with permission of Jadoun Science Publishing Group India.

## A. Microemulsion in pharmaceuticals

Microemulsions deliver drugs through several routes of entry, which are discussed as follows:

### 1. *Transdermal route*

Over the years topical application of microemulsion has been an area of increasing interest. There are a number of reasons that topical administration of drugs can be advantageous over other approaches. Topical administration has the capacity to enable the avoidance of hepatic first pass metabolism and thus reducing toxicity effects. Another advantage is the ability to directly deliver the drug to the affected area, such as the skin or eyes.[113]

The three main properties that determine transdermal permeation of drugs include, the mobility of drug within the vehicle, the release of drug from the vehicle, and permeation of the drug into the skin. Microemulsion represents an approach, which has the ability to improve transdermal delivery of numerous drugs over standard formulations such as emulsions[114] and gels.[115] Microemulsions also have more facile drug mobility,[116] in comparison with gel former, which increases the viscosity and thus will decrease the permeation in the skin.[117] The high solubilization properties of lipophilic and hydrophilic drugs give microemulsions superior transdermal delivery over many other approaches. Microemulsions are also able to generate increased thermodynamic activity towards the skin[114,116] and microemulsions may affect drug permeability in the skin. Furthermore, the components of microemulsions can serve as permeation enhancers. It is known that several compounds commonly used for microemulsion preparation have the ability to alter the structure of the stratum corneum thereby improving transdermal permeation. For instance, short chain alkanols have been widely used as permeation enhancers.[118]

### 2. *Nasal route*

Sheetal *et al.*[119] demonstrated in 2013 that microemulsions can enhance uptake across the nasal mucosa, and characterized carbamazepine as a trans-nasal microemulsion. In order to accomplish this oleic acid was selected to use as oil, and Tween 80 and propylene glycol were used as surfactant and co-surfactant respectively based on their known solubilities. Pseudoternary phase diagrams were used to determine the optimized ratio of Tween 80: propylene glycol for microemulsion preparations. These microemulsions were evaluated based on globule size, viscosity, pH, conductivity, and percent transmittance. Additionally, an *ex vivo* diffusion study to determine the optimized microemulsion was performed on sheep nasal mucosa and was able to identify the diffusion flux and permeability coefficients. Pharmacodynamics performance was also identified in rats using electrically induced seizures. The nasal route has also proven useful for the administration of diazepam, because of its rapid onset of action during emergency treatment of status epilepticus.[120]

### 3. *Oral route*

The oral administration route of microemulsion preparations offer numerous benefits in comparison with conventional oral formulation including increased absorption, improved clinical potency, and decreased drug toxicity. Thus, microemulsion is reported as an ideal delivery tool for drugs such as steroids, hormones, diuretic, and antibiotics.[121] Development of effective oral DDS is critical for drug performance, because drug efficacy requires stability upon administration and solubility in the gastrointestinal fluid. Thus, the performance of the drug is dependent on the successful formulation of the finished product. Microemulsions can provide the solution to challenges that other approaches cannot address such as enhanced solubilization of poorly soluble drugs and the ability to overcome the dissolution related to bioavailability issues. Microemulsions are able to act as super solvents for drugs and can also be optimized to enable consistent bioavailability. Additionally, they have utility for the delivery of hydrophilic drugs such as proteins and peptides due to polar, nonpolar, and interfacial domains, which allow for the encapsulation of drugs with varying solubility.[10]

### 4. *Parenteral route*

Parenteral administration of drugs with limited solubility is a major challenge within the pharmaceutical industry, because this leads to a low efficiency of drug reaching the target site. Microemulsion formulations have several distinct advantages over conventional emulsion delivery systems when administered through the parenteral route, because microemulsions are cleared slower resulting in increased circulation times. Furthermore, both O/W and W/O microemulsions are used for parenteral delivery.[102] For example, Rhee *et al.* was able to formulate itraconazole that contained parenteral microemulsion, using an o/w micro emulsion system. The average droplet size of the prepared microemulsions was found to be less than 150 nm, and the results of the hemolysis test demonstrated no toxicity from this formulation to red blood cells. Other experiments investigated the pharmacokinetic profiles of the ITZ-microemulsion for itraconazole and its major metabolite, hydroxyitraconazole, and compared them to PEG 400 solution and cyclodextrin formulations in rats. In summary, these results exploit the potential of an ITZ-microemulsion formulation system for delivery using the parenteral route.[122]

### 5. *Ocular delivery*

Both ocular and pulmonary delivery approaches are often used for the treatment of eye diseases, with many drugs being delivered topically as well. The o/w microemulsion has previously been investigated for ocular administration, and was found to be able to dissolve poorly soluble drugs, increase absorption and improve release profile.[113] Some of the drawbacks of these systems are low corneal bioavailability and lack of efficiency in the posterior segment of ocular tissue. Since they have emerged microemulsions have been giving rise to promising formulations for ocular administration and have achieved improved patient compliance making them favorable for ophthalmological use.[123] Okur *et al.*[124] through the investigation of the effect of ofloxacin-loaded microemulsion formulations using ocular delivery determined that OFX microemulsions represent a promising strategy for ocular drug delivery.

### 6. *Tumor targeting*

Microemulsions have demonstrated utility as delivery vehicles for chemotherapeutic or diagnostic agents to specifically treat neoplastic cells.[125] One method for treating neoplasms is to increase treatment specificity, being that neoplastic cells have an increased number of low density lipoprotein (LDL) receptors on their cell surface as compared to normal cells. In order to take advantage of this, a microemulsion was created comprised of a nucleus of cholesterol esters, ~20% triglycerides, surrounded by a core of phospholipids, free cholesterol and containing a chemotherapeutic drug. Microemulsions were then incorporated into cells using the receptors for LDL to target and deliver the incorporated molecules. This resulted in a higher concentration of anticancer drugs being delivered into the neoplastic cells, which have overexpression of the LDL receptors. Using the LDL targeting system promoted specificity, which was able to reduce toxic effects on normal tissues and organs. Additionally, Shiokawa *et al.*[126] was able to demonstrate targeted ACM delivery using Folate-linked microemulsions. This work showed that this folate modification, if a sufficiently long PEG chain on emulsions is used, promotes selective targeting of tumor cells.

### 7. *Cancer treatment*

Paclitaxel microemulsion was the first microemulsion cancer treatment reported by He *et al.*[127] Of particular interest, paclitaxel microemulsion had significantly less hypersensitivity reactions in comparison to Taxol® as a result of the very low Cremophor® EL content. Therefore, if appropriately designed microemulsions systems are employed, there is improvement in both tolerability and efficacy for therapeutic agents.[128]

O/W emulsions have numerous advantages for formulating lipophilic anti-cancer drugs in the sub-micron range. Since the oil phase solubilizes the lipophilic compounds, lipophilic drugs can be enhanced in this emulsion system, which leads to smaller administration volumes in comparison with the aqueous solution. Emulsion systems have demonstrated promise in cancer chemotherapy, as vehicles for prolonging drug release post intramuscular or intratumoral injection.[129]

## B. Microemulsion challenges

There are multiple challenges that arise from using a microemulsion system. Microemulsion formulation is a very costly process, because the size reduction of droplets is extremely difficult and requires the use of special instruments and process methods.[130] For example, instruments such as homogenizers (an instrument required for the micro-emulsion formulation) and processes such as micro fluidization and ultrasonication (manufacturing process) require significant investment. Microemulsions are also known to be quite unstable, which is problematic for the long-term storage of formulations. The stability problem of micro-emulsions stems from a process called Ostwald ripening. This process occurs as a result of the high rate of curvature of the small droplet, which yields a greater solubility than large droplets with a low radius of curvature.[130]

# V. Nanotechnology for Drug Delivery Systems

NPs represent a particularly attractive drug delivery approach for drug delivery for several reasons including composition, structure, and surface characteristics.[131] The NP composition can be widely varied, which allows the carriers to be modified for specific applications and target specificity. Polymeric and inorganic NPs are the most common base design for targeted drug delivery applications. Polymeric NPs can be sub-categorized into liposomes, micelles, dendrimers, nanospheres, and nanocapsules. This section will highlight the benefits and weaknesses of these different NP systems and their use as drug delivery vehicles.

## A. Polymers in drug delivery

The ability to engineer polymeric nanostructures such as hyperbranched polymers, dendrimers, and polymeric micelles is a rapidly growing area of biomaterials science, due to their unique characteristics and vast potential for drug delivery.[132–134] An important component of any polymer to be used in drug delivery, is that the polymer must be biocompatible. Moreover, biocompatible polymers used for drug delivery are often biodegradable and have byproducts, which are non-harmful, such as non-toxic alcohols, acids, and other relatively easily eliminated products. These byproducts can then contribute to drug release properties as a result of their erosion/degradation, not to mention drug diffusion through polymeric material. A biodegradable polymer to be used for the development of DDS must meet very specific requirements such as:

- Biocompatible backbone of the polymer and its degradation products
- Mechanical strength sufficient to meet the needs of specific applications
- Degradability with degradation kinetics matching a biological process
- Processability using available equipment
- Solubility in various solvents
- Chemical, structural and application versatility
- Economically acceptable shelf life
- European Medicine Evaluation Agency (EMEA) or FDA

Polymers that respond to stimulus, such as "intelligent", "smart" or "environmentally sensitive" polymers, are systems that have the ability to exhibit marked changes in response to physical stimuli (such as temperature, solvents or light) or to chemical stimuli (such as reactants, pH, ions in solution or chemical recognition). The responses may vary depending on the particular stimulus applied, which could be changes in shape, volume, mechanical properties or permeation rates as well as others. These response systems have various applications including encapsulation, controlled delivery or as intelligent switches.[135] Some common polymers used for drug delivery include polymeric micelles, polymeric NPs and dendrimers. This section is focused on polymeric NPs.

## 1. *Liposomes*

The amphiphilic molecules of liposomes are indicated in Figure 9a and are discussed in previous sections. The self-assembly that liposomes produce are well characterized and can be described as spherical in structure, with the polar components of the molecule in

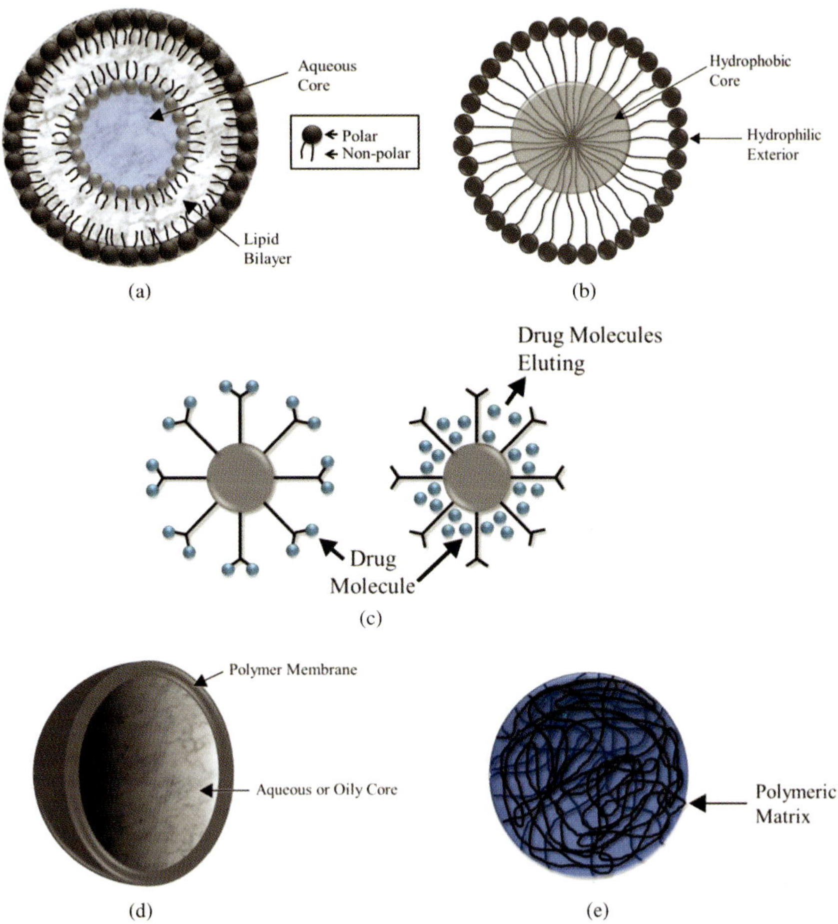


Figure 9. Particle schematics. (a) Liposome, (b) micelle, (c) dendrimers functionalized with complexed (left) and encapsulated (right) drug molecules, (d) nanosphere, and (e) nanocapsule. Taken from Ref. [140] with permission of Elsevier.

contact with the polar environment and the nonpolar components in contact with the nonpolar environment.[40] As previously discussed in other sections, liposomes are very attractive for drug delivery applications for a number of reasons. A thorough discussion on the topic of liposomal DDS can be found in previous sections of this chapter.

### 2. *Micelles*

A micelle can be defined as a set of amphiphilic molecules, which self-assemble into a structure containing a hydrophobic core and a hydrophilic exterior (Figure 9b).[131] The structure of the micelle is particularly useful for drug delivery applications for a number of reasons. Micelles generally will have diameters in the range of less than 100 nm, which permits extravasation through the fenestrations of tumor vessels and reduces uptake by the MPS/RES system. Micelles are also shielded from recognition during circulation as a result of their hydrophilic surface, which results in increased circulation time in the body.[136] Additionally, hydrophobic drugs can be loaded as the core of the micelle, which will now be protected by the hydrophilic corona during transport to the tumor site.[137]

### 3. *Dendrimers*

A dendrimer is characterized as a highly branched molecule that displays a well-defined structure and a high degree of monodispersity.[138] Dendrimers are very stable with surfaces that are readily functionalized with targeting ligands such as folic acid.[139] Drug molecules that are encapsulated in the dendrimer's multifunctional core and are thereby protected by the extensive branching. Drugs such as paclitaxel have also been attached to the exterior of the dendrimer (Figure 9c).[139]

### 4. *Nanospheres and nanocapsules*

Nanospheres are described as a spherical polymeric matrix, which encapsulates a drug (Figure 9d). The drug is first distributed evenly throughout the matrix and subsequently released into the environment through the mechanism of diffusion. The polymer matrix composition as well as its ability to imbibe fluids will control the drug release rate.[141] Nanocapsules have often been referred to as reservoir systems, since they contain the active ingredient at the core, which is protected from the environment by a polymeric membrane (Figure 9e).[15] Once the core has become saturated, active ingredient(s) can diffuse through the membrane at an approximately constant release rate.[142] The release behavior of these particles is particularly attractive for many drug delivery applications. These NP systems have been extensively investigated for diffusion-driven drug release using their large surface-to-volume ratios, which enables drug release at clinically relevant time scales. However, there has been a surge in development for NP systems, which use alternative mechanisms of drug release other than diffusion. Rather, this new class of NP has the ability to respond to environmental, chemical, thermal or biological cues.[143,144] These "smart materials" release their therapeutic payload only upon being triggered by specific stimulus.

## B. Inorganic nanoparticles

Inorganic NPs are described as particles of metal oxide or metallic composition with at least one length scale in the nanometer range. These nanostructures generally exhibit significantly distinct chemical, physical, biological and functional properties due to their nanoscale size. However, the preparation of inorganic NPs comes with several challenges. The production process can be very challenging for NPs and most procedures tend to differ substantially between institutions or industrial scale laboratories.[145,146] One of the most common methods for NP synthesis is to use the sol–gel route in which the preparation of a solution of inorganic precursor has its particle growth determined by the thermal or pH conditions of the solution. Typical inorganic precursors include metal salts, metal halides and inorganic alkoxides. These precursors are then synthesized by hydrolysis and condensation reactions to produce the desired metal oxide species. Using mineralizers (acids or bases) allows for the control of both hydrolysis and condensation rates independently, switching from kinetic-based to equilibrium based particle growth mechanisms, and ultimately enables controlled growth rates of various facets versus others.[147] The use of the spray-pyrolysis process is an example of a scalable technique.[148] In this method, a homogenized precursor solution composed of the inorganic compounds and relevant additives are sprayed and brought to temperatures at or about the boiling point of the solvent within a specially designed chamber. Precursor solutions are then atomized through a nozzle into droplets with the use of flowing gas. Microemulsion processing is another method used for the preparation of NPs. Additionally, microemulsions have also been used for the production of magnetic NPs[12] and superconductor NPs.[149] Microemulsions can be produced spontaneously, that is without the need for significant mechanical agitation, making it a mostly straightforward technique. However, in order for a type of NP to be utilized clinically, concerns associated with long-term tissue damage, toxicity, immunogenicity, carcinogenesis, and inflammation of inorganic NPs need to be elucidated. Inorganic NPs need to be designed such that stability, circulation times and localization can be modified without compromising theranostic efficacies.[150] Furthermore, gold NPs, quantum dots, carbon nanotubes, magnetic NPs, and ceramic NPs are widely used for cancer therapy and imaging.[150] In particular, among inorganic NPs, superparamagnetic iron oxide (SPIONs) magnetic materials have received ever increasing attention and utilization.

### 1. *SPIONs*[151]

The discovery of new or improved methods for the synthesis of SPIONs is a major focus in this field in order to obtain reliable/reproducible SPIONs with optimum surface charge, shape, size, colloidal stability in a biological environment, biocompatibility, and saturation magnetization. One important issue that always needs careful attention when selecting SPIONs for drug delivery or other purposes is the fate of the SPIONs after the drug delivery.[151] For example, poly(methyl methacrylate) is biocompatible, but its byproducts, such as methacrylate monomer, could be reactive and toxic. Additionally, the selection of SPIONs for drug specific loading should be determined based on how the drug and shell materials complement each other; otherwise the combination of drug and shell materials could result in a burst effect producing toxic chemicals. A comprehensive understanding of various interactions of SPIONs with biomolecules within the body (i.e. protein–NP interactions) is required in order to design SPIONs using optimum surface properties, which overcome problems relating to rejection of magnetic NPs, biocompatibility and toxicity in future studies, and treatments. Ultimately SPIONs in biomedicine can be used to improve patient care by increasing efficacy while avoiding adverse effects. These approaches can also be used to treat various stages of cancer, especially early stage metastatic events (Figure 10).

As an approach to increase the targeting efficacy of SPIONs to brain tissue, a number of polymers could be useful (e.g. n-hexadecylcyanoacrylate and polyethylene glycol), and they have demonstrated the potential for crossing the BBB as coating materials. The ability to traverse the BBB offers great enhancement in delivery efficiency and utility for transferring drugs (e.g. antidepressants) into deep areas of the brain, such as the striatum, hippocampus, and hypothalamus.[151]

## C. Nanotechnology challenges

Nanotechnology will seemingly assume an essential role in drug delivery and greatly improve human therapeutics. While the development of DDS is just beginning, it has an extremely promising future. Novel nanotechnology based approaches provide opportunities for scientists to develop systems that are becoming ever closer to the sophistication and precision of naturally occurring biological structures.

Nanotechnology has been an emerging field that can potentially change the way diseases are treated through enhanced drug delivery. However, this technology suffers from the following challenges[152]:

- Lack of knowledge regarding NP components and their characteristics
- Lack of uniformity of toxicity

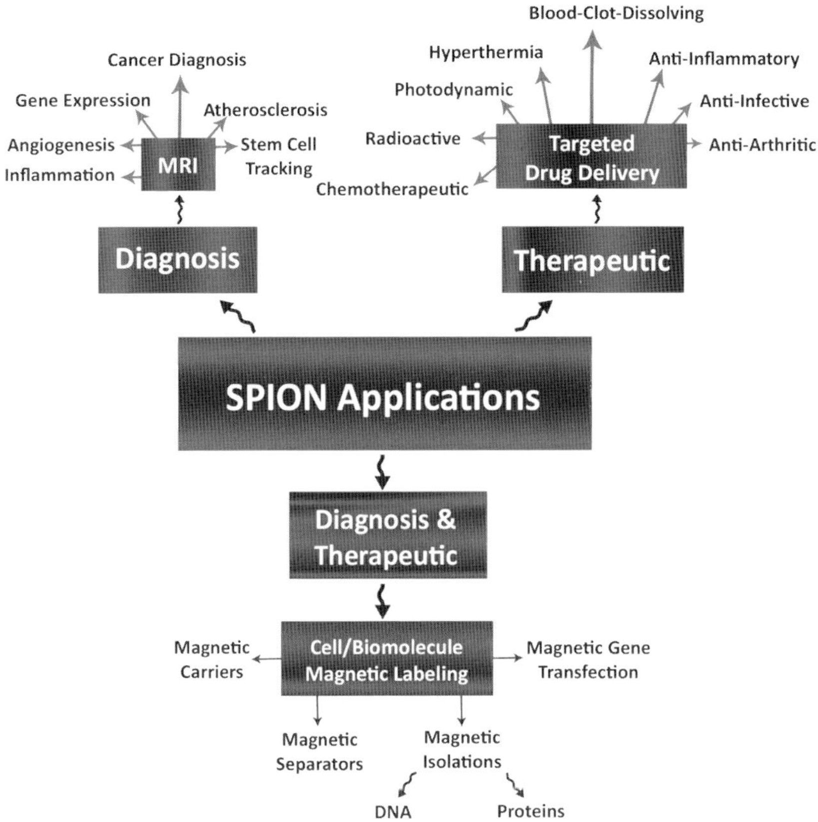


Figure 10. Recent diagnostic and therapeutic applications of SPIONs. Taken from Ref. [151] with permission of Elsevier.

- Lack of standardization in model systems and assays
- Lack of standard synthesis protocols
- Lack of efficient analytical tools
- Lack of understanding of the impact on biological systems
- Lack of *in vivo* monitoring systems
- Lack of standardized safety guidelines
- Lack of well-trained workforce

Significant challenges will need to be addressed before this field can progress into developing clinically viable therapies. One of the current barriers to translating these technologies is the inability to standardize the design and testing of how these nanomaterials will behave once they enter the body. Other methods that require refinement is the ability to target nanomaterials to specific sites of the body while avoiding nonspecific interactions with organs such as the liver and spleen.

## VI. Concluding Remarks

Liposomes have many applications as DDS; their usefulness stems from the fact that the liposomes are exceptionally versatile in use. Liposomes have many distinct advantages including ease of chemical modifications, the ability to carry drugs/genes of different kinds and the potential administration through different routes. The ability to safely transport drugs to the target site and even trigger controlled release represents an exceptionally powerful therapeutic system. Novel applications in diagnosis and even theranostics have been realized. Tremendous advancements in liposome technology have been demonstrated since their discovery in 1965. However, there is still much work to do in order to continue the advancement of liposomes as there has only been limited demonstration of translational success from bench to bedside.[153]

The transdermal drug delivery route offers many advantages and represents a hugely important method of delivering drugs through the skin. The skin having such large surface area and accessibility provides for a convenient and patient-friendly target for drug delivery. Circumvention of the gastrointestinal tract, and limited effects of hepatic first-pass metabolism and bioavailability may be improved with this approach. Additionally, many of the side effects, which occur during other delivery methods, may be reduced. In contrast to an oral tablet or injectable, the external nature of the delivery system allows for greater control over the administered dosage. This drives the motivation to investigating the possibilities of the transdermal route. While considerable progress has been made towards novel enhancement methodologies, there is still room to grow in fully exploiting the commercial potential of the transdermal drug delivery market. To ensure continued progression using the transdermal route for drug delivery, the most promising solution is the ability to develop improved technologies to overcome the limiting barriers of the skin. There has been an abundance of evidence to support enhanced skin permeability using these techniques for a variety of drug compounds and now moving forward the task will be to translate this into successful marketed products. There are many lessons to be learned from the few approvals granted by the FDA over the years. Particular attention should be given to the eventual fate of these products, compared to product ideas that didn't progress to the market, whether that be for financial reasons, safety concerns or technical complexity challenges.[95]

Microemulsion systems provide novel solutions, which overcome many challenges in the field. However, some of the barriers to market entry such as poor aqueous solubility of highly lipophilic drug compounds and more consistent and reproducible bioavailability will need to be resolved. Microemulsions have proven efficacy for oral administration of cyclosporine (Neoral), however microemulsion based technologies need not be confined to oral routes of administration. Additionally, microemulsions can be easily used for developing novel cargoes for phytoactives and extracts as well. There is great potential in developing effective targeting therapeutics using multiple aspects of microemulsions using various targeting molecules and due to their tremendous capacity for carrying lipophilic drugs. Several targeting cargoes have been developed in recent years using the concept of microemulsions as the foundation. It appears feasible that in the near future microemulsions will become a highly versatile system for standard and specialized purposes such as phytoactive applications. It is projected that the potential of microemulsions as novel compartmentalized liquids will become even more significant in the future.[10]

Nanotechnology offers tremendous potential for the development of novel products and incredibly powerful nanomedicines and nanodevices. These development processes are heavily interfaced with biotechnology and information technology, making its scope broad reaching. Nanotechnology based products have the capability to overcome many of the limitation of conventional methods, however, several major challenges remain including toxicity, environmental hazards, production cost and accessibility.[152]

## References


1. Verma, R. K.; Garg, S.; Directions, F. *Pharm. Technol.* **2001**, *25* (February), 1.
2. Almeida, A. J.; Souto, E. *Adv. Drug Deliv. Rev.* **2007**, *59* (6), 478.
3. Allen, T. M.; Cullis, P. R. *Science* **2004**, *303* (5665), 1818.
4. Zhang, Y.; Chan, H. F.; Leong, K. W. *Adv. Drug Deliv. Rev.* **2013**, *65* (1), 104.
5. Torchilin, V. P. *Eur. J. Pharm. Sci.* **2000**, *11 (2)*, S81.
6. Tiwari, G.; Tiwari, R.; Sriwastawa, B.; Bhati, L.; Pandey, S.; Pandey, P.; Bannerjee, S. K. *Int. J. Pharm. Investig.* **2012**, *2* (1), 2.
7. Langer, M. R. P. R. *Nat. Biotechno.* **2009**, *26* (11), 1261.
8. Gaudana, R.; Ananthula, H. K.; Parenky, A.; Mitra, A. K. *AAPS J.* **2010**, *12* (3), 348.
9. Illum, L. *Drug Discov. Today* **2002**, *7* (23), 1184.
10. Sahu G. K.; Sharma, H.; Gupta, A.; Kaur, C. D. *Int. J. Pharm. Sci. Dev. Res.* **2015**, *1*, 8.
11. Kreuter, J. *Adv. Drug Deliv. Rev.* **2014**, *71*, 2.
12. Jafari, T.; Simchi, A.; Khakpash, N. *J. Colloid Interface Sci.* **2010**, *345* (1), 64.
13. Schipper, M. L.; Iyer, G.; Koh, A. L.; Cheng, Z.; Ebenstein, Y.; Aharoni, A.; Keren, S.; Bentolila, L. A.; Li, J.; Rao, J.; Chen, X.; Banin, U.; Wu, A. M.; Sinclair, R.; Weiss, S.; Gambhir, S. S. *Small* **2009**, *5* (1), 126.
14. Lasic, D. D. *Trends Biotechnol.* **1998**, *16* (7), 307.
15. Haley, B.; Frenkel, E. *Urol. Oncol. Semin. Orig. Investig.* **2008**, *26* (1), 57.
16. Jesorka, A.; Orwar, O. *Annu. Rev. Anal. Chem.* **2008**, *1*, 801.
17. Klibanov, A. L.; Maruyama, K.; Torchilin, V. P.; Huang, L. *FEBS Lett.* **1990**, *268* (1), 235.
18. Bangham, A. D.; Standish, M. M.; Watkins, J. C. *J. Mol. Biol.* **1965**, *13* (1), 238.
19. Deamer, D. W. *FASEB J.* **2010**, *24* (5), 1308.
20. Gregoriadis, G. *FEBS Lett.* **1973**, *36* (3), 292.
21. Gregoriadis, G.; Ryman, B. E. *Biochem. J.* **1971**, *124* (5), 58.
22. Poste, G.; Papahadjopoulos, D. *Proc. Natl. Acad. Sci.* **1976**, *73* (5), 1603.

23. Lian, T.; Ho, R. J. Y. *J. Pharm. Sci.* **2001**, *90* (6), 667.
24. Zhigaltsev, I. V.; Belliveau, N.; Hafez, I.; Leung, A. K. K.; Huft, J.; Hansen, C.; Cullis, P. R. *Langmuir* **2012**, *28* (7), 3633.
25. Allen, T. M.; Everest, J. M. *J. Pharmacol. Exp. Ther.* **1983**, *226* (2), 539.
26. Maeda, H.; Wu, J.; Sawa, T.; Matsumura, Y.; Hori, K. *J. Control. Release* **2000**, *65* (1–2), 271.
27. Nagayasu, A.; Uchiyama, K.; Kiwada, H. *Adv. Drug Deliv. Rev.* **1999**, *40* (1–2), 75.
28. Gill, P. S.; Wernz, J.; Scadden, D. T.; Cohen, P.; Mukwaya, G. M.; von Roenn, J. H.; Jacobs, M.; Kempin, S.; Silverberg, I.; Gonzales, G.; Rarick, M. U.; Myers, A. M.; Shepherd, F.; Sawka, C.; Pike, M. C.; Ross, M. E. *J. Clin. Oncol.* **1996**, *14* (8), 2353.
29. Gabizon, A.; Papahadjopoulos, D. *Proc. Natl. Acad. Sci. U.S.A.* **1988**, *85* (18), 6949.
30. Torchilin, V. *Adv. Drug Deliv. Rev.* **2011**, *63* (3), 131.
31. Papisov, M. I. *Adv. Drug Deliv. Rev.* **1998**, *32* (1–2), 119.
32. Blume, G.; Cevc, G. *BBA — Biomembr.* **1993**, *1146* (2), 157.
33. Connor, J.; Huang, L. *Cancer Res.* **1986**, *46* (7), 3431.
34. Carter, G.; White, P.; Fernie, M.; King, S.; McLean, G.; Titball, R.; Carr, F. J. *Int. J. Oncol.* **1998**, *13* (4), 819.
35. Gabizon, A. *J. Liposome Res.* **2003**, *13* (1), 17.
36. Bakker-Woudenberg, IAJM, *Int. J. Antimicrob. Ag.* **2002**, *19* (4), 299.
37. Moghimi, S. M.; Szebeni, J. *Prog. Lipid Res.* **2003**, *42* (6), 463.
38. Dwivedi, C.; Sahu, R.; Tiwari, S. P.; Satapathy, T.; Roy, A. *J. Drug Deliv. Ther.* **2009**, *4* (2), 116.
39. Ebrahim, S.; Peyman, G. A.; Lee, P. J. *Surv. Ophthalmol.* **2005**, *50* (2), 167.
40. Lasic, D. D. *Trends Biotechnol.* **1998**, *16* (7), 307.
41. Sercombe, L.; Veerati, T.; Moheimani, F.; Wu, S. Y.; Sood, A. K.; Hua, S. *Front. Pharmacol.* **2015**, *6* (DEC), 1.
42. Jain, R. K. *Annu. Rev. Biomed. Eng.* **1999**, *1* (96), 241.
43. Hobbs, S. K.; Monsky, W. L.; Yuan, F.; Roberts, W. G.; Griffith, L.; Torchilin, V. P.; Jain, R. K. *Proc. Natl. Acad. Sci. U.S.A.* **1998**, *95* (8), 4607.
44. Torchilin, V. P. *Nat. Rev. Drug Discov.* **2005**, *4* (2), 145.

45. Sawant, R. R.; Torchilin, V. P. *AAPS J.* **2012**, *14* (2), 303.
46. Fang, J.; Nakamura, H.; Maeda, H. *Adv. Drug Deliv. Rev.* **2011**, *63* (3), 136.
47. Crielaard, B. J.; Lammers, T.; Schiffelers, R. M.; Storm, G. *J. Control. Release* **2012**, *161* (2), 225.
48. Hong, R.; Huang, C.; Tseng, Y. *Clin. Cancer Res.* **1999**, *5* (November), 3645.
49. Yokoi, K.; Tanei, T.; Godin, B.; van de Ven, A.; Alexander, J.; Ferrari, M. *Cancer Res.* **2014**, *73* (8), 4973.
50. Bae, Y. *J. Control. Release Off. J.* **2009**, *133* (1), 2.
51. Patel, N. R.; Pattni, B. S.; Abouzeid, A. H.; Torchilin, V. P. *Adv. Drug Deliv. Rev.* **2013**, *65* (13–14), 1748.
52. Heldin, C.-H.; Rubin, K.; Pietras, K.; Ostman, A. *Nat. Rev. Cancer* **2004**, *4* (10), 806.
53. Ferrari, M. *Nat. Rev. Cancer* **2005**, *5* (3), 161.
54. Drummond, D. C.; Meyer, O.; Hong, K.; Kirpotin, D. B.; Papahadjopoulos, D. *Pharmacol. Rev.* **1999**, *51* (4), 691.
55. Van Rooy, I.; Hennink, W. E.; Storm, G.; Schiffelers, R. M.; Mastrobattista, E. *Eur. J. Pharm. Sci.* **2012**, *45* (3), 330.
56. Kirpotin, D.; Park, J. W.; Hong, K.; Zalipsky, S.; Li, W. L.; Carter, P.; Benz, C. C.; Papahadjopoulos, D. *Biochemistry* **1997**, *36* (1), 66.
57. Sapra, P.; Allenz, T. M. *Cancer Res.* **2002**, *62* (24), 7190.
58. Liebert, M.; Wedemeyer, G. A; Stein, J. A.; Cells, C.; Washington, R. V.; Flint, A.; Ren, L.; Grossman, H. B. *Cancer Res.* **1989**, *49*, 6720.
59. Park, J. W.; Hong, K.; Kirpotin, D. B.; Colbern, G.; Shalaby, R.; Baselga, J.; Shao, Y.; Nielsen, U. B.; Marks, J. D.; Moore, D.; Papahadjopoulos, D.; Benz, C. C. *Clin. Cancer Res.* **2002**, *8* (April), 1172.
60. Kirpotin, D. B.; Drummond, D. C.; Shao, Y.; Shalaby, M. R.; Hong, K.; Nielsen, U. B.; Marks, J. D.; Benz, C. C.; Park, J. W. *Cancer Res.* **2006**, *66* (13), 6732.
61. Sofou, S.; Sgouros, G. *Expert Opin. Drug Deliv.* **2008**, *5* (2), 189.
62. Munster, P.; Krop, I.; Miller, K.; Dhindsa, N.; Niyijiza, C.; Nielsen, U.; Odueyungbo, A.; Rajarethinam, A.; Marande, M.; Campbell, K.; Geretti, E.; Reynolds, J.; Hendriks, B.; Wickham, T.; Moyo, V.; LoRusso, P. *Cancer Res.* **2014**, *73* (24 Supplement), P4.

63. Elbayoumi, T. A.; Torchilin, V. P. *Clin. Cancer Res.* **2009**, *15* (6), 1973.
64. Matsumura, Y.; Gotoh, M.; Muro, K.; Yamada, Y.; Shirao, K.; Shimada, Y.; Okuwa, M.; Matsumoto, S.; Miyata, Y.; Ohkura, H.; Chin, K.; Baba, S.; Yamao, T.; Kannami, A.; Takamatsu, Y.; Ito, K.; Takahashi, K. *Ann. Oncol.* **2004**, *15* (3), 517.
65. Senzer, N. N.; Matsuno, K.; Yamagata, N.; Fujisawa, T.; Wasserman, E.; Sutherland, W.; Sharma, S.; Phan, A. *Mol. Cancer Ther.* **2009**, *8* (1 Supplement), C36.
66. Riviere, K.; Huang, Z.; Jerger, K.; Macaraeg, N.; Szoka, F. C. *J. Drug Target.* **2011**, *19* (1), 14.
67. Forma, I.; Ejemplo, C. US Patent 2013/0071321 A1, **2013**.
68. Copolovici, D. M.; Langel, K.; Eriste, E.; Langel, Ü. *ACS Nano* **2014**, *8* (3), 1972.
69. Pappalardo, J. S.; Quattrocchi, V.; Langellotti, C.; Di Giacomo, S.; Gnazzo, V.; Olivera, V.; Calamante, G.; Zamorano, P. I.; Levchenko, T. S.; Torchilin, V. P. *J. Control. Release* **2009**, *134* (1), 41.
70. Biswas, S.; Deshpande, P. P.; Perche, F.; Dodwadkar, N. S.; Sane, S. D.; Torchilin, V. P. *Cancer Lett.* **2013**, *335* (1), 191.
71. Goren, D.; Horowitz, A. T.; Zalipsky, S.; Woodle, M. C.; Yarden, Y.; Gabizon, A. *Br. J. Cancer* **1996**, *74* (11), 1749.
72. Smistad, G.; Jacobsen, J.; Sande, S. A. *Int. J. Pharm.* **2007**, *330* (1–2), 14.
73. Dokka, S.; Toledo, D.; Shi, X.; Castranova, V.; Rojanasakul, Y. *Pharm. Res.* **2000**, *17* (5), 521.
74. Sastry, S. V.; Nyshadham, J. R.; Fix, J. A. *Pharm. Sci. Technolo. Today* **2000**, *3* (4), 138.
75. Wokovich, A. M.; Prodduturi, S.; Doub, W. H.; Hussain, A. S.; Buhse, L. F. *Eur. J. Pharm. Biopharm.* **2006**, *64* (1), 1.
76. Wu, X. Y.; Zhou, Y. *J. Control. Release* **1998**, *51*, 57.
77. Patel, M. P.; Gupta, M. M. *Pharma Innov. J.* **2013**, *2* (3), 149.
78. Benson, H. *Curr. Drug Deliv.* **2005**, *2* (1), 23.
79. Prausnitz, M. R. *Adv. Drug Deliv. Rev.* **2004**, *56* (5), 581.
80. Kogan, A.; Garti, N. *Adv. Colloid Interface Sci.* **2006**, *123–126* (SPEC. ISS.), 369.
81. Gentile, L.; Behrens, M. A.; Porcar, L.; Butler, P.; Wagner, N. J.; Olsson, U. *Langmuir* **2014**, *30* (28), 8316.

82. Pommella, A.; Caserta, S.; Guida, V.; Guido, S. *Phys. Rev. Lett.* **2012**, *108* (13), 1.
83. Schreier, H.; Bouwstra, J. *J. Control. Release* **1994**, *30* (1), 1.
84. Förster, M.; Bolzinger, M.-A.; Fessi, H.; Briançon, S. *Eur. J. Dermatol.* **2009**, *19* (4), 309.
85. Godin, B.; Touitou, E. *Crit. Rev. Ther. Drug Carrier Syst.* **2003**, *20* (1), 63.
86. Muzzalupo, R.; Tavano, L.; Cassano, R.; Trombino, S.; Ferrarelli, T.; Picci, N. *Eur. J. Pharm. Biopharm.* **2011**, *79* (1), 28.
87. Rawat, M.; Singh, D.; Saraf, S.; Saraf, S. *Biol. Pharm. Bull.* **2006**, *29* (9), 1790.
88. Shim, J.; Kang, H. S.; Park, W. S.; Han, S. H.; Kim, J.; Chang, I. S. *J. Control. Release* **2004**, *97* (3), 477.
89. Dragicevic, N., Maibach, H. I., Eds.; Springer Berlin Heidelberg: Berlin, Heidelberg, **2015**.
90. Knorr, F.; Lademann, J.; Patzelt, A.; Sterry, W.; Blume-Peytavi, U.; Vogt, A. *Eur. J. Pharm. Biopharm.* **2009**, *71* (2), 173.
91. Pathak, A.; Shrivasatava, N.; Baghel, S. *World J. Pharm. Sci.*, **2014**, *3* (i), 422.
92. Geethu, V. S.; Manikandan, P.; Nethaji, R.; Surenderan, N.S.; Babu, G. *Pharma Innov. J.* **2015**, *4* (1), 37.
93. Singh, P.; Roberts, M. S. *J. Pharm. Exp. Ther.* **1994**, 144.
94. Boccia, R. V.; Gordan, L. N.; Clark, G.; Howell, J. D.; Grunberg, S. M. *Support. Care Cancer* **2011**, *19* (10), 1609.
95. Watkinson, A. C.; Kearney, M.-C.; Quinn, H. L.; Courtenay, A. J.; Donnelly, R. F. *Expert Opin. Drug Deliv.* **2016**, *5247* (January).
96. Walter, J. R.; Xu, S. *Drug Discov. Today* **2015**, *20* (11), 1293.
97. Wiedersberg, S.; Guy, R. H. *J. Control. Release* **2014**, *190*, 150.
98. Donnelly, R. F.; Moffatt, K.; Alkilani, A. Z. A.; Vicente-Pérez, E. M.; Barry, J.; McCrudden, M. T. C.; Woolfson, A. D. *Pharm. Res.* **2014**, *31* (8), 1989.
99. Sharma, N.; Agarwal, G.; RANA, A.C.; Alibhat, Z.; Kumar, D. *Int. J. Drug Dev. Res.* **2011**, *3* (3), 189.
100. Vyas, S.; Khar, R. *vallabh prakashan* Delhi (India) **2002**.
101. Patel, D.; Chaudhary, S. A; Parmar, B.; Bhura, N. *Pharma Innov.* **2012**, *1* (4), 78.

102. Hoar, T. P.; Schulman, J. H. *Nature.* **1943**, *152*, 102.
103. Schulman, J. H.; Stoeckenius, W.; Prince, L. M. *J. Phys. Chem.* **1959**, *63* (10), 1677.
104. Shinoda, K.; Lindman, B. *Langmuir* **1987**, *3* (2), 135.
105. Lawrence, M. J. *Adv. Colloid Interface Sci.* **2006**, *123–126* (SPEC. ISS.), 369.
106. Sarciaux, J. M.; Acar, L.; Sado, P. A. *Int. J. Pharm.* **1995**, *120* (2), 127.
107. Gasco, M. R.; Pattarino, F.; Lattanzi, F. *Int. J. Pharm.* **1990**, *62* (2–3), 119.
108. Lee, V. H. *Crit. Rev. Ther. Drug Carrier Syst.* **1988**, *5* (2), 69.
109. Constantinides, P. P. *Pharm. Res.* **1995**, *12* (11), 1561.
110. Malcolmson, C.; Satra, C.; Kantaria, S.; Sidhu, A.; Lawrence, M. J. *J. Pharm. Sci.* **1998**, *87* (1), 109.
111. Thakur, A.; Walia, M. K.; Kumar, S. L. H. *Pharmacophore* **2013**, *4* (1), 15.
112. Lawrence, M. J.; Rees, G. D. *Adv. Drug Deliv. Rev.* **2012**, *64* (Suppl.), 175.
113. Kumar, P.; Mittal, K. L. *Handbook of Microemulsion Science and Technology*; CRC Press, **1999**.
114. Kreilgaard, M.; Pedersen, E. J.; Jaroszewski, J. W. *J. Control. Release* **2000**, *69* (3), 421.
115 Kriwet, K.; Muller Goymann, C. C. *Int. J. Pharm.* **1995**, *125* (2), 231.
116. Alvarez-Figueroa, M. J.; Blanco-Méndez, J. *Int. J. Pharm.* **2001**, *215* (1–2), 57.
117. Gasco, M. R.; Gallarate, M.; Pattarino, F. *Int. J. Pharm.* **1991**, *69* (3), 193.
118. Kim, Y.-H.; Ghanem, A.-H.; Mahmoud, H.; Higuchi, W. I. *Int. J. Pharm.* **1992**, *80* (1–3), 17.
119. Acharya, S. P.; Pundarikakshudu, K.; Panchal, A.; Lalwani, A. *Drug Deliv. Transl. Res.* **2013**, *3* (3), 252.
120. Li, L.; Nandi, I.; Kim, K. H. *Int. J. Pharm.* **2002**, *237* (1–2), 77.
121. Agrawal, O. P.; Agrawal, S. *Asian J. Pharm. Sci. Technol.* **2012**, *2* (1), 5.
122. Rhee, Y.-S.; Park, C.-W.; Nam, T.-Y.; Shin, Y.-S.; Chi, S.-C.; Park, E.-S. *Arch. Pharm. Res.* **2007**, *30* (1), 114.
123. Ligório Fialho, S.; da Silva-Cunha, A. *Clin. Exp. Ophthalmol.* **2004**, *32* (6), 626.

124. Üstündag-Okur, N.; Gökçe, E. H.; Eğrilmez, S.; Özer, Ö.; Ertan, G. *J. Ocul. Pharmacol. Ther.* **2014**, *30* (4), 319.
125. Maranhao, R. C.; Zerbini, F. E. J. Microemulsions used as vehicles for carrying chemotherapeutic agents to neoplastic cells. US Patent 5578583 A, Nov 26, **1996**.
126. Shiokawa, T.; Hattori, Y.; Kawano, K.; Ohguchi, Y. **2005**, *11*, 2018.
127. He, L.; Wang, G. L.; Zhang, Q. *Int. J. Pharm.* **2003**, *250* (1), 45.
128. Date, A. A.; Nagarsenker, M. S. *Int. J. Pharm.* **2008**, *355* (1–2), 19.
129. Joy, J.; Krishnakumar, K.; John, A.; Dineshkumar, B. **2015**, *5* (1), 1.
130. Muzaffar, F.; Singh, U. K.; Chauhan, L. *Int. J. Pharm. Pharm. Sci.* **2013**, *5* (3), 39.
131. Liechty, W. B.; Peppas, N. a. *Eur. J. Pharm. Biopharm.* **2012**, *80* (2), 241.
132. Xu, Q.; Liu, Y.; Su, S.; Li, W.; Chen, C.; Wu, Y. *Biomaterials* **2012**, *33* (5), 1627.
133. Kim, J. M.; Shin, D. A.; Byun, D. H.; Kim, H. S.; Kim, S.; Kim, H. I. *J. Korean Neurosurg. Soc.* **2012**, *52* (5), 435.
134. Lim, J.; Simanek, E. E. *Adv. Drug Deliv. Rev.* **2012**, *64* (9), 826.
135. York, A. W.; Kirkland, S. E.; McCormick, C. L. *Adv. Drug Deliv. Rev.* **2008**, *60* (9), 1018.
136. Lavasanifar, A.; Samuel, J.; Kwon, G. S. *Adv. Drug Deliv. Rev.* **2002**, *54* (2), 169.
137. Kataoka, K.; Kwon, G. S.; Yokoyama, M.; Okano, T.; Sakurai, S. *J. Control. Release* **1993**, *24* (1–3), 119.
138. Hughes, G. A. *Nanomedicine Nanotechnology, Biol. Med.* **2005**, *1* (1), 22.
139. Majoros, J.; Myc, A.; Thomas, T.; Mehta, C. B.; Baker, J. R. **2006**, No. Mwco 3500, 572.
140. Steichen, S. D.; Caldorera-Moore, M.; Peppas, N. A. *Eur. J. Pharm. Sci.* **2013**, *48* (3), 416.
141. Brigger, I.; Dubernet, C.; Couvreur, P. *Adv. Drug Deliv. Rev.* **2012**, *64*, 24.
142. Ratner, B. D.; Hoffman, A. S.; Schoen, F. J.; Lemons, J. E. *Biomaterials Science: An Introduction to Materials in Medicine*; Academic Press, **2004**.

143. Schoener, C. A.; Hutson, H. N.; Peppas, N. A. *J. Biomed. Mater. Res. — Part A* **2013**, *101 A* (8), 2229.
144. Caldorera-Moore, M.; Guimard, N.; Shi, L.; Roy, K. *Expert Opin. Drug Deliv.* **2010**, *7* (4), 479.
145. Khakpash, N.; Simchi, A.; Jafari, T. *J. Mater. Sci. Mater. Electron.* **2011**, *23* (3), 659.
146. Brinker, C. J.; Scherer, G. W. *Sol-Gel Science: The Physics and Chemistry of Sol-Gel Processing*; Elsevier Science, **2013**.
147. Lee, E. J. H.; Ribeiro, C.; Longo, E.; Leite, E. R. *Chem. Phys.* **2006**, *328* (1–3), 229.
148. Guild, C.; Biswas, S.; Meng, Y.; Jafari, T.; Gaffney, A. M.; Suib, S. L. *Catal. Today* **2014**, *238*, 87.
149. Zhao, C.; Krall, A.; Zhao, H.; Zhang, Q.; Li, Y. *Int. J. Hydrogen Energy* **2012**, *37* (13), 9967.
150. Huang, H.-C.; Barua, S.; Sharma, G.; Dey, S. K.; Rege, K. *J. Control. Release* **2011**, *155* (3), 344.
151. Mahmoudi, M.; Sant, S.; Wang, B.; Laurent, S.; Sen, T. *Adv. Drug Deliv. Rev.* **2011**, *63* (1–2), 24.
152. Pate, S. *Med. Chem.* **2015**, *5* (12), 528.
153. Pattni, B. S.; Chupin, V. V.; Torchilin, V. P. *Chem. Rev.* **2015**, *115* (19), 10938.

# Chapter 2

# Nanoparticles in Circulation: Blood Stability

Saeid Zanganeh*, Tahereh Jafari†, Nasser Khakpash‡, Mohsen Erfanzadeh§, and Jim Q. Ho¶

*Department of Radiology, Molecular Imaging Program at Stanford, Stanford University, CA, USA

†Institute of Materials Science, University of Connecticut, 97 North Eagleville Road, Storrs, CT, USA

‡Department of Chemical and Biomolecular Engineering, University of Connecticut, 191 Auditorium Road, Storrs, CT, USA

§Department of Biomedical Engineering, University of Connecticut, Storrs, CT, USA

¶Middlebury College, Middlebury, VT, USA

## Abbreviations

| | |
|---|---|
| FDA | Food and Drug Administration |
| ITIMs | immunoreceptor tyrosine-based inhibitory motifs |
| MPS | mononuclear phagocytic system |
| MRI | magnetic resonance imaging |
| PEG | polyethylene glycol |
| PCL | poly(-caprolactone) |
| RBCs | red blood cells |
| RES | reticuloendothelial system |
| SIRP$\alpha$ | signal regulatory protein $\alpha$ |
| SHP-1 | Src homology phosphatases 1 |
| SIRPa | signal regulatory protein alpha |

## I. Introduction

Nanoparticles have made a pivotal impact on biomedical applications including imaging, diagnostics, and drug delivery.[1,2] Liposomes, polymeric nanoparticles, micelles, and metallic colloids are various types of nanoparticles used to enhance the therapeutic efficacy and patient compliance of medications. The ability of nanoparticles to circulate in blood, target specific tissues, internalize into target cells, and release drugs can be specifically modified via engineering their material, size, shape, and surface chemistry.[3,4] The therapeutic potential of nanoparticles is strongly dependent on their ability to remain in the circulation. However, it is particularly challenging to maintain the nanoparticles in the circulation long enough to reach their therapeutic target. The reticuloendothelial system (RES), also known as mononuclear phagocytic system (MPS), rapidly removes most nanoparticles from the circulation.[5] Once the plasma proteins (opsonins) on nanoparticles are adsorbed, macrophages in liver and spleen recognize the nanoparticles, which results in phagocytosis and elimination of the particles.[6] Such rapid clearance of nanoparticles results in their short half-life and limited accumulation at the target site and limits their applications.

Various schemes have been investigated in order to (i) escape immune clearance to avoid non-target uptake and (ii) target specific tissues to deliver the appropriate amount of therapeutics to the pathological site.[7] These schemes include chemical modification of nanoparticles such as antibody functionalization for targeting or polyethylene glycol (PEG) functionalization for stealth,[8,9] physical modification of nanoparticles such as shape, size, and elasticity modifications,[10,11] and biologically-inspired modifications including mimicking attributes of natural carriers, such as blood cells.[12,13] However, in spite of the aforementioned advances, targeting and long-circulating nanoparticles with clinical applications are yet to be developed.[7,14,15]

## II. Nanoparticles for Drug Delivery

### A. Systemic delivery

For systemic applications, the circulation time of nanoparticles directly influences the drug delivery duration. Low stability, poor solubility, and undesirable pharmacokinetics and biodistribution are limitations of various types of drugs.[16] A variety of drugs can be encapsulated in nanoparticles to be protected from degradation within the body. Furthermore, nanoparticles can provide a controlled manner of releasing drugs in the blood compartment, hence improving the pharmacokinetics.[17,18]

### B. Targeted drug delivery

Similar to systemic delivery, the circulation time of nanoparticles directly affects their ability to accumulate at the target site for targeted delivery. This is because longer circulation time provides more contact time between the nanoparticles and the target tissue. The ultimate goal of most nanoparticle based therapeutics has been the targeted delivery. Providing focused accumulation in the target tissue, targeted nanoparticles reduce the side effects caused by toxic therapeutics. Both passive and active mechanisms have been developed for targeting nanoparticles. Passive targeting deploys natural physiological differences between different tissues and organs such as the difference between cancerous and healthy sites. Moreover, leaky vasculatures in tumors facilitate the nanoparticle delivery to the cancerous sites.[19,20] Conversely, active targeting utilizes ligands such as peptides and antibodies to direct nanoparticles to particular tissues. The efficacy of both passive and active targeting mechanisms is highly dependent on the prolonged circulation time of nanoparticles.

Variations in vascular physiology occur as a result of various tissue and pathological conditions and provide the possibility of deploying targeted long-circulating nanoparticles. Physical entrapment of nanoparticles can be best performed in tissues with leaky vasculature or organs, such as spleen, which are naturally responsible for clearing nanoparticles.[21] The altered vascular physiology of tumors has directed most efforts on passive targeting to the accumulation of nanoparticles in tumors.[22] The permeability of the vasculature around the tumors is significantly increased due to the abnormal blood flow pattern and insufficient lymphatic drainage in the tumor site. Therefore nanoparticles and liposomes can be trapped and accumulated in tumors.[21,23] Unfortunately, the organs of the RES and, in some cases, the bone marrow are the accumulation sites for a large percentage of nanoparticles. Therefore considerable tissue damage to these organs and serious side effects, including bacterial infection and anemia have happened upon the usage of such nanoparticles.[24]

There have been outstanding advances in discovering peptides and antibodies that target particular organs. The biochemical differences in different tissue vasculatures are the basis for the action of these biomolecules.[25,26] As an example, the overexpression of new peptidases, proteases, and integrins on the endothelium and in the extracellular matrix make significant changes to the tumor vasculature.[26] Many different ligands, including various antibodies, peptides and aptamers, are ligands which are found to specifically recognize tumors.[26–30]

Attaching these ligands to nanoparticles has enabled researchers to develop several tumor-targeting therapies.[31–33] Liposomes made by direct attachment of targeting ligands to lipids have also been investigated.[26,34] Nanoparticles with long circulation times can enhance the targeting efficiency due to the higher likelihood of interaction between targeting ligands and their respective receptors. However, possible interference of targeting ligands on particle circulation has raised concerns about reducing the benefits of active targeting.[35–37] To maximize therapeutic benefits, the balance between prolonged circulation and high targeting specificity should be optimized.

## III. Nanoparticle Clearance from Blood

The complete elimination of substances which are externally introduced to a living organism is called *clearance*. The RES performs renal and hepatic clearance, destabilization, aggregation, opsonization and clearance of the nanoparticles in the blood (Figure 1).[38] Summation of the renal clearance, hepatic clearance, and lung clearance makes up the total body clearance. Glomerular filtration, secretion from the peritubular capillaries to the nephron, and re-absorption from the nephron back to the peritubular capillaries are the determinant factors for clearance.

Whether the therapeutic objective of the nanoparticles is systemic delivery, passive targeting, or active targeting, their function is highly dependent on the circulation time in blood. Nanoparticle concentration in blood has a direct relationship with the circulation time for systemic delivery. There is a direct relationship between the circulation time of nanoparticles and the amount of drug they deliver. It has been shown that the accumulation of nanoparticles at the target site in targeted delivery is also directly linked to the circulation time.[39,40] The rapid clearance by the RES deprives most nanoparticles of the opportunity to interact with the target tissue. Nanoparticles may pass through the target tissue as few times as only once depending on their location.[5] For nanoparticles that do reach the target tissue, it is still possible that they cannot contact, adhere to, and permeate through the vascular wall.[41] To remain in blood circulation, nanoparticles should be capable of avoiding capture by the RES phagocytic cells,

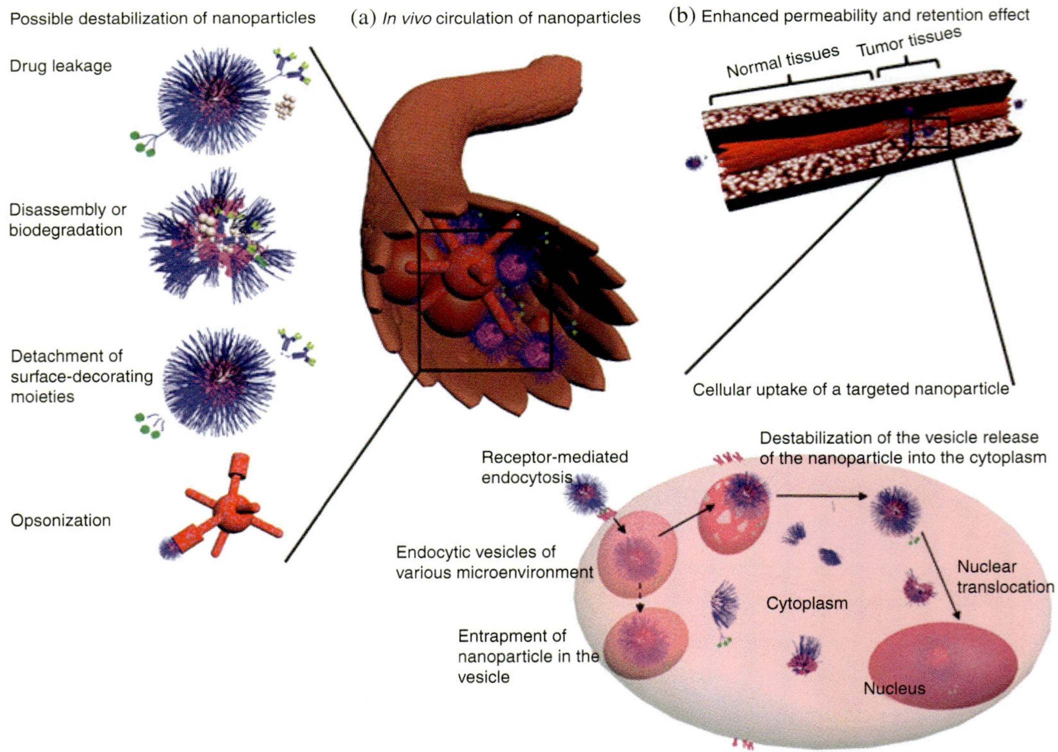


Figure 1. Possible destabilization and degradation pathways of nanoparticles during *in vivo* circulation (a) the EPR effect and intracellular fate of nanoparticles. (b) Drug-leakage, disassembly or degradation, detachment of surface decorating moieties, opsonization, and clearance of nanoparticles during circulation can adversely affect the efficiency of nanoparticles. Taken from Ref. [38] with permission of Royal Society of Chemistry.

which are responsible for removing foreign extracellular materials. Circulating monocytes and macrophages of the liver, spleen, lung, and bone marrow make up the RES.[42] Nanoparticles of any composition are usually cleared from the circulation in 10 minutes, unless they are specifically designed to avoid clearance.[42–44] Opsonization is the conciliator of the phagocytic clearance of nanoparticles.[45–47] Opsonization is the process in which plasma components (i.e. opsonins) are adsorbed onto the surface of a foreign particle in order to facilitate phagocytosis. Immunoglobulins, blood-clotting factors including thrombospondin and fibronectin, and components of the complement system, such as C3, C4, and C5 are different examples of opsonins.[21,48] Moreover, the opsonization process is strongly influenced by electrostatics and hydrogen bonding.[49–51] Higher hydrophobicity of the nanoparticles results in higher opsonin adsorption because the binding of opsonins and nanoparticles happens by hydrophobic interactions.[52,53] Depending on the size and surface characteristics of the nanoparticles, opsonins that adsorb onto them can vary.[21,43–46]

Significant alterations of biodistribution was observed when neutral, sterically stabilized liposomes, were injected repeatedly.[54] The long-circulating characteristics of the PEGylated liposomes were altered and extensive accumulation in the liver was observed.[55] Accelerated blood clearance is the term assigned to this phenomenon. Splenic B-cells produce anti-PEG IgM after the first dose of PEG-coated liposomes is introduced. As a result of repeated injection in a few days, the anti-PEG IgM binds to the PEG on the surface and leads to the activation of the complement system. Therefore, C3 fragments of the complement system cause opsonization leading to an increase in Kupffer cells uptake of the nanoparticle.[56–58] The similar condition was reported for stabilized liposomes that contain nucleic acids and the nucleic acid content demonstrated a strong adjuvant effect.[59] Therefore deployment of non-immunostimulatory siRNA is essential for avoiding the accelerated blood clearance of siRNA/liposome complexes. In order to stop the production of PEG-antibodies and therefore the enhanced blood clearance of the PEG-liposomes after re-administration, siRNA/liposome complexes that contain sheddable PEG-lipids, such as rapidly diffusible PEG-lipids[59] or serum esterase sensitive PEG-lipids[60] shall be developed.

# IV. Designing Long-Circulating Nanoparticles

## A. Size

Modification of the particle size can significantly change the biological function of the particles injected intravenously.[10,23,61] The circulation time, clearance, selective tissue distribution, and the intracellular fate of the nanoparticles can be strongly altered by tailoring their size. Kidneys quickly filter out the nanoparticles smaller than 10 nm in diameter.[62,63] It has been shown that a renal filtration threshold exists for quantum dots. Quantum dots with a final hydrodynamic diameter below 5.5 nm are totally filtered out by the kidneys and therefore are detected in the urine.[61,64] Nanomaterials larger than 10 nm in diameter can only be eliminated by the kidneys if they comprise degradable materials such as polymers, lipids, or hydrogels. Among the nanoparticles that have sizes larger than a certain threshold necessary to avoid renal clearance, smaller particles are more resistant to other types of clearance. It has been shown that PEGylated nanoparticles with diameters of 220 nm were cleared more rapidly than the same nanoparticles with 60 nm diameter.[65] It was demonstrated that only 6% opsonization occurred for particles with a diameter of 80 nm while larger particles were prone to higher opsonization (23% for 170 nm and 34% for 240 nm). Another study focused on the comparison of serum protein adsorption and phagocytosis for nanoparticles with three different sizes (80, 170 and 240 nm).[66] The blood filtration system in the spleen is made of a tight reticular mesh with interendothelial cell slits that are 200 nm in width.[21,67] Therefore, the spleen takes up rigid nanoparticles larger than 200 nm more effectively.[64] It has been shown that during 24 hours after injection of poloxamine 908-coated polystyrene nanoparticles, which are 250 nm in diameter, up to 50% of the nanoparticles are accumulated in the spleen.[68,69] Large capillary beds in the lungs result in the easy uptake of particles larger than 7 $\mu$m.[70,71] The toxicity of silver nanoparticles has been shown to vary with the nanoparticles' size. Particles with the smallest size (10–20 nm in diameter) are usually more cytotoxic as opposed to larger particles (50–100 nm in diameter).[72–74] Generally, nanoparticles (<500 nm) have a better performance than microparticles (>500 nm) in intravenous drug delivery due to their stronger capability to circulate, remain in blood, and target pathological tissues (such as tumors).[7]

The structural limitations of the target tissue limit the ability of siRNA/liposome complexes to reach the target cells. The size of siRNA/liposome complexes need to be carefully designed to overcome this limitation. Blood cells and endothelial cells are accessible to siRNA/liposome complexes upon systemic injection. The size of siRNA/liposome complexes must be higher than a certain threshold in order to be able to reach these cells.[75–77] Furthermore, small (10–30 nm) polymeric micelles that contain a PEG shell do not exhibit the accelerated blood clearance, hence the size-dependency of this effect.[78]

To achieve long circulation times for nanoparticles, their size should be larger than a certain threshold to avoid renal filtration but it should not be too large to increase the chance of opsonization and RES clearance. Upon proper design, polymeric nanoparticles with sizes between 20 nm and 100 nm are most capable of circulating in the blood for a long time; therefore they possess the highest potential for *in vivo* applications. As mentioned above, these nanoparticles have the intermediate size suitable for minimizing renal and lymphatic clearance and opsonization. Furthermore, cells can internalize nanoparticles of such sizes more easily than larger or smaller particles.[79] These nanoparticles also have the ability of entering certain tissues like liver and spleen as well as passing through the leaky or highly permeable vasculature of particular tissues like tumors and inflammation sites.[38] It is worth emphasizing that the aforementioned size limits are approximate values and depend on various factors including the physiological conditions such as the cell type and status and location of the disease as well as the composition, size, geometry, charge, and surface chemistry of the nanoparticles. For instance, the permeation through skin or plasma membranes is faster and easier for small and hydrophobic molecules than hydrophilic macromolecules.[38]

## B. Morphology

The most widespread shape for drug delivery carriers is a sphere. Nevertheless, particles with different shapes have been introduced for drug delivery purposes in the past few years.[54] Shape is one of the most prominent tunable parameters of the nanoparticles for improving their biological functions.[81–84] Targeting, circulation, internalization, immune cell association, adhesion, cellular uptake, and intracellular trafficking can be modified via shape tuning.[82,83,85–90] Shape modification can result in improved margination and adhesion to the endothelium under flow conditions.[81,82] It has been reported that the circulation time of rod-shaped micelles can be as much as 10 times longer than that of spherical micelles.[91] Theoretical models have been proposed to explain the advantages of non-spherical shapes in targeting and internalization.[64] It was experimentally shown by Gratton *et al.* that the increase in the aspect ratio of the particles enhances the cellular uptake in cancer cells.[85] However, the cellular uptake of gold nanoparticles has been shown to decrease with increase in their aspect ratio.[93] A larger increase in lung/liver ratio is observed for ICAM-antibody-coated rods as opposed to ICAM-antibody-coated spheres. Therefore the uptake in lung targeting is considerably higher for rod-shaped particles with ICAM-antibody functionalization and the uptake in both the liver and spleen is reduced. On the contrary, only lung targeting and spleen avoidance are improved in spherical particles with ICAM-antibody functionalization.[7]

It has been demonstrated that phagocytosis by macrophages is significantly affected by the geometry of the nanoparticles. The circulation time in blood is enhanced for worm-like flexible micelles.[91] Biodegradable PEG-b-poly(e-caprolactone) and non-biodegradable PEG-b-polyethylethylene filamentous micelles (filomicelles) were compared with spheres of similar chemistry by Discher *et al.*[91] They demonstrated that 18 mm-long filomicelles with diameters of 20–60 nm possess 10 times longer circulation times as opposed to their spherical counterparts as a result of the decrease in phagocytosis rate and clearance by the RES. It was shown in another study that the interaction between macrophages and elliptical disk shaped particles varies depending on the contact point of the particle and macrophages while spherical particles are easily phagocytosed by macrophages.[89] As shown in Figure 2, when the initial contact point

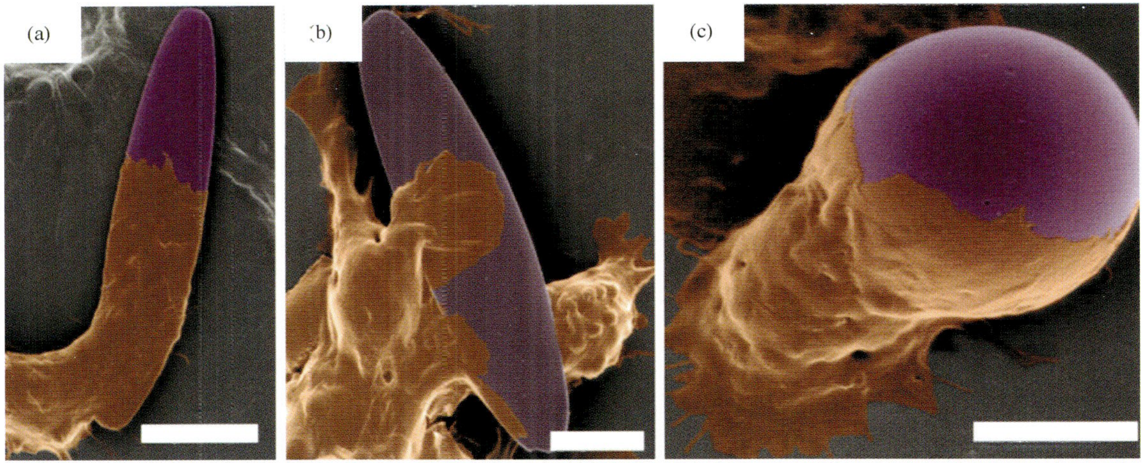


Figure 2. Phagocytosis of stretched and spherical shape particles by macrophages and the effect of the initial contact point (a) The initial contact point is the pointed end point of the elliptical disk shaped particle (Scale bar: 10 $\mu$m). (b) The initial contact point is the flat side of the disk (Scale bar: 5 $\mu$m). (c) Phagocytosis is easily progressed for a spherical particle (Scale bar: 5 $\mu$m). Taken from Ref. [89] with permission of National Academy of Sciences.

of macrophages and the elliptical disk-shaped particles was the flat side of the elliptical disk, macrophages were not successful in phagocytosis, whereas phagocytosis occurred when the initial contact point was the pointed end of the particle. Being highly elongated, worm-shaped particles are almost completely resistant to phagocytosis by macrophages.[83] In a related study, polymer particles were coated with anti-ICAM by Muro *et al.* and were intravenously injected in mice.[86] Longer circulation times and higher targeting capability were observed for elliptical disk-shaped particles as opposed to spherical particles. Furthermore, slower internalization and unique intracellular trafficking were demonstrated by elliptical disk-shaped PLGA particles in comparison to spherical PLGA particles.[94]

The influence of particle shape on *in vivo* biodistribution was studied by Decuzzi *et al.*, where silica particles without PEGylation formed in four different shapes (sphere, hemisphere, discoidal, and cylindrical shapes) were examined in mice.[10] Among these groups, the discoidal particles demonstrated the highest likelihood of avoiding accumulation in the liver, which is in agreement with the previously mentioned findings for anti-ICAM-coated elliptical disk particles.[86] Liver accumulation was approximately similar for the other three shapes. Another study investigated the *in vivo* accumulation of different shapes of polymer lipid nanostructures (LIPOMER) in the liver and spleen.[95] It was shown that LIPOMERs with irregular shapes are less accumulated in the liver than spherical LIPOMERs and can avoid liver macrophages more effectively. It has also been shown that for the nanoparticles that have the same shape (i.e. cylindrical), those with a higher aspect ratio (height divided by diameter) undergo a more rapid cellular uptake.[85,89] This result is in coordination with the findings reported by Champion and Mitragotri for polymer microparticles.[89] Recognizing the importance of particle shape in drug delivery, several studies have been attempted to determine the optimal shape for long circulation times in blood. However, further *in vivo* circulation studies using particles with well-defined sizes and shapes are needed to consolidate the optimal shapes for long-circulating nanoparticles.

## C. Elasticity

Tuning the elasticity of nanoparticles can potentially result in improved circulation, reduced uptake by the immune system, and enhanced targeting, hence improving the biological fate of the nanoparticles. However, the effect of particle elasticity modification on drug delivery has not been studied to a large degree.[11,82,96–98] Increased stiffness in polyacrylamide particles increases the stimulation of phagocytosis.[99] Geng *et al.* demonstrated a very long circulation time of about 1 week for flexible block co-polymer worm-like micelles while their rigid counterparts that were prepared by core crosslinking had a much lower circulation time of several hours.[91] In another study, the influence of elasticity was investigated on the circulation and biodistribution of red blood cell shaped microparticles over a range of 10–63.8 kPa.[11] The elimination half-life of softer microparticles were increased more than 30 times when the particle modulus was increased 8 times. The fundamental role of nanoparticles' elasticity in facilitating cellular uptake, circulation, biodistribution, and tissue targeting was investigated by Asselmo *et al.*[100] They managed to synthetize hydrogel nanoparticles with elasticity values ranging from 0.255 kPa to 3000 kPa in order to study the role of elasticity on circulation and targeting for small spherical nanoparticles (~200 nm). An identical number of radio-labeled soft and hard hydrogel nanoparticles were intravenously injected into mice for *in vivo* evaluation of circulation time. Soft nanoparticles had a considerably longer circulation time in blood compared to hard nanoparticles at short times. Longer distribution half-life (R1) and elimination half-life ($\beta 1$) were observed for softer nanoparticles. It can be assumed that because softer particles avoid permanent clearance by the RES more than harder particles, they stay longer in the circulation at short time points. As mentioned before, phagocytosis is less effective on elongated or stretched nanoparticles.[89] Furthermore, when the amount of cell membrane tension that should be endured for completing the phagocytosis increases, the phagocytosis is reduced. It has also been shown for spherical particles that phagocytosis is reduced when the curvature of the particles decreases.[100,101]

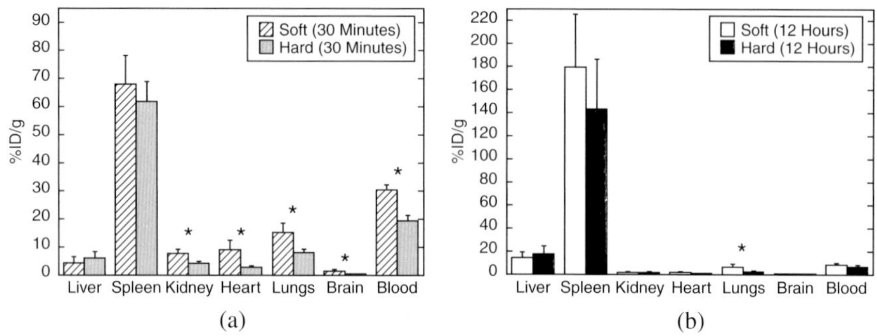


Figure 3. Biodistribution of soft and hard hydrogel nanoparticles. (a) Biodistribution in different organs of soft (hatched bars) and hard (light gray bars) hydrogel nanoparticles at 30 minutes. (b) Biodistribution in different organs of soft (white bars) and hard (dark gray bars) hydrogel nanoparticles at 12 hours. Taken from Ref. [10] with permission of Elsevier.

Soft and hard hydrogel nanoparticles have been studied for their biodistribution in 30-minutes and 12-hour periods. Biodistribution was significantly different in kidneys, heart, brain, and lungs between soft and hard particles at 30 minutes (Figure 3a). However, at 12 hours the only organs that showed a statistically different biodistribution between soft and hard particles were lungs (Figure 3b). It can be assumed that the difference in the circulation time of the particles causes the difference in their biodistribution. Particularly, because the softer nanoparticles are more capable of remaining in blood for short periods (less than 2 hours), they are more likely to be present in organs that receive a large amount of circulation.[100]

Although the precise role of mechanical properties in determining circulation times is yet to be investigated, the data clearly demonstrate that the particle's mechanical properties can be best utilized in tandem with tuning of other physical (e.g. shape) and chemical (e.g. targeting ligands) parameters to create more advanced nanoparticle delivery systems.

## D. Surface modification

Coating the nanoparticles with a dense brush layer of neutral hydrophilic flexible polymers has been a successful method to avoid adsorption of plasma proteins and hide the particles from the RES and increase the circulation time. Coating of the quantum dots affects their colloidal stability and decreases the toxicity because otherwise quantum dots will cover the cell surface and reduce their growth.[102] The main goal of surface modification for opsonin adsorption prevention is to reduce the hydrophobicity and surface charge density of the particle. High hydrophobicity and surface charge density increase opsonization. Polysaccharides, PEG and PEG-derived polymer (poloxamines and poloxamers) are the most common coating materials used for nanoparticle surface tuning.

The biomimetic properties of polysaccharides obtained from natural sources make them suitable candidates for surface modifying agents.[103–105] The opsonization process has been successfully affected by various polysaccharides such as dextran, heparin, hyaluronic acid, and chitosan.[106] The *in vivo* circulation time of poly(methyl methacrylate) nanoparticles is significantly enhanced when they are coated with heparin and dextran.[107] Reduction in the complement activation for particles with modified surfaces may be the cause of the enhancement in the circulation time.[108] Furthermore, it has been observed that the cell toxicity and protein adsorption of poly(-caprolactone) (PCL) was reduced due to dextran coating. On the other hand the complement activation was increased as a result of dextran-coating on nanoparticles.[105,106] Iron oxide nanoparticles that are coated with dextran and have a long circulation time are common magnetic resonance imaging (MRI) agents in order to image blood pools, assess lymph node metastases, and investigate hepatosplenic tumors.[109,110] Basically, immunogenic responses are the main obstacles for deployment of polysaccharides. Polysaccharides are recognized by the immune system with the help of circulating antibodies.[21]

PEG is a biocompatible, non-ionic, flexible, and hydrophilic polymer made by repeating units of ethylene oxide. When used as a surface modifying agent, PEG forms a steric layer around the nanoparticles and stops flocculation and opsonization, leading to the reduction of complement activation.[111] PEG is the most prevalent nanoparticle coating polymer and the US Food and Drug Administration (FDA) has approved it for certain internal applications. Immordino *et al.* have comprehensively reviewed the utilization of PEG to coat liposomes in order to increase their circulation time.[112] Circulation time and clearance of polymeric nanoparticles are significantly influenced by PEGylation, the length of PEG chains, and the number of PEG chains per nanoparticle.[113,114] Furthermore, for nanoparticles with neutral net surface charge, PEG forms a neutral stabilizing interface, reduces their aggregation, and limits their non-specific interactions with cells. Reduction of protein binding to the surface of the liposome is the basic advantage of PEG coating and will result in lower opsonization and lower uptake by RES macrophages. The flexibility of PEG on the surface of nanoparticles is a key determinant of how well the nanoparticles can be hidden from the macrophages. Therefore the density, thickness, and configuration (mushroom or brush-like configuration) of the PEG layer need to be carefully studied.[115] The increase in the density of PEG on the surface of the liposome changes the condition of the regimen of the PEG polymers. As depicted in Figure 4 the regimen transforms from more heavily coiled, flat pancake structures to intermediate mushroom structures to low-coiled, extended brush structures.[116,117] It is proposed by Buyens *et al.* that in order to have good stealth liposomes, a high — but not too high — concentration of PEG-lipids is needed.[115] The cytoplasmic deposition of the siRNA cargo is hampered when the PEG is strongly concentrated. As a result, a sheddable PEG coat is essential for PEGylated siRNA/liposome complexes.[117]

The length and surface density of the PEG chains are vital factors for a successful PEG coating.[118–120] Without the full coverage by PEG, the particles are still prone to opsonins' adsorption and only a low stealth efficiency is obtained. A prolonged circulation time is achieved by high PEG coverage. On the other hand, the mobility and flexibility of the PEG may reduce as a result of excessive PEG coverage and can decrease the circulation time.[121] The optimal thicknesses of the PEG layer have been suggested by various guidelines in order to reach the needed stealth properties.[47]

Although PEGylation is believed to reduce the opsonization, its effect on complement activation is not completely clear.[122,123] PEGylation results in the decrease in protein adsorption, however it is possible that some components of the complement system are still

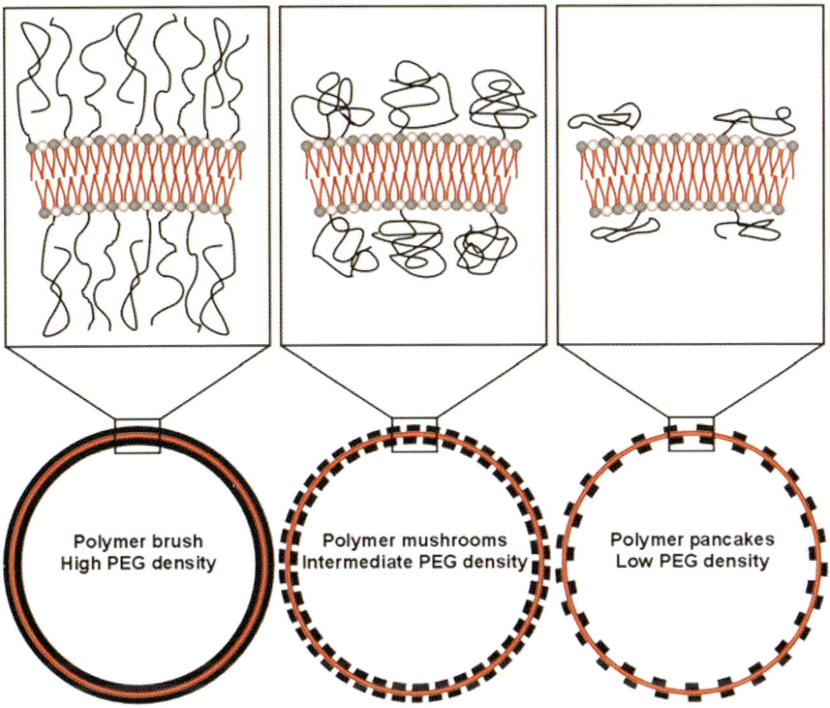


Figure 4. Possible regimens of the PEG polymers on the surface of liposomes. Schematic representation (bottom: general and top: detailed zoom) of the brush (left), mushroom (middle) and pancake (right) regimen of the PEG polymer. The surface coverage of the lipid bilayer by PEG is complete, intermediate, and incomplete for brush, mushroom, and pancake, respectively. Taken from Ref. [115] with permission of Elsevier.

attached to them. Furthermore, despite the achieved improvements these strategies have not been able to provide circulation times higher than 8–10 hours.[21] Moreover, the accelerated blood clearance limits the possibility of repeated administration of surface-modified particles. For instance, the circulation time was significantly decreased for a second dose of poloxamine 908 coated particles that was administered 3–10 days after an initial dose.[124,125] Anti-PEG antibody facilitates complement activation when the PEGylated nanoparticles are repeatedly injected and enhances phagocytosis by the RES.[126,127]

The stealth properties of PEGs are improved upon their combination with polysaccharides. Coating PLA nanoparticles with a combination of PEG and a water-soluble form of chitosan has been reported to significantly decrease the *in vitro* macrophage uptake, hence leading to a considerably high blood circulation half-life of up to 63.5 hours.[128] In spite of the improvements in preventing opsonization and achieving prolonged circulation times via surface modifications with PEG or polysaccharides, it has been suggested that increasing the stealth properties of the nanoparticles could have adverse effects on particle-target interactions and diminish the therapeutic benefits.[129,130]

## E. CD47 as a "self" marker

The balance between prophagocytic ("eat me") and antiphagocytic ("don't eat me") signals on target cells is a determinant factor for phagocytosis.[13,131–134] CD47, which was first seen on stem cells, is a transmembrane protein that constrains phagocytosis by binding to its receptor, signal regulatory protein $\alpha$ (SIRP$\alpha$), which is expressed on phagocytes.[135–139] Expeditious clearance of erythrocytes, platelets, and lymphohematopoetic cells by macrophages has been observed due to the lack of CD47 on their surface which leads to the elimination of the CD47-SIRP$\alpha$–mediated antiphagocytic signal.[137,140,141] When CD47 binds to SIRP$\alpha$, the immunoreceptor tyrosine-based inhibitory motifs (ITIMs) on SIRP$\alpha$ are phosphorylated and Src homology phosphatases 1 and 2 (SHP-1 and SHP-2) are recruited. Both of these conditions hinder the accumulation of myosin-IIA at the phagocytic synapse.[142,143]

A CD47 analog is expressed by a number of viruses including variola and vaccine viruses to circumvent macrophages.[144,145] In order to evade macrophages, hematopoietic stem cells and leukemia cells upregulate CD47 during circulation.[146,147] Therefore, the importance of CD47-SIPR interaction in phagocytosis of colloidal carriers has been studied by Hsu *et al.*[148] Their results indicated that antagonizing SIRP on macrophages via pretreatment with water soluble CD47, results in a drastically decreased internalization of perfluorocarbons

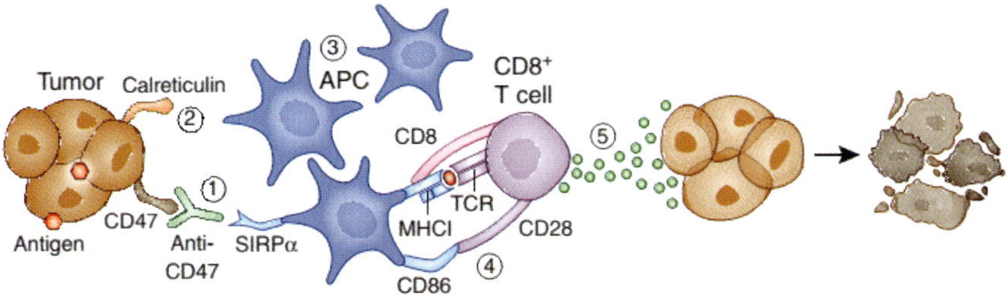


Figure 5. The CD47 immune checkpoint. CD47 blocking antibodies (1) and pro-phagocytic molecules such as calreticulin (2) work together to increase tumor engulfment by APCs, (3) enabling the processing and presentation of tumor antigens as peptides in the groove of MHC class I molecules (4) and the priming of CD8+ T cells specific for tumor antigens. (5) MHC, major histocompatibility complex; TCR, T cell receptor. Taken from Ref. [136] with permission of Nature Publishing Group.

emulsions (340 nm in diameter) as opposed to the case with normal macrophages. Moreover, coating polystyrene particles with recombinant CD47 prevented phagocytosis by macrophages to a large degree.[142] Therefore, CD47-SIRP interaction can be utilized to design a novel approach for the design of macrophage-evading nanoparticles for prolonged blood circulation. It was also demonstrated by Liu *et al.*[143] that the therapeutic effects of CD47 blockade are facilitated by two other types of immune cells — CD8+ T cells and dendritic cells (DCs). They particularly showed that innate immune DC activation and T cell priming are bridged by CD47 blockade (Figure 5).

CD47 acts as a signaling receptor for the secreted matricellular protein thrombospondin-1 and the counter-receptor for signal-regulatory protein as a marker of self.[149,150] Lack of CD47 or thrombospondin-1 strongly increases the resistance of mice to tissue stress on signal regulatory protein alpha (SIRPa), which on phagocytic cells recognizes CD47 involvement related to ischemia, ischemia/reperfusion, and high dose irradiation.[151–153] Radioresistance related to CD47 blockade is cell-independent and uncontrolled by NO signaling. This demonstrates that further pro-survival signaling pathways are regulated by CD47.[154,155]

In most cases, increased expression of CD47 results in an indirect survival advantage *in vivo*. CD47 inhibits phagocytic clearance by engaging SIRPa on macrophages.[137,149] Similarly, killing of several types of cancer cells by macrophages or NK cells has been hampered due to the enhanced expression of CD47.[147,156,157] On the contrary, CD47 antibodies block SIRPa binding and reinforce tumor clearance by macrophages.[158,159] It should be mentioned that some groups have indicated that such clearance can occur regardless of the inhibitory SIRPa signaling.[160–162] Consequently, two conflicting roles have been suggested for CD47 in cell survival. A balance should be maintained between the benefits of decreasing CD47 expression that results in cell autonomy and less inhibitory CD47 signaling against maintaining sufficient CD47 levels to inhibit phagocytic clearance *in vivo*. High CD47 expression is demonstrated by hematopoietic stem cells. It has been suggested that elevated CD47 expression in the stem cell niche is crucial to protect stem cells from innate immune surveillance.[146]

## F. RBC-hitchhiking nanoparticles

The phenomenon of RBC-hitchhiking occurs for a number of pathogens including haemobartonella and eperythrozoon. In this phenomenon bacteria attach to the outer surface of the RBC and the pathogens can have a circulation time of up to 60 days in some cases.[163] The prolonged circulation time of the pathogens bound on RBCs has stimulated the study of RBC-hitchhiking polymeric nanoparticles. According to Chambers and Mitragotri, polymeric nanoparticles that were adhered to rat RBCs via non-covalent adhesion exhibit circulation times over 10 hours, while unattached particles are cleared within minutes.[164,165] RBCs can potentially be used as natural carriers for vascular drug delivery. They can be of particular interest when surface loading is performed and there is no requirement for encapsulation and entering the cargoes into the inner volume of the erythrocyte by damaging the encapsulation.[166–173] Improvements in drug delivery and therapeutic effects as a result of coupling the drugs to a RBC surface have been observed in preclinical studies of various animal species.[174,175] The behavior of nanoparticles in the circulation can be significantly changed when they are coupled to RBCs. The circulation time of RBCs were not significantly altered ($p > 0.05$) when low doses of nanoparticles were attached to them. On the contrary, the circulation time of RBCs was decreased when higher doses of nanoparticles (100:1) were attached to them, which can be due to the accelerated removal of RBCs that carry multiple particles.

Nanoparticle circulation time is distinctly increased when they are carried by RBCs. When evaluating nanoparticle uptake in the organs, the significant difference in blood level between free nanoparticles and those bound to RBCs should be considered. Such consideration is useful in detecting nanoparticles that are kept in the tissue compared to those in the residual blood pool. It has been observed that attachment to RBCs increases the lung/liver and lung/spleen accumulation ratio of nanoparticles, while the accumulation values of free nanoparticles did not alter considerably over a period of 24 hours. Accumulation in kidneys and the heart were also enhanced for RBC-bound nanoparticles, however the RBC-hitchhiking improvement is more significant in lungs than in kidneys and the heart.[172]

It has been demonstrated that nanoparticles bound to RBCs can target and accumulate in organs like brain and lung that are rich in endothelium while defying liver and spleen as two of the main RES clearance organs.[172] Carboxylated polystyrene (PS) nanoparticles affixed to the surface of purified red blood cells (RBCs) at 100:1 particle to cell incubation ratio were prepared by Anselmo *et al.* As shown in Figure 6b rod-shaped nanoparticles made by stretching 200 nm diameter spheres (Figure 6a) were adhered to the surface of RBCs and were used for *in vivo* experiments.[7] It was demonstrated that the lung accumulation of RBC-bound nanoparticles coated with IgG was higher in comparison to free nanoparticles coated with IgG, even without chemical targeting via ICAM-antibody. Further enhancement in lung accumulation was observed for nanoparticles with both ICAM-antibody and RBC-hitchhiking. Whether hitchhiked or non-hitchhiked, nanoparticles coated with ICAM-antibody exhibit higher lung/liver and lung/spleen ratios as compared to IgG-coated particles. This demonstrates how the surface chemistry, particularly ICAM-antibody coating, of nanoparticles attached to RBCs

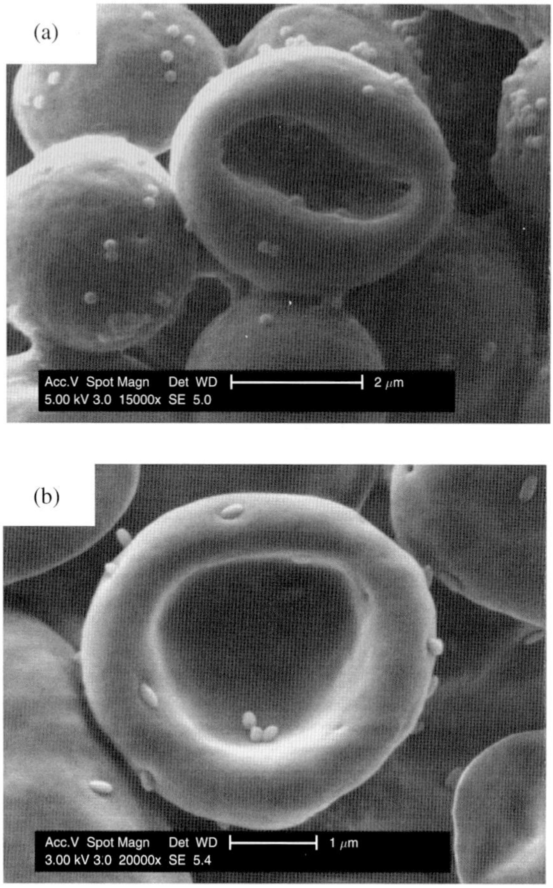


Figure 6. Attachment of nanoparticles to RBCs (a) 200 nm spheres affixed to RBCs, and (b) rods stretched from 200 nm spheres affixed to RBCs. Taken from Ref. [7] with permission of Elsevier.

improves accumulation in lungs while simultaneously limiting RES uptake in clearance organs.

A highly complicated relationship exists between size, geometry and plasticity of the carrier, ligand's affinity, spatial freedom and density, and nature of the target determinants — accessibility, clustering, and surface density.[12,19,20,176] Relative contributions of different nanoparticle features can be investigated by comparing the uptake in lungs and the ability of the nanoparticle to avoid liver and spleen for all groups under investigation. It is clear that surface chemistry, shape, and RBC hitchhiking all play a distinct role in facilitating lung uptake and RES avoidance. Spherical particles coated with ICAM-antibody offer significant benefits in terms of lung targeting while avoiding both the liver and the spleen. Modification of the particles' shape from spheres to rods as well as hitchhiking onto the surface of RBCs further enhance these advantages. Evading both liver and spleen is only achieved when combining either hitchhiking and ICAM-antibody, or rod-shaped particles and ICAM-antibody. Combining the three proposed modifications results in hitchhiked rod-shaped ICAM-antibody-coated particles that allow the highest level of lung targeting and RES organ avoidance. In some cases, the increase in lung/liver and lung/spleen ratios may only stem from the increase in accumulation at target sites and not necessarily avoidance of liver or spleen. For RBC hitchhiking rods coated with ICAM-antibody, although high lung accumulation does contribute to the increase in target organ to clearance organ ratios, the main contributor is the actual evasion of the RES organs as is evident from the extremely low total accumulation of hitchhiked rods coated with ICAM-antibody in the liver and spleen. Such a low uptake in both liver and spleen is unique to the hitchhiked rods coated with ICAM-antibody while high lung targeting values have been achieved for other groups as well.

## V. Conclusions

Urgent needs for maintaining long blood circulation times for nanoparticles have propelled significant advances in particle design strategies. Although macrophage-evading surface modifications, especially PEGylation, are effective and widely popular nanoparticle tuning methods, other approaches including CD47-coating, flexibility, non-spherical shapes, and hitchhiking on RBCs have shown great potential in avoiding clearance. Combining physical, chemical, and biological modifications provides the highest targeting capability and the lowest RES organ uptake. The shape-engineered and RBC hitchhiking nanoparticle formulations can be used to design more sophisticated carrier systems that not only target specific tissues but also avoid association with clearance organs that typically cause toxicity issues. Antibody targeting, shape effect and cellular hitchhiking have shown their applicability in targeting nanoparticles to specific tissues as well as avoiding nanoparticle uptake by the immune system, to a certain degree. Integration of all the aforementioned strategies leads to the development of a single formulation that can accomplish the required goals more effectively than any individual method alone.

Designing long-circulating nanoparticles is not a facile but complex task requiring detailed appreciation of the interactions between nanoparticles and the biological milieu. Therefore, carefully tailored combinations of available strategies including the modification of size, surface shielding, mechanical properties, and shape as well as the utilization of natural cells are required to achieve dramatic advances in the circulation time.

## References

1. Zhang, L.; Gu, F. X.; Chan, J. M.; Wang, A. Z.; Langer, R. S.; Farokhzad, O. C. *Clin. Pharmacol. Ther.* **2008**, *83* (5), 761.
2. Sanvicens, N.; Marco, M. P. *Trends Biotechnol.* **2008**, *26* (8), 425.
3. Doshi, N.; Mitragotri, S. *Adv. Funct. Mater.* **2009**, *19* (24), 3843.
4. Mitragotri, S.; Lahann, J. *Nat. Mater.* **2009**, *8* (1), 15.
5. Yoo, J.-W.; Chambers, E.; Mitragotri, S. *Curr. Pharm. Des.* **2010**, *16* (21), 2298.
6. Gref, R.; Minamitake, Y.; Peracchia, M. T.; Trubetskoy, V.; Torchilin, V.; Langer, R. *Science* **1994**, *263* (5153), 1600.
7. Anselmo, A. C.; Kumar, S.; Gupta, V.; Pearce, A. M.; Ragusa, A.; Muzykantov, V.; Mitragotri, S. *Biomaterials* **2015**, *68*, 1.
8. Torchilin, V. P.; Trubetskoy, V. S. *Adv. Drug Deliv. Rev.* **1995**, *16* (2–3), 141.
9. Farokhzad, O. C.; Cheng, J.; Teply, B. A.; Sherifi, I.; Jon, S.; Kantoff, P. W.; Richie, J. P.; Langer, R. *Proc. Natl. Acad. Sci. U. S. A.* **2006**, *103* (16), 6315.
10. Decuzzi, P.; Godin, B.; Tanaka, T.; Lee, S. Y.; Chiappini, C.; Liu, X.; Ferrari, M. *J. Control. Release* **2010**, *141* (3), 320.

11. Merkel, T. J.; Jones, S. W.; Herlihy, K. P.; Kersey, F. R.; Shields, A. R.; Napier, M.; Luft, J. C.; Wu, H.; Zamboni, W. C.; Wang, A. Z.; Bear, J. E.; DeSimone, J. M. *Proc. Natl. Acad. Sci.* **2011**, *108* (2), 586.
12. Anselmo, A. C.; Mitragotri, S. *J. Control. Release* **2014**, *190*, 531.
13. Grimsley, C.; Ravichandran, K. S. *Trends Cell Biol.* **2003**, *13* (12), 648.
14. Anselmo, A. C.; Mitragotri, S. *AAPS J.* **2015**, *17* (5), 1041.
15. Anselmo, A. C.; Mitragotri, S. *J. Control. Release* **2014**, *190*, 15.
16. Allen, T. M.; Cullis, P. R. *Science* **2004**, *303* (5665), 1818.
17. Wang, A. Z.; Langer, R.; Farokhzad, O. C. *Annu. Rev. Med.* **2012**, *63*, 185.
18. Faraji, A. H.; Wipf, P. *Bioorg. Med. Chem.* **2009**, *17* (8), 2950.
19. Howard, M.; Zern, B. J.; Anselmo, A. C.; Shuvaev, V. V.; Mitragotri, S.; Muzykantov, V. *ACS Nano* **2014**, *8* (5), 4100.
20. Pillai, J. D.; Dunn, S. S.; Napier, M. E.; Desimone, J. M. *IUBMB Life* **2011**, *63* (8), 596.
21. Moghimi, S. M.; Hunter, A. C.; Murray, J. C. *Pharmacol. Rev.* **2001**, *53* (2), 283.
22. Folkman, J. *Nat. Med.* **1995**, *1* (1), 27.
23. Moghimi, S. M. *FASEB J.* **2005**, *19* (3), 311.
24. Brigger, I.; Dubernet, C.; Couvreur, P. *Adv. Drug Deliv. Rev.* **2012**, *64* (Suppl.), 24.
25. Bessette, P. H.; Rice, J. J.; Daugherty, P. S. *Protein Eng. Des. Sel.* **2004**, *17* (10), 731.
26. Ruoslahti, E. *Drug Discov. Today* **2002**, *7* (22), 1138.
27. Brannon-Peppas, L.; Blanchette, J. O. *Adv. Drug Deliv. Rev.* **2012**, *64* (Suppl.), 206.
28. Farokhzad, O. C.; Jon, S.; Khademhosseini, A.; Tran, T. T.; Lavan, D. A; Langer, R. *Cancer Res.* **2004**, 7668.
29. Bibby, D. C.; Talmadge, J. E.; Dalal, M. K.; Kurz, S. G.; Chytil, K. M.; Barry, S. E.; Shand, D. G.; Steiert, M. *Int. J. Pharm.* **2005**, *293* (1–2), 281.
30. Torchilin, V. P. *Eur. J. Pharm. Sci.* **2000**, *11* (Suppl. 2), S81.
31. Byrne, J. D.; Betancourt, T.; Brannon-Peppas, L. *Adv. Drug Deliv. Rev.* **2008**, *60* (15), 1615.

32. Cho, K.; Wang, X.; Nie, S.; Chen, Z.; Shin, D. M. *Clin. Cancer Res.* **2008**, *14* (5), 1310.
33. Nobs, L.; Buchegger, F.; Gurny, R.; Allémann, E. *J. Pharm. Sci.* **2004**, *93* (8), 1980.
34. Ferrari, M. *Nat. Rev. Cancer* **2005**, *5* (3), 161.
35. McNeeley, K. M.; Karathanasis, E.; Annapragada, A. V.; Bellamkonda, R. V. *Biomaterials* **2009**, *30* (23–24), 3986.
36. McNeeley, K. M.; Annapragada, A.; Bellamkonda, R. V. *Nanotechnology* **2007**, *18* (38), 385101.
37. Pirollo, K. F.; Chang, E. H. *Trends Biotechnol.* **2008**, *26* (10), 552.
38. Elsabahy, M.; Wooley, K. L. *Chem. Soc. Rev.* **2012**, *41* (7), 2545.
39. Gabizon, A.; Catane, R.; Uziely, B.; Kaufman, B.; Safra, T.; Cohen, R.; Martin, F. *Cancer Res.* **1994**, *54* (4), 987.
40. von Maltzahn, G.; Park, J.-H.; Agrawal, A.; Bandaru, N. K.; Das, S. K.; Sailor, M. J.; Bhatia, S. N. *Cancer Res.* **2009**, *69* (9), 3892.
41. Decuzzi, P.; Pasqualini, R.; Arap, W.; Ferrari, M. *Pharm. Res.* **2009**, *26* (1), 235.
42. Brain, J. D.; Molina, R. M.; DeCamp, M. M.; Warner, A. E. *Am. J. Physiol.* **1999**, *276* (1 Pt 1), L146.
43. Ogawara, K.; Yoshida, M.; Higaki, K.; Kimura, T.; Shiraishi, K.; Nishikawa, M.; Takakura, Y.; Hashida, M. *J. Control. Release* **1999**, *59*, 15.
44. Ogawara, K. ichi; Furumoto, K.; Takakura, Y.; Hashida, M.; Higaki, K.; Kimura, T. *J. Control. Release* **2001**, *77* (3), 191.
45. Ogawara, K. I.; Yoshida, M.; Takakura, Y.; Hashida, M.; Higaki, K.; Kimura, T. *Biochim. Biophys. Acta — Gen. Subj.* **1999**, *1472* (1–2), 165.
46. Ogawara, K. ichi; Yoshida, M.; Kubo, J. ichi; Nishikawa, M.; Takakura, Y.; Hashida, M.; Higaki, K.; Kimura, T. *J. Control. Release* **1999**, *61* (3), 241.
47. Stolnik, S.; Illum, L.; Davis, S. S. *Adv. Drug Deliv. Rev.* **2012**, *64* (Suppl.), 290.
48. Dobrovolskaia, M. A.; McNeil, S. E. *Nat. Nanotechnol.* **2007**, *2* (8), 469.
49. Gessner, A.; Lieske, A.; Paulke, B. R.; Müller, R. H. *Eur. J. Pharm. Biopharm.* **2002**, *54* (2), 165.

50. Gessner, A.; Lieske, A.; Paulke, B.-R.; Müller, R. H. *J. Biomed. Mater. Res. A* **2003**, *65* (3), 319.
51. Yoon, J. Y.; Kim, J. H.; Kim, W. S. *Colloids Surfaces B Biointerfaces* **1998**, *12* (1), 15.
52. Gessner, A.; Waicz, R.; Lieske, A.; Paulke, B. R.; Mader, K.; Muller, R. H. *Int. J. Pharm.* **2000**, *196* (2), 245.
53. Yoon, J. Y.; Kim, J. H.; Kim, W. S. *Colloids Surfaces A Physicochem. Eng. Asp.* **1999**, *153* (1–3), 413.
54. Dams, E. T. M.; Laverman, P.; Oyen, W. J. G.; Storm, G.; Scherphof, G. L.; van der Meer, J. W. M.; Corstens, F. H. M.; Boerman, O. C. *J. Pharmacol. Exp. Ther.* **2000**, *292* (3), 1071.
55. Laverman, P.; Carstens, M. G.; Boerman, O. C.; Dams, E. T.; Oyen, W. J.; van Rooijen, N.; Corstens, F. H.; Storm, G. *J. Pharmacol. Exp. Ther.* **2001**, *298* (2), 607.
56. Ishida, T.; Ichihara, M.; Wang, X.; Kiwada, H. *J. Control. Release* **2006**, *115* (3), 243.
57. Ishida, T.; Ichihara, M.; Wang, X.; Yamamoto, K.; Kimura, J.; Majima, E.; Kiwada, H. *J. Control. Release* **2006**, *112* (1), 15.
58. Ishida, T.; Wang, X.; Shimizu, T.; Nawata, K.; Kiwada, H. *J. Control. Release* **2007**, *122* (3), 349.
59. Judge, A. D.; Bola, G.; Lee, A. C. H.; MacLachlan, I. *Mol. Ther.* **2006**, *13* (3), 494.
60. Xu, H.; Wang, K. Q.; Deng, Y. H.; Chen, D. W. *Biomaterials* **2010**, *31* (17), 4757.
61. Choi, H. S.; Liu, W.; Misra, P.; Tanaka, W.; Zimmer, J. P.; Ipe, B. I.; Bawendi, M. G.; Frangioni, J. V. *Nat. Biotechnol.* **2007**, *25* (10), 1165.
62. Kircher, M. F.; Mahmood, U.; King, R. S.; Weissleder, R.; Josephson, L. *Cancer Res.* **2003**, *63* (23), 8122.
63. Rao, J. *ACS Nano* **2008**, *2* (10), 1984.
64. Albanese, A.; Tang, P. S.; Chan, W. C. W. *Annu. Rev. Biomed. Eng.* **2012**, *14* (1), 1.
65. Moghimi, S. M.; Hedeman, H.; Muir, I. S.; Illum, L.; Davis, S. S. *BBA — Gen. Subj.* **1993**, *1157* (2), 233.
66. Fang, C.; Shi, B.; Pei, Y. Y.; Hong, M. H.; Wu, J.; Chen, H. Z. *Eur. J. Pharm. Sci.* **2006**, *27* (1), 27.

67. Moghimi, S. M.; Hunter, A. C. *Trends Biotechnol.* **2000**, *18* (10), 412.
68. Moghimi, S. M.; Muir, I. S.; Illum, L.; Davis, S. S.; Kolb-Bachofen, V. *BBA — Mol. Cell Res.* **1993**, *1179* (2), 157.
69. Moghimi, S. M.; Porter, C. J. H.; Muir, I. S.; Illum, L.; Davis, S. S. *Biochem. Biophys. Res. Commun.* **1991**, *177* (2), 861.
70. Azarmi, S.; Roa, W. H.; Löbenberg, R. *Adv. Drug Deliv. Rev.* **2008**, *60* (8), 863.
71. Santhi, K.; Dhanaraj, S. A; Koshy, M.; Ponnusankar, S.; Suresh, B. *Drug Dev. Ind. Pharm.* **2000**, *26* (12), 1293.
72. Park, M. V. D. Z.; Neigh, A. M.; Vermeulen, J. P.; de la Fonteyne, L. J. J.; Verharen, H. W.; Briedé, J. J.; van Loveren, H.; de Jong, W. H. *Biomaterials* **2011**, *32* (36), 9810.
73. Soenen, S. J.; Parak, W. J.; Rejman, J.; Manshian, B. *Chem. Rev.* **2015**, *115* (5), 2109.
74. Kim, T. H.; Kim, M.; Park, H. S.; Shin, U. S.; Gong, M. S.; Kim, H. W. *J. Biomed. Mater. Res. — Part A* **2012**, *100 A* (4), 1033.
75. Santel, A.; Aleku, M.; Keil, O.; Endruschat, J.; Esche, V.; Fisch, G.; Dames, S.; Löffler, K.; Fechtner, M.; Arnold, W.; Giese, K.; Klippel, A.; Kaufmann, J. *Gene Ther.* **2006**, *13* (16), 1222.
76. Santel, A.; Aleku, M.; Keil, O.; Endruschat, J.; Esche, V.; Durieux, B.; Löffler, K.; Fechtner, M.; Röhl, T.; Fisch, G.; Dames, S.; Arnold, W.; Giese, K.; Klippel, A.; Kaufmann, J. *Gene Ther.* **2006**, *13* (18), 1360.
77. Kaufmann, J.; Ahrens, K.; Santel, A. *Microvasc. Res.* **2010**, *80* (2), 286.
78. Koide, H.; Asai, T.; Hatanaka, K.; Urakami, T.; Ishii, T.; Kenjo, E.; Nishihara, M.; Yokoyama, M.; Ishida, T.; Kiwada, H.; Oku, N. *Int. J. Pharm.* **2008**, *362* (1–2), 197.
79. Iversen, T.-G.; Skotland, T.; Sandvig, K. *Nano Today* **2011**, *6* (2), 176.
80. Champion, J. A; Katare, Y. K.; Mitragotri, S. *Proc. Natl. Acad. Sci. U. S. A.* **2007**, *104* (29), 11901.
81. Doshi, N.; Prabhakarpandian, B.; Rea-Ramsey, A.; Pant, K.; Sundaram, S.; Mitragotri, S. *J. Control. Release* **2010**, *146* (2), 196.
82. Anselmo, A. C.; Modery-Pawlowski, C. L.; Menegatti, S.; Kumar, S.; Vogus, D. R.; Tian, L. L.; Chen, M.; Squires, T. M.; Sen Gupta, A.; Mitragotri, S. *ACS Nano* **2014**, *8* (11), 11243.
83. Champion, J. A.; Mitragotri, S. *Pharm. Res.* **2009**, *26* (1), 244.

84. Stoehr, L. C.; Gonzalez, E.; Stampfl, A.; Casals, E.; Duschl, A.; Puntes, V.; Oostingh, G. J. *Part. Fibre Toxicol.* **2011**, *8*, 36.
85. Gratton, S. E. A; Ropp, P. A; Pohlhaus, P. D.; Luft, J. C.; Madden, V. J.; Napier, M. E.; DeSimone, J. M. *Proc. Natl. Acad. Sci. U. S. A.* **2008**, *105* (33), 11613.
86. Muro, S.; Garnacho, C.; Champion, J. A.; Leferovich, J.; Gajewski, C.; Schuchman, E. H.; Mitragotri, S.; Muzykantov, V. R. *Mol. Ther.* **2008**, *16* (8), 1450.
87. Kolhar, P.; Anselmo, A. C.; Gupta, V.; Pant, K.; Prabhakarpandian, B.; Ruoslahti, E.; Mitragotri, S. *Proc. Natl. Acad. Sci.* **2013**, *110* (26), 10753.
88. Decuzzi, P.; Ferrari, M. *Biomaterials* **2006**, *27* (30), 5307.
89. Champion, J. A; Mitragotri, S. *Proc. Natl. Acad. Sci. U. S. A.* **2006**, *103*, 4930.
90. Gilbert, J. B.; O'Brien, J. S.; Suresh, H. S.; Cohen, R. E.; Rubner, M. F. *Adv. Mater.* **2013**, *25* (41), 5948.
91. Geng, Y.; Dalhaimer, P.; Cai, S.; Tsai, R.; Tewari, M.; Minko, T.; Discher, D. E. *Nat. Nanotechnol.* **2007**, *2* (4), 249.
92. Decuzzi, P.; Ferrari, M. *Biophys. J.* **2008**, *94* (10), 3790.
93. Chithrani, B. D.; Ghazani, A. A.; Chan, W. C. W. *Nano Lett.* **2006**, *6* (4), 662.
94. Yoo, J. W.; Doshi, N.; Mitragotri, S. *Macromol. Rapid Commun.* **2010**, *31* (2), 142.
95. Devarajan, P. V; Jindal, A. B.; Patil, R. R.; Mulla, F.; Gaikwad, R. V.; Samad, A. *J. Pharm. Sci.* **2010**, *99* (6), 2576.
96. Discher, D. E.; Janmey, P.; Wang, Y.-L. *Science* **2005**, *310* (5751), 1139.
97. Merkel, T. J.; Chen, K.; Jones, S. W.; Pandya, A. A.; Tian, S.; Napier, M. E.; Zamboni, W. E.; Desimone, J. M. *J. Control. Release* **2012**, *162* (1), 37.
98. Haghgooie, R.; Toner, M.; Doyle, P. S. *Macromol. Rapid Commun.* **2010**, *31* (2), 128.
99. Beningo, K. A.; Wang, Y. *J. Cell Sci.* **2002**, *115*, 849.
100. Anselmo, A. C.; Zhang, M.; Kumar, S.; Vogus, D. R.; Menegatti, S.; Helgeson, M. E.; Mitragotri, S. *ACS Nano* **2015**, *9* (3), 3169.

101. Herant, M.; Heinrich, V.; Dembo, M. *J. Cell Sci.* **2005**, *118* (Pt 9), 1789.
102. Kirchner, C.; Liedl, T.; Kudera, S.; Pellegrino, T.; Javier, A. M.; Gaub, H. E.; Stölzle, S.; Fertig, N.; Parak, W. J. *Nano Lett.* **2005**, *5* (2), 331.
103. Labarre, D.; Vauthier, C.; Chauvierre, C.; Petri, B.; Müller, R.; Chehimi, M. M. *Biomaterials* **2005**, *26* (24), 5075.
104. Lemarchand, C.; Gref, R.; Couvreur, P. *Eur. J. Pharm. Biopharm. Off. J. Arbeitsgemeinschaft für Pharm. Verfahrenstechnik e.V* **2004**, *58* (2), 327.
105. Lemarchand, C.; Gref, R.; Passirani, C.; Garcion, E.; Petri, B.; Müller, R.; Costantini, D.; Couvreur, P. *Biomaterials* **2006**, *27* (1), 108.
106. Lemarchand, C.; Gref, R.; Lesieur, S.; Hommel, H.; Vacher, B.; Besheer, A.; Maeder, K.; Couvreur, P. *J. Control. Release* **2005**, *108* (1), 97.
107. Passirani, C.; Barratt, G.; Devissaguet, J. P.; Labarre, D. *Pharm. Res.* **1998**, *15* (7), 1046.
108. Passirani, C.; Barratt, G.; Devissaguet, J.-P.; Labarre, D. *Life Sci.* **1998**, *62* (8), 775.
109. Harisinghani, M. G.; Saini, S.; Weissleder, R.; Halpern, E. F.; Schima, W.; Rubin, D. L.; Stillman, A. E.; Sica, G. T.; Small, W. C.; Hahn, P. F. *Radiology* **1997**, *202* (3), 687.
110. Hahn, P. F.; Saini, S. *Radiol. Clin. North Am.* **1998**, *36* (2), 287.
111. Harper, G. R.; Davis, S. S.; Davies, M. C.; Norman, M. E.; Tadros, T. F.; Taylor, D. C.; Irving, M. P.; Waters, J. A.; Watts, J. F. *Biomaterials* **1995**, *16* (6), 427.
112. Immordino, M. L.; Dosio, F.; Cattel, L. *Int. J. Nanomedicine* **2006**, *1* (3), 297.
113. Sun, G.; Hagooly, A.; Xu, J.; Nyström, A. M.; Li, Z.; Rossin, R.; Moore, D. A.; Wooley, K. L.; Welch, M. J. *Biomacromolecules* **2008**, *9* (7), 1997.
114. Pressly, E. D.; Rossin, R.; Hagooly, A.; Fukukawa, K.-I.; Messmore, B. W.; Welch, M. J.; Wooley, K. L.; Lamm, M. S.; Hule, R. A.; Pochan, D. J.; Hawker, C. J. *Biomacromolecules* **2007**, *8* (10), 3126.
115. Buyens, K.; De Smedt, S. C.; Braeckmans, K.; Demeester, J.; Peeters, L.; Van Grunsven, L. A.; De Mollerat Du Jeu, X.; Sawant, R.;

Torchilin, V.; Farkasova, K.; Ogris, M.; Sanders, N. N. *J. Control. Release* **2012**, *158* (3), 362.

116. Allen, C.; Dos Santos, N.; Gallagher, R.; Chiu, G. N. C.; Shu, Y.; Li, W. M.; Johnstone, S. A.; Janoff, A. S.; Mayer, L. D.; Webb, M. S.; Bally, M. B. *Biosci. Rep.* **2002**, *22* (2), 225.
117. Li, S.-D.; Huang, L. *J. Control. Release* **2010**, *145* (3), 178.
118. Avgoustakis, K.; Beletsi, A.; Panagi, Z.; Klepetsanis, P.; Livaniou, E.; Evangelatos, G.; Ithakissios, D. S. *Int. J. Pharm.* **2003**, *259* (1–2), 115.
119. Mosqueira, V. C. F.; Legrand, P.; Morgat, J. L.; Vert, M.; Mysiakine, E.; Gref, R.; Devissaguet, J. P.; Barratt, G. *Pharm. Res.* **2001**, *18* (10), 1411.
120. Beletsi, A.; Panagi, Z.; Avgoustakis, K. *Int. J. Pharm.* **2005**, *298* (1), 233.
121. Storm, G.; Belliot, S. O.; Daemen, T.; Lasic, D. D. *Adv. Drug Deliv. Rev.* **1995**, *17* (1), 31.
122. Meng, F.; Engbers, G. H. M.; Gessner, A.; Müller, R. H.; Feijen, J. *J. Biomed. Mater. Res. A* **2004**, *70* (1), 97.
123. Moghimi, S. M.; Szebeni, J. *Prog. Lipid Res.* **2003**, *42* (6), 463.
124. Moghimi, S. M.; Gray, T. *Clin. Sci. (Lond).* **1997**, *93* (4), 371.
125. Armstrong, T. I.; Moghimi, S. M.; Davis, S. S.; Illum, L. *Pharmaceutical Research.* 1997, pp 1629–1633.
126. Armstrong, J. K.; Hempel, G., Koling, S.; Chan, L. S.; Fisher, T.; Meiselman, H. J.; Garratty, G. *Cancer* **2007**, *110* (1), 103.
127. Wang, X.; Ishida, T.; Kiwada, H. *J. Control. Release* **2007**, *119* (2), 236.
128. Sheng, Y.; Liu, C.; Yuan, Y.; Tao, X.; Yang, F.; Shan, X.; Zhou, H.; Xu, F. *Biomaterials* **2009**, *30* (12), 2340.
129. Erbacher, P.; Bettinger, T.; Belguise-Valladier, P.; Zou, S.; Coll, J. L.; Behr, J. P.; Remy, J. S. *J. Gene Med.* **1999**, *1* (3), 210.
130. Hong, R.; Huang, C.; Tseng, Y. *Clin. Cancer Res.* **1999**, *5* (November), 3645.
131. McCracken, M. N.; Cha, A. C.; Weissman, I. L. *Clin. Cancer Res.* **2015**, *21* (16), 3597.
132. Gardai, S. J.; McPhillips, K. A.; Frasch, S. C.; Janssen, W. J.; Starefeldt, A.; Murphy-Ullrich, J. E.; Bratton, D. L.; Oldenborg, P. A.; Michalak, M.; Henson, P. M. *Cell* **2005**, *123* (2), 321.

133. Chao, M. P.; Jaiswal, S.; Weissman-Tsukamoto, R.; Alizadeh, A. A; Gentles, A. J.; Volkmer, J.; Weiskopf, K.; Willingham, S. B.; Raveh, T.; Park, C. Y.; Majeti, R.; Weissman, I. L. *Sci. Transl. Med.* **2010**, *2* (63), 63ra94.
134. Chao, M. P.; Majeti, R.; Weissman, I. L. *Nat. Rev. Cancer* **2011**, *12* (1), 58.
135. Willingham, S. B.; Volkmer, J.-P.; Gentles, a. J.; Sahoo, D.; Dalerba, P.; Mitra, S. S.; Wang, J.; Contreras-Trujillo, H.; Martin, R.; Cohen, J. D.; Lovelace, P.; Scheeren, F. A.; Chao, M. P.; Weiskopf, K.; Tang, C.; Volkmer, A. K.; Naik, T. J.; Storm, T. A.; Mosley, A. R.; Edris, B.; Schmid, S. M.; Sun, C. K.; Chua, M.-S.; Murillo, O.; Rajendran, P.; Cha, A. C.; Chin, R. K.; Kim, D.; Adorno, M.; Raveh, T.; Tseng, D.; Jaiswal, S.; Enger, P. O.; Steinberg, G. K.; Li, G.; So, S. K.; Majeti, R.; Harsh, G. R.; van de Rijn, M.; Teng, N. N. H.; Sunwoo, J. B.; Alizadeh, A. A.; Clarke, M. F.; Weissman, I. L. *Proc. Natl. Acad. Sci.* **2012**, *109* (17), 6662.
136. Vonderheide, R. H. *Nat. Med.* **2015**, *21* (10), 1122.
137. Oldenborg, P. a; Zheleznyak, A.; Fang, Y. F.; Lagenaur, C. F.; Gresham, H. D.; Lindberg, F. P. *Science* **2000**, *288* (5473), 2051.
138. Blazar, B. R.; Lindberg, F. P.; Ingulli, E.; Panoskaltsis-Mortari, A.; Oldenborg, P.-A.; Iizuka, K.; Yokoyama, W. M.; Taylor, P. A. *J. Exp. Med.* **2001**, *194* (4), 541.
139. Barclay, A. N.; van den Berg, T. K. *Annu. Rev. Immunol.* **2014**, *32* (1), 25.
140. Zheng, Q.; Van Die, I.; Cummings, R. D. *J. Biol. Chem.* **2002**, *277* (42), 39833.
141. Olsson, M.; Bruhns, P.; Frazier, W. A.; Ravetch, J. V.; Oldenborg, P. A. *Blood* **2005**, *105* (9), 3577.
142. Tsai, R. K.; Discher, D. E. *J. Cell Biol.* **2008**, *180* (5), 989.
143. Liu, X.; Pu, Y.; Cron, K.; Deng, L.; Kline, J.; Frazier, W. A.; Xu, H.; Peng, H.; Fu, Y.-X.; Xu, M. M. *Nat. Med.* **2015**, *21* (10), 1209.
144. Lindberg, F. P.; Lublin, D. M.; Telen, M. J.; Veile, R. A.; Miller, Y. E.; Donis-Keller, H.; Brown, E. J. *J. Biol. Chem.* **1994**, *269* (3), 1567.
145. Parkinson, J. E.; Sanderson, C. M.; Smith, G. L. *Virology* **1995**, *214* (1), 177.

146. Jaiswal, S.; Jamieson, C. H. M.; Pang, W. W.; Park, C. Y.; Chao, M. P.; Majeti, R.; Traver, D.; van Rooijen, N.; Weissman, I. L. *Cell* **2009**, *138* (2), 271.
147. Majeti, R.; Chao, M. P.; Alizadeh, A. A.; Pang, W. W.; Jaiswal, S.; Gibbs, K. D.; van Rooijen, N.; Weissman, I. L. *Cell* **2009**, *138* (2), 286.
148. Hsu, Y. C.; Acuña, M.; Tahara, S. M.; Peng, C. A. *Pharm. Res.* **2003**, *20* (10), 1539.
149. Matozaki, T.; Murata, Y.; Okazawa, H.; Ohnishi, H. *Trends Cell Biol.* **2009**, *19* (2), 72.
150. Roberts, D. D.; Miller, T. W.; Rogers, N. M.; Yao, M.; Isenberg, J. S. *Matrix Biol.* **2012**, *31* (3), 162.
151. Isenberg, J. S.; Maxhimer, J. B.; Powers, P.; Tsokos, M.; Frazier, W. A.; Roberts, D. D. *Surgery* **2008**, *144* (5), 752.
152. Isenberg, J. S.; Maxhimer, J. B.; Hyodo, F.; Pendrak, M. L.; Ridnour, L. A.; DeGraff, W. G.; Tsokos, M.; Wink, D. A.; Roberts, D. D. *Am. J. Pathol.* **2008**, *173* (4), 1100.
153. Isenberg, J. S.; Hyodo, F.; Matsumoto, K.-I.; Romeo, M. J.; Abu-Asab, M.; Tsokos, M.; Kuppusamy, P.; Wink, D. A.; Krishna, M. C.; Roberts, D. D. *Blood* **2007**, *109* (5), 1945.
154. Maxhimer, J. B.; Soto-Pantoja, D. R.; Ridnour, L. A.; Shih, H. B.; Degraff, W. G.; Tsokos, M.; Wink, D. A.; Isenberg, J. S.; Roberts, D. D. *Sci. Transl. Med.* **2009**, *1* (3), 3ra7.
155. Kaur, S.; Soto-Pantoja, D. R.; Stein, E. V; Liu, C.; Elkahloun, A. G.; Pendrak, M. L.; Nicolae, A.; Singh, S. P.; Nie, Z.; Levens, D.; Isenberg, J. S.; Roberts, D. D. *Sci. Rep.* **2013**, *3*, 1673.
156. Chan, K. S.; Espinosa, I.; Chao, M.; Wong, D.; Ailles, L.; Diehn, M.; Gill, H.; Presti, J.; Chang, H. Y.; van de Rijn, M.; Shortliffe, L.; Weissman, I. L. *Proc. Natl. Acad. Sci. U. S. A.* **2009**, *106* (33), 14016.
157. Kim, M. J.; Lee, J.-C.; Lee, J.-J.; Kim, S.; Lee, S. G.; Park, S.-W.; Sung, M. W.; Heo, D. S. *Tumour Biol.* **2008**, *29* (1), 28.
158. Chao, M. P.; Alizadeh, A. A.; Tang, C.; Jan, M.; Weissman-Tsukamoto, R.; Zhao, F.; Park, C. Y.; Weissman, I. L.; Majeti, R. *Cancer Res.* **2011**, *71* (4), 1374.

159. Chao, M. P.; Alizadeh, A. A.; Tang, C.; Myklebust, J. H.; Varghese, B.; Gill, S.; Jan, M.; Cha, A. C.; Chan, C. K.; Tan, B. T.; Park, C. Y.; Zhao, F.; Kohrt, H. E.; Malumbres, R.; Briones, J.; Gascoyne, R. D.; Lossos, I. S.; Levy, R.; Weissman, I. L.; Majeti, R. *Cell* **2010**, *142* (5), 699.
160. Zhao, X. W.; van Beek, E. M.; Schornagel, K.; Van der Maaden, H.; Van Houdt, M.; Otten, M. A.; Finetti, P.; Van Egmond, M.; Matozaki, T.; Kraal, G.; Birnbaum, D.; van Elsas, A.; Kuijpers, T. W.; Bertucci, F.; van den Berg, T. K. *Proc. Natl. Acad. Sci. U. S. A.* **2011**, *108* (45), 18342.
161. Soto-Pantoja, D. R.; Miller, T. W.; Frazier, W. A.; Roberts, D. D. *Proc. Natl. Acad. Sci. U. S. A.* **2012**, *109* (42), E2842; author reply E2844.
162. Zhao, X. W.; Kuijpers, T. W.; van den Berg, T. K. *Blood* **2012**, *119* (18), 4333.
163. Mohandas, N.; Chasis, J. A. *Semin. Hematol.* **1993**, *30* (3), 171.
164. Chambers, E.; Mitragotri, S. *Exp. Biol. Med.* (*Maywood*). **2007**, *232* (7), 958.
165. Chambers, E.; Mitragotri, S. *J. Control. Release* **2004**, *100* (1), 111.
166. Muzykantov, V. R. *Expert Opin. Drug Deliv.* **2013**, *10* (1), 1.
167. Muzykantov, V. R.; Sakharov, D. V.; Smirnov, M. D.; Samokhin, G. P.; Smirnov, V. N. *BBA — Gen. Subj.* **1986**, *884* (2), 355.
168. Shi, G.; Mukthavaram, R.; Kesari, S.; Simberg, D. *Adv. Healthc. Mater.* **2014**, *3* (1), 142.
169. Zaitsev, S.; Kowalska, M. A.; Neyman, M.; Carnemolla, R.; Tliba, S.; Ding, B. Sen; Stonestrom, A.; Spitzer, D.; Atkinson, J. P.; Poncz, M.; Cines, D. B.; Esmon, C. T.; Muzykantov, V. R. *Blood* **2012**, *119* (20), 4779.
170. Krantz, A. *Blood Cells. Mol. Dis.* **1997**, *23* (1), 58.
171. Hamidi, M.; Zarrin, A.; Foroozesh, M.; Mohammadi-Samani, S. *J. Control. Release* **2007**, *118* (2), 145.
172. Anselmo, A. C.; Gupta, V.; Zern, B. J.; Pan, D.; Zakrewsky, M.; Muzykantov, V. **2013**, No. 12, 11129.
173. Muzykantov, V. R. *Expert Opin. Drug Deliv.* **2010**, *7* (4), 403.
174. Zaitsev, S.; Spitzer, D.; Murciano, J. C.; Ding, B. Sen; Tliba, S.; Kowalska, M. A.; Marcos-Contreras, O. A.; Kuo, A.; Stepanova, V.;

Atkinson, J. P.; Poncz, M.; Cines, D. B.; Muzykantov, V. R. *Blood* **2010**, *115* (25), 5241.

175. Danielyan, K.; Ganguly, K.; Ding, B. Sen; Atochin, D.; Zaitsev, S.; Murciano, J. C.; Huang, P. L.; Kasner, S. E.; Cines, D. B.; Muzykantov, V. R. *Circulation* **2008**, *118* (14), 1442.
176. Shuvaev, V. V.; Brenner, J. S.; Muzykantov, V. R. *J. Control. Release* **2015**, *219*, 576.

# Chapter 3

# How do Nanoparticles (NPs) Pass Barriers?

Saeid Zanganeh*, Ryan Spitler*,
Najme Javdani†, and Jim Q. Ho‡

* *Department of Radiology, Molecular Imaging Program at Stanford, Stanford University, CA, USA*
† *Institut De Recherche Clinique De Montreal, Montreal, Canada*
‡ *Middlebury College, Middlebury, VT, USA*

## Abbreviations

| | |
|---|---|
| BBBs | brain blood barriers |
| CME | clathrin-mediated endocytosis |
| CvME | caveolin-mediated endocytosis |
| IV | intravenous administration |
| NPs | nanoparticles |
| SWNT | single-walled carbon nanotubes |
| RES | reticuloendothelial system |

## I. Introduction

One of the essential requirements of any delivery system is its ability to move about freely in available environments and have the ability to cross various barriers that are encountered. Nanoparticles (NPs) possess several key advantages as therapeutic materials, which allow some or all of these obstacles to be overcome. For example, the small size of some NPs allows them to pass through biological barriers and extend bioavailability in tumor tissues.[1] Another attractive characteristic of NPs is that the properties of NPs can be engineered through synthesis including size, shape, and surface characteristics. For instance, the addition of ligands or receptors onto the surface of NPs can enhance their target-specificity.[1]

The body has many natural biological barriers, which create challenges for drug administration. However, much work is underway to address this issue in an effort to bypass these barriers to better treat different disease. By deeply studying the various obstacles that NPs have to pass, it will enable engineers to design more applicable and efficient nanomaterials for drug delivery. Some of the most challenging barriers include blood vessels barriers, cell barriers, and brain blood barriers (BBBs).

Within different cell types there are specific mechanisms that select which materials can pass. One such mechanism is the endocytosis system in which only particular particle types can be transported through the cell barriers. NPs, based on their composition and cell targeting characteristics, have the advantage that they can use the endocytosis machinery to typically enter almost any cell type. However, several other physicochemical characteristics of NPs such as NP size, shape surface functional groups that give rise to charge and surface chemistry influence their cellular uptake.

Of particular importance, it should be noted that the NP properties are not necessarily applicable in biological mediums. In biological environments the surface of the NP will become coated with a layer of biomolecules especially proteins, which will form a protein corona. The composition and molecular properties of the protein corona that forms generally change the behavior of the NPs especially for the cellular uptake.

In this chapter, an understanding of how NPs pass barriers will be described; first the biological barriers and cell uptake mechanisms will be introduced, followed by NPs properties, which affect their transport and uptake.

## II. Biological Barriers

Some of the major challenges in the NP drug delivery field is to improve transport of therapeutics across biological barriers including the small intestine, nasal, skin, ocular, the mouth mucosa, BBB and cell membrane. Many natural defense mechanisms of the human body represent significant barriers for drug delivery.

Under normal conditions these barriers are designed to keep foreign material out of the body and only allow specific small molecules across. Thus, any obstacle that prevents a drug from reaching its target site of action is to be considered a barrier to drug delivery. A strong fundamental understanding of how these barriers function, will allow for the design of novel drug delivery systems based on nanotechnology.[2]

### A. Blood vessel barriers

A primary barrier for systemic NP delivery are blood vessel barriers, which normally transport materials in the body. As blood moves through the vessels, they become thinner and thinner until they are finally converted to capillaries through extensive branching and narrowing. These very narrow capillaries are in close vicinity to the individual cells. Once reaching their narrowest sizes, the capillaries begin to merge, forming veins. The blood from the veins then takes the contents back to the heart where they are recirculated.[2] However, simply traveling along the vessels does not accomplish efficient drug delivery. Most importantly, the delivery system must reach the target destination. This initially requires crossing of the blood capillary wall in order to reach the extracellular fluid of the tissue and then once again crossing of other non-target cells, which can also represent a barrier until finally entering the target cell. It is clear that there are major barriers for NPs during transit, especially in going through capillaries to reach the target site.

The two routes that NPs use in order to traverse blood capillaries and other cell layers are the transcellular and paracellular routes. In passing through the transcellular route, a particulate system will enter the cell from one side and leave the cell from the opposite side in order to reach a tissue. On the other hand, in the paracellular system the NPs move between the cells using the junctions. These junctions have the advantage of not requiring a system to avoid being destroyed by the cell. Some common moieties using the paracellular pathway include ions, larger molecules, and leukocytes. This process is controlled by the association of tight and adherence junctions to the cytoskeleton, which is referred to as the apical junction complex. Tight junctions act as a barrier for regulation, while adherence junctions control the development and stabilization of tight junctions. Due to differences in structure and presence of tight junctions, epithelial and endothelial barriers have different permeability values. Thus, tight junctions determine what is transported by the paracellular pathway transport. For example, the diffusion of larger molecules will likely not be feasible, however the migration of certain cell types could be permitted. Certain regions of the body are particularly stringent, especially the epithelia and brain capillary endothelium, while the vascular endothelium in various other tissues tends to have much greater permeability. A strong fundamental understanding of this regulation mechanism is essential in order to enable the development of NPs that reach their intended target cells with minimal off target effects.

## B. Blood brain barriers (BBB)

The blood-brain barrier (BBB) is a highly specialized and stringent barrier system consisting of endothelial cells that separates the blood from the body and underlying brain cells, which in turn provides protection to brain cells and overall brain homeostasis. The brain restricts the paracellular pathway as well using an endothelium containing a complex arrangement of tight junctions that restricts the passage of certain molecules. The glial processes that surround the brain serve as an additional physical barrier. The BBB is permeable to small and lipophilic molecules (up to 800 atomic mass units), however larger molecules will not be transported across this barrier without the use of an active transport system.[3]

### 1. *Nanoparticles used to cross the BBB*

One important consideration for the application of nanobiotechnology in the brain is the ability to facilitate the delivery of therapeutics across the BBB. Some NPs known to pass the BBB include lipid NPs, liposome, and polymeric NPs.

A lipid NP is composed of nanometer-sized spheres that are surrounded by a lipid bilayer. The lipophilic features on the surface of these NPs facilitate traversing the BBB to enter the brain using the endocytosis pathway.[4] In order to achieve crossing the BBB with liposomes, properties such as lipid composition, size, surface charge, and the preparation method must be considered. Generally small liposomes or nanoliposomes usually measure between 25–50 nm in diameter. Additionally, liposomes can be functionalized using ApoE-derived peptides, which can facilitate cellular uptake and transport across the BBB.[5] Polymeric NPs can be engineered for targeted drug delivery across the BBB, controlled release at the target site and can be biodegradable.[2]

## C. The biology of a tumor and therapeutic barriers

Delivering drugs within tumor cells can be very challenging due to the structure and environment of the tumor mass. However, there are some distinct characteristics of tumor cells such as rapid angiogenesis and local necrosis that can be used for targeted delivery. Due to this rapid angiogenesis, vessel formation is abnormal leading to vessels that are branched, short-circuited and of uneven diameter, which causes poorly constructed leaky walls. Depending on the type of tumor, the gaps between the endothelial cells are abnormally large,[2] 200 nm to 1.2 $\mu$m in comparison with the gap in healthy vessels that are ~5–10 nm.[2] Poor vascularization, as a result of abnormal vessels in these tumors leads to a lack of oxygen and nutrients driving anaerobic metabolism.[2] During anaerobic metabolism lactic acid is produced as a result of the low pH environment, which is typical in tumors. For comparison, the pH of blood is ~7.4, while tumors can have pH values as low as 5.6.[2]

Factors such as angiogenesis, leakiness of vessels and acidic environments are properties that can be exploited for targeted delivery. However there are also a number of tumor properties that can make drug delivery extremely challenging. Tumor cells can be hard to reach, with some cells surviving even under marginal conditions and away from blood vessels. New methods in NP design in combination with an understanding of the biological systems involved will be necessary in order to design successful treatments that are effective for the eradication of the entire tumor.[2]

## III. Cellular Uptake Mechanism

Prior to engineering NPs, a fundamental understanding of the mechanisms involved such as internalization mechanisms and the intracellular routing is essential. The cell membrane is an important and essential component of every cell. This essential barrier is required in order to separate the internal from the external environment of the cell.[2] Some small molecules such as ions, some small proteins, and molecules (glucose and amino acids) can be transported through specialized transporter proteins, while larger substances such as most nutrients, antigens, antibodies, signaling, growth factors, and NPs cannot diffuse through the membrane and are thus internalized through endocytosis.[2]

Endocytosis requires energy and involves folding the plasma membrane into vesicles, which are then released from the membrane, and its contents transported into the cell.[6] There are two general categories of the endocytosis mechanisms: phagocytosis and pinocytosis.

### A. Phagocytosis

Phagocytosis is a common mechanism for the uptake of particles typically larger than 500 nm in diameter and is essential for the elimination of viruses, bacteria and infected cells by the immune system using macrophages, dendritic cells, and neutrophils. Initially a particle is enclosed within the phagocytic pocket; the cell membrane is then closed forming a sealed vacuole with the particle inside, which is referred to as the phagosome.[7,8]

There are three stages of phagocytosis as shown in Figure 1) opsonization and reorganization of the particles 2) particle adhesion to the cell membrane 3) particle ingestion by the cell. Opsonins coat foreign particles, making them visible to phagocytic cells.[2] The opsonin-coated foreign particles are first recognized by receptors on the cell surface, followed by receptor binding which induces a cascade of intracellular signaling events triggering the ingestion of the

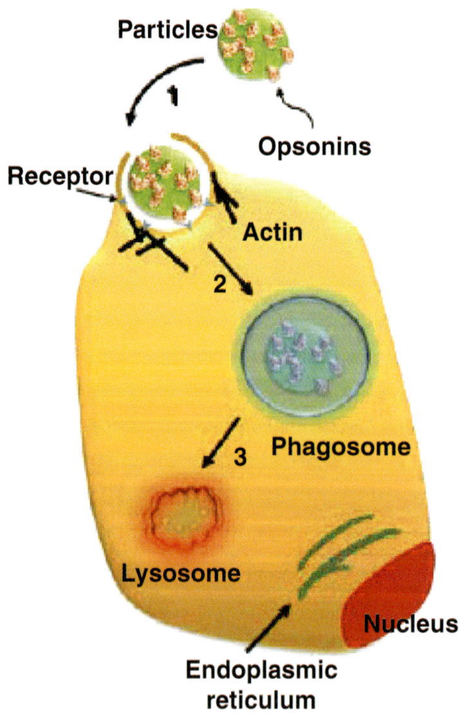


Figure 1. Simplified schematic of the phagocytosis of particles in three steps. Phagocytosis is an actin process dependent process. Taken from Ref. [23] with permission of Elsevier.

foreign particles.[8,9] The particle shape at the point of where cell attachment occurs can also trigger phagocytosis.[2]

## B. Pinocytosis

Pinocytosis is defined as the uptake of fluids and molecules within small vesicles during phagocytosis, and generally engulfing particles of less than approximately 500 nm in diameter, (e.g. microorganisms and cell debris). Pinocytosis is known to occur during several different processes including macropinocytosis, clathrin-mediated endocytosis (CME), caveolin-mediated endocytosis (CvME), and clathrin- and caveolin-independent endocytosis.[8,10] The five endocytic pathways, phagocytosis included, are illustrated in Figure 2. As depicted, macropinocytosis like phagocytosis is an actin dependent process.[11,12] All the above mechanisms are described in detail as follow:

### 1. *Macropinocytosis*

Macropinocytosis is a non-specific mechanism where fluid contents are taken up at the same concentration as the surrounding environment. In contrast to CvME and CME, uptake by macropinocytosis is non-specific and is not receptor mediated,[2] with the vesicle size typically being 100 nm to 5 mm.

Macropinocytosis is an actin dependent process and forms protrusions at the outer cell membrane, which then can go on to

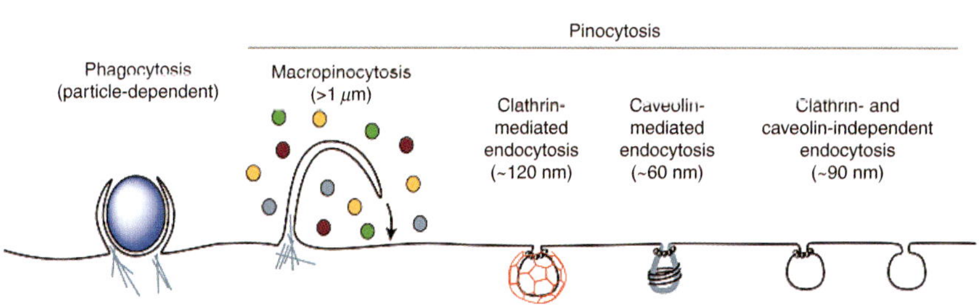


Figure 2. The different endocytic pathways in a mammal cell. The pathways differ in size and mechanism of vesicle formation. Taken from Ref. [10] with permission of Nature Publishing Group.

fuse with the cell membrane when larger fragments or debris are taken up.[2]

2. *Clathrin-mediated endocytosis (CME)*

CME is an example of receptor-mediated endocytosis having vesicle sizes of ~120 nm.[2] This type of endocytosis is well characterized and similar to other pinocytotic pathways, representing a form of receptor-mediated endocytosis. This pathway is crucial for the uptake of several molecules including low-density lipoprotein and transferrin.[2,13]

3. *Caveolin-mediated endocytosis (CvME)*

CvME is the primary mechanisms of endocytosis, which occurs after CME. This endocytosis mechanism captures small particles typically in the diameter range 60 nm.[2] Numerous cell types including the capillary endothelium, type I epithelial cells, muscle cells and fibroblasts exhibit CvME, which occurs where there are lipid rafts.[10,14]

4. *Clathrin- and caveolin-independent endocytosis*

Clathrin- and caveolin-independent endocytosis produces vesicles of ~50–100 nm and is not a selective mechanism (i.e. it is not dependent on receptors or stimuli from the internalized material).[23]

Yet, not all cell types contain the required machinery in order to perform the entire full spectrum of endocytic pathways. Therefore, these pathways are cell type specific and determine the trafficking and intracellular fate of the particles that are encountered.[2] For example, red blood cells are considered non-phagocytic cells, because they do not have any phagocytic receptors on their surface and no actin–myosin system. The non-phagocytic character of red blood cells allows them to serve as a good non-phagocytic model for how NPs penetrate through cell membrane.[2]

## C. Mechanisms for the uptakes of NPs into cells

There have been numerous reported studies on NP uptake mechanisms and each study was reported to be either phagocytosis or pinocytosis driven. It is unclear whether non-endocytic uptake mechanisms, such as diffusion or active transport, can be utilized by NPs.[2] The primary factors that determine if NPs will be taken up are the size and charge of their surface. Generally, particles above 200 nm in diameter are not internalized through a single clathrin-coated vesicle and are expected to have limited internalization through classical endocytosis.[2] The uptake of larger particles occurs through phagocytosis with liquid internalization occurring by macropinocytosis.

### 1. *Phagocytic uptake of NPs*

The local shape and geometry of NPs determine their initial attachment to the cell membrane in phagocytosis; however binding of opsonin proteins to the NP surface is mainly what triggers phagocytic uptake of NPs. It is known that opsonins will bind to most NPs, targeting them for rapid elimination through the reticuloendothelial system (RES) when injected *in vivo*.[7,15,16] However, the most important characteristics for NP opsonization are the ionic and hydrophobic interactions that occur with the NP surface.[8,9,17] The surface hydrophobicity and negative charge are seemingly key factors for opsonization.[2] Special coatings have been specifically designed for NPs to limit phagocytic uptake by cells associated with the immune system in order to avoid removal and sequestration into the RES.[2] For example, NPs coated with PEG delays immune clearance of foreign particles.[18,19]

### 2. *Clathrin-mediated endocytosis (CME) pathway for uptake of NPs*

CME pathway can be utilized for NP uptake; using low density lipoprotein and transferrin receptors in combination with drug delivery devices can result in rapid and selective cellular uptake.[6,20]

Internalization via CME is dependent on NP size (50–200 nm) and cell type.[21]

### 3. *Caveolae-mediated endocytosis (CvME) pathway for uptake of NPs*

The second major trafficking pathway used is caveolae-mediated endocytosis (CvME), which facilitates the uptake of biomolecules and NPs.[2] There are also several classes of NPs that enter cells through caveolae.[6] Based on *in vitro* studies, the CvME pathway is postulated to have a slower uptake mechanism.[2]

## IV. Macropinocytosis for Uptake of NPs

Nanoparticle uptake by endocytosis is a highly complex process involving the interplay of the energy release by the number of bound receptors, the number of vesicles that form, receptor diffusion toward the attached particle and time to complete the process. Being that for large NPs, only one particle is taken up per vesicle, attachment occurs to numerous receptors simultaneously, resulting in a substantial change in free energy. Also, larger particles require more time for membrane wrapping, with larger contact areas with the cell membrane resulting in a larger decrease in free energy.[21,22] In contrast, several smaller NPs can be taken up in one vacuole; however, only single attachment sites are used resulting in less free energy change.

### A. NPs properties influencing their uptake

There are many specific physical and chemical properties that will influence cellular internalization of NPs, not to mention the surrounding environment (i.e. cell medium) (Figure 3). Some of the key properties of NPs to consider are size, shape, and surface functional groups. Therefore, great effort has been put forth towards the development of NPs with well-defined characteristics. The effect of the core composition is not discussed here, because the surface

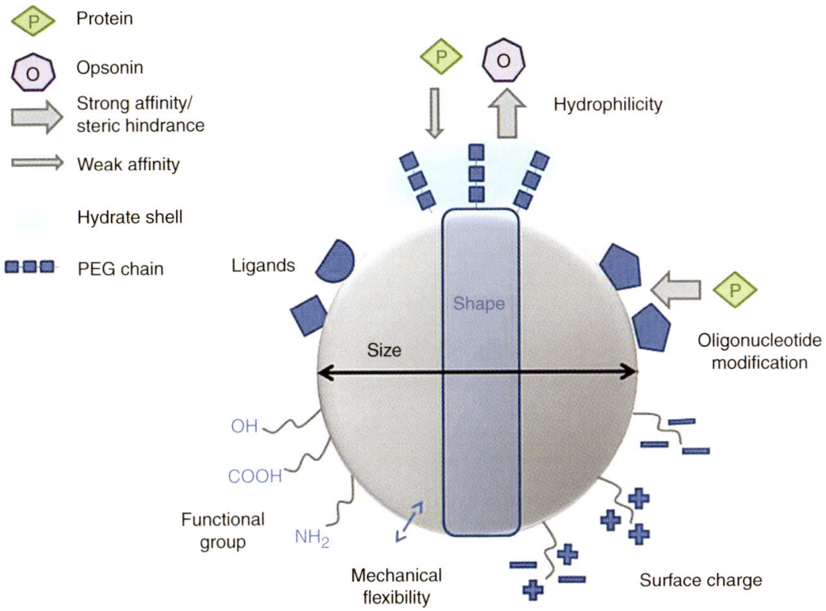


Figure 3. Determinants of nanoparticle interactions with cells as determined by experimental conditions (presence of protein and opsonins): shape (spheres, short–long rods, cubes, and triangles with different aspect ratios), cell type, size, surface chemistry, and addition of ligands. Taken from Ref. [2] with permission of Wiley & Sons.

characteristics are greatest, as the surface properties determine the protein corona and hence likely the biological impacts.[2]

## B. Effect of size

Some of the most critical general issues to consider in the field of nanotechnology is the size and effect on the biodistribution, the kinetics of release and cellular uptake of NPs.[2] In particular, spherical NPs up to 500 nm in diameter have been shown to be taken up by non-phagocytic cells, whereas none of the 1000 nm NPs were taken up by these cells.[2] Small NPs have the advantage of often not being recognized as foreign agents by macrophages and can enter macrophage cells through membrane pores; however, microparticles are easily taken up by RES.[23] Diffusion of NPs into cells can be facilitated by decreasing the size and increasing the surface area of the NPs.[2] For example, more efficient cellular and nuclear internalization was observed with 30 nm sized single-walled carbon nanotubes (SWNT) compared to 50 nm SWNTs.[2] The optimum NP size for internalization was determined to be 50 nm, and this was found to be true for a number of cell types.[2]

According to previous findings, the NP diameter was a determinant in the internalization pathway engaged. For example, clathrin coated pits control the penetration of microspheres having a diameter of less than 200 nm; however if the size is increased to 500 nm, the caveolae-mediated process is predominant for internalization.[24] Additionally, NPs with diameters of 50 nm are able to efficiently target human mesenchymal stem cells using clathrin, which is known to act independently from endocytosis.[2] Interestingly, in cancer cells, NP uptake increases as the size of the NP is decreased, which occurs regardless of surface modifications on the NPs.[25] For instance, submicron multi walls have demonstrated better cell internalization capacity than larger ones.[2]

However, the greatest efficiency for surface attachment has been demonstrated using particles with dimensions on the scale of 2–3 mm.[2] In summary, NP uptake efficiency by non-phagocytic cells illustrates a clear trend: cellular uptake initially increases as a function of NP size until reaching the optimum limit of ~50 nm and then uptake decreases when this size is exceeded. In contrast, for phagocytic cells, the uptake is not necessarily as related to just NP size.[22,24,26]

## C. Effect of shape and geometry

It is well known that the shape and geometry of NPs significantly impacts their ability to pass cellular barriers and become internalized. The optimal shape can confer several advantages such as improved circulation time, biodistribution and residency time of NPs once internalized inside cells.[27,28] Additionally, the shape of NPs can also influence cell targeting. For example, receptor-mediated endocytosis of nano-carriers is directly affected by their shape.[2] Elongated NPs have demonstrated increased efficiency for cell adherence in comparison to spherical NPs. The curve shape of spherical particles is known to limit the number of possible exposed binding sites, which are available to interact with the receptors on the target cell. In contrast, elongated NPs have more surface area available for multivalent interaction with the cell surface.[2] There, the higher aspect ratio of the elongated allow these NPs to be taken up by cells more frequently than spherical NP counterparts of the same size. For example, rod-shaped NPs which have higher aspect ratios will be taken up more efficiently by Hela and caco-2 cells, than lower aspect ratio NPs or larger 1 mm sized particles.[2]

Furthermore, rod-shaped NPs have a high affinity and internalization efficiency to endothelial cells, which allow for improved efficacy both *in vitro* and *in vivo* experiments.[2] It has also been shown, that discoid NPs are taken up better by Hela cells than rod-shaped NPs of the same volume.[2]

It should also be noted that NPs possessing sharp shapes tend to penetrate endosomal membranes and localize to the cytoplasm, leading to reduced exocytosis of these NPs compared to spherical NPs.[2] Other studies have shown differential cellular uptake of spherical NPs having different aspect ratios and resulting in a faster rate of absorption for those NPs with a high aspect ratio.[2] In addition, recent investigations have identified that ellipsoidal NPs are taken up less frequently than spherical NPs.[2]

## D. Effect of surface functional group

Surface modifications can change the various physical, chemical or biological characteristics at the surface of NPs, which can be used to manipulate the charge of the NPs. Understanding the interactions between the negative membrane of cells and NPs with varying charges is essential for the design of adaptable and useful NPs.[2]

### 1. *Neutral modification of particle surface*

Nonionic polymers are able to increase particle stability using steric stabilization, which limits the direct interaction between cells and the phagocytic system. PEG is a commonly used neutral and hydrophilic material that forms a barrier on the surface of NPs, which can lead to the steric stabilization, low opsonization, high blood circulation time and escape from reorganization by the phagocytosis process.[2] Another use of PEG is to modify the hydrophilicity of polyester or polyalkylcyanoacrylate NPs as needed.[2] Also it is known that reducing NP aggregation has been shown to improve cellular uptake. Additionally, both PEG and modified PLGA NPs have demonstrated as much as five-fold higher cellular uptake than unmodified NPs.[2] Another advantage of PEG or folic acid NP surface modifications is that they have been shown to reduce the adsorption of proteins and thus recognition by macrophages. In terms of possible applications, these modified NPs can be easily taken up by cancer cells for efficient and selective cancer therapies.[2] As an alternative approach to the PEGylation method, polysaccharides can also be used and are known to interact with specific receptors, which allows for targeted delivery.[2]

### 2. *Positive modification of particles surface*

While neutral modification results in less particle aggregation, neutrally charged NPs have led to low cellular uptake.[2] An explanation for this is the low affinity of the neutral NPs toward the negatively charged cell membrane. However, positive modification of NPs causes electrostatic interactions to be formed between the surface of the NPs and the negative charge on the cell membrane.[29,30] Thus, positively charged NPs have demonstrated more efficient uptake than negatively charged NPs using the endocytosis pathways. The negatively charged proteins found on the outer surface of gastrointestinal epithelial cells allowed for increased cell permeability with the use of positively charged drug carriers.[2]

### 3. *Negative modification of particles surface*

The negative modification of NPs has also been shown to affect their cellular uptake. However, cellular internalization of negatively charged NPs by RME (receptor-mediated endocytosis) has yielded some opposition. Negatively charged NPs have been reported to both increase cellular uptake efficiency as a function of increased charge as well as inhibit cellular uptake as reported by He *et al.*[2] Another surface property that affects cellular uptake is hydrophobicity.

## V. NPs and Biological Entities Interaction

Immediately after intravenous administration (IV), NPs come in contact with blood as part of the physiological environment. The blood plasma represents a highly complex environment containing hundreds of thousands of different proteins ranging up to 12 orders of magnitude in concentration based on the particular protein.[2] The blood plasma has a full range of proteins at different concentrations. Thus, upon IV injection of NPs entering the bloodstream, there is binding competition between various biological molecules, which determine what is adsorbed onto the NPs surface. Initially the most abundant proteins predominate adsorption on the NPs surface, however over longer periods of time these proteins will become replaced by higher affinity proteins.[2]

### A. Protein corona

When NPs are introduced into biological environments, a protein-rich layer is adsorbed onto their surface, which is known as the protein corona. Understanding the process of protein corona formation is critical as subsequent downstream interactions will be mediated based on the nature of this corona. However, understanding the relationship between the original functionality of NPs and protein corona formation can be very challenging and is an area that still remains elusive.[31,32] There are several factors that contribute to how the protein corona is formed; NP surface chemistry, size[33] and charge[2] are pivotal factors in determining the formation and nature of the resultant protein corona. The composition and molecular properties of the protein corona are known to be influential factors for the cellular uptake of NPs,[34,35] however, any direct links between structure and downstream effect still remain to be elucidated. Therefore, it is necessary to understand the adsorption of the protein corona on the NP surface to be able to reliably predict how they will then interact with cells after this has occurred.

It is important to note that there are a number of diverse biomarkers that bind and cover NPs almost immediately after they come in contact with biological fluids.[2] Protein coronas can be classified into two types: those proteins that bind to NPs with high affinity, known as a hard corona, or those with low affinity, known as a soft corona. The initial corona that is formed is considered as a "soft" protein corona, which allows for a more highly dynamic exchange of proteins, which will coat the surface of the NPs. However, over time the composition of the "soft" corona will slowly transition into a "hard", which has less dynamic exchange (Figure 4).[36,37]

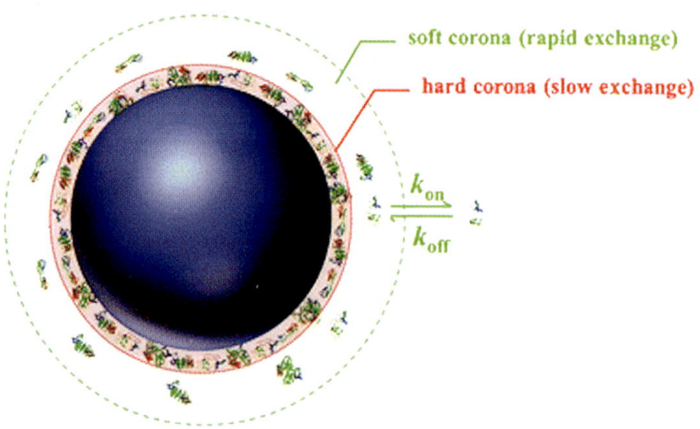


Figure 4. Illustration showing soft and hard corona on NPs in a biological environment. Taken from Ref. [45] with permission of Royal Society of Chemistry.

### 1. *Effect of corona on NP characteristics (size)*

While a NP may start out with at a particular size after synthesis, this size can vary after NPs are exposed to the biological environment. More specifically, the protein corona modifies several properties of the original NPs such as size and surface composition, which determine downstream events including physiological responses, kinetics, transport, accumulation, exclusion and toxicity of the NPs.

### 2. *Effect of corona on NP uptake*

As mentioned previously, the NP protein corona complex can influence cellular uptake. For example, changes to the adsorbed protein structure can even inhibit cellular uptake of NPs. Additionally, should the adsorbed protein corona become unfolded or undergo some other structural rearrangement, it may change accessibility of NPs to cell surface receptors.[43] However, it has been demonstrated that complete serum media and serum depletion of several abundant proteins do not influence the NP interaction with endothelial cells. There, the cellular association of NP protein complexes is a non-specific process and is mostly dependent upon the total amount of adsorbed proteins rather than the particular of type of protein coating on the NP surface.[2]

In general, even after addition of serum, the efficiency of NP uptake by phagocytes is unaffected, and contingent on the kind of protein present.[2] While many contradictory effects occur, this can be explained by a differential population of proteins in the medium, and by dynamic changes in protein corona composition over time, driven by protein abundance, affinity[38] and by opsonization. It has also been found that increasing the amount of serum present in the culture medium of non-phagocytic cells results in a reduction in cellular NP uptake. The extent of NP hydrophobicity decreases the effect of NPs while increasing hydrophobicity increases the effect. It is known that the hydrophobicity of NPs determines binding affinity to hydrophobic pockets of bovine serum albumin. Differential NP uptake by non-phagocytic cell lines is explained by a number of properties of the cells including different binding proteins present on the cell surface, charge differences among cells and the rate of receptor turnover.[39,40]

As previously discussed, NP uptake is a multi-step process, in which NPs initially adhere to the cell membrane and then are subsequently internalized into the cell by energy dependent pathways. Typically, the protein corona strongly reduces NP adhesion to cellular membranes in contrast to bare nano-materials and thus results in a concomitant decrease in nanoparticle uptake efficiency.[2]

## 3. *Prevention strategies to inhibit corona formation*

The biomolecular protein corona is still largely an unpredictable complex factor, which in some cases can trigger desired outcomes, but in others can result in toxic biological responses.[44] Therefore, there is currently and has been much effort put forth in order to synthetically prevent protein adsorption. One possible solution was to functionalize NPs with polymer chains such as PEG, making the NPs' surface highly biocompatible, which limits non-specific interactions with biological components.[2] While PEGylation is able to partially mitigate protein adsorption to NPs, it has still been unable to completely prevent protein/biomolecule corona formation.[37,41] An alternative solution that could be employed is to functionalize the NP surface with zwitterions. There have been NPs designed such that hydrophobicity is tunable in which the NPs don't adsorb proteins at moderate serum levels and do not form hard coronas at physiological serum concentrations. Novel strategies such as the one just mentioned have the capacity to lead to new analytical methods to investigate the interaction of nano materials in biological environments without compounding variables such as interference from protein binding. However, even with recent gain, further development in producing "corona-free" NPs will be required and is currently underway.[42]

## VI. Concluding Remarks

NPs have several unique properties such as small size and high surface area, which are distinctly advantageous in biomedicine. The small size of NPs allows them to enhance drug transport across biological barriers (leading to an increased bioavailability of the entrapped drug) or improve intracellular drug delivery which can then be applied to gene or cancer therapy. NPs can be designed to transport different substances across various barriers depending on the properties they possess. With so many possibilities it is important to understand the system one wishes to exploit and then design a nanocarrier to accomplish or exceed previous limits of already characterized systems.

In general, the following trends can be concluded from this chapter: based on the morphology of NPs, uptake by non-phagocytic cells is optimum at a NP size of ~50 nm, and for phagocytes, the results are still inconclusive. The effect of increased surface charge modification, either positive or negative, tends to favor cellular uptake in either cell type. However, when all other factors are comparable, positively charged particles are taken up more extensively.

NPs in biological medium exhibit different properties than the original synthetic NPs; in order to reduce unpredicted NP behavior in the body, surface enhancements such as PEGylation should be employed.

## References

1. Albanese, A.; Tang, P. S.; Chan, W. C. *Annu. Rev. Biomed. Eng.* **2012**, *14*, 1.
2. Kettler, K.; Veltman, K.; van de Meent, D.; van Wezel, A.; Hendriks, A. J. *Environ. Toxicol. Chem.* **2014**, *33* (3), 481–492.
3. Vick, N. A.; Khandekar, J. D.; Bigner, D. D. *Arch. Neurol.* **1977**, *34*, 523.
4. Bondi, M. L.; Di Gesu, R.; Craparo, E. F. *Methods Enzymol.* **2012**, *508*, 229.
5. Re, F.; Cambianica, I.; Zona, C.; Sesana, S.; Gregori, M.; Rigolio, R.; La Ferla, B.; Nicotra, F.; Forloni, G.; Cagnotto, A.; Salmona, M.; Masserini, M.; Sancini, G. *Nanomedicine* **2011**, *7*, 551.

6. Bareford, L. M.; Swaan, P. W. *Adv. Drug Deliv. Rev.* **2007**, *59*, 748.
7. May, R. C.; Machesky, L. M. *J. Cell Sci.* **2001**, *114*, 1061.
8. Hillaireau, H.; Couvreur, P. *Cell Mol. Life Sci.* **2009**, *66*, 2873.
9. Groves, E.; Dart, A. E.; Covarelli, V.; Caron, E. *Cell Mol. Life Sci.* **2008**, *65*, 1957.
10. Conner, S. D.; Schmid, S. L. *Nature* **2003**, *422*, 37–44.
11. Aderem, A.; Underhill, D. M. *Annu. Rev. Immunol.* **1999**, *17*, 593.
12. Kumari, S.; Mg, S.; Mayor, S. *Cell Res.* **2010**, *20*, 256.
13. Schmid, S. L. *Annu. Rev. Biochem.* **1997**, *66*, 511.
14. Gehr, P.; Clift, M. J.; Brandenberger, C.; Lehmann, A.; Herzog, F.; Rothen-Rutishauser, B. *Compr. Physiol.* **2011**, *1*, 1159.
15. Khalil, I. A.; Kogure, K.; Akita, H.; Harashima, H. *Pharmacol. Rev.* **2006**, *58*, 32.
16. Sahay, G.; Alakhova, D. Y.; Kabanov, A. V. *J. Control. Release* **2010**, *145*, 182.
17. Dramsi, S.; Cossart, P. *Annu. Rev. Cell Dev. Biol.* **1998**, *14*, 137.
18. Fontana, G.; Licciardi, M.; Mansueto, S.; Schillaci, D.; Giammona, G. *Biomaterials* **2001**, *22*, 2857.
19. Tyagi, R.; Lala, S.; Verma, A. K.; Nandy, A. K.; Mahato, S. B.; Maitra, A.; Basu, M. K. *J. Drug Target* **2005**, *13*, 161.
20. McMahon, H. T.; Boucrot, E. *Nat. Rev. Mol. Cell Biol.* **2011**, *12*, 517.
21. Gao, H.; Shi, W.; Freund, L. B. *Proc. Natl. Acad Sci. U.S.A.* **2005**, *102*, 9469.
22. Chithrani, B. D.; Chan, W. C. *Nano Lett.* **2007**, *7*, 1542.
23. Choi, J.; Zhang, Q.; Reipa, V.; Wang, N. S.; Stratmeyer, M. E.; Hitchins, V. M.; Goering, P. L. *J. Appl. Toxicol.* **2009**, *29*, 52.
24. Rejman, J.; Oberle, V.; Zuhorn, I. S.; Hoekstra, D. *Biochem. J.* **2004**, *377*, 159.
25. Cho, E. C.; Au, L.; Zhang, Q.; Xia, Y. *Small* **2010**, *6*, 517.
26. Chithrani, B. D.; Ghazani, A. A.; Chan, W. C. *Nano Lett.* **2006**, *6*, 662.
27. Dasgupta, S.; Auth, T.; Gompper, G. *Nano Lett.* **2014**, *14*, 687.
28. Truong, N. P.; Whittaker, M. R.; Mak, C. W.; Davis, T. P. *Expert Opin. Drug. Deliv.* **2015**, *12*, 129.
29. He, C.; Hu, Y.; Yin, L.; Tang, C.; Yin, C. *Biomaterials* **2010**, *31*, 3657.
30. Lankoff, A.; Arabski, M.; Wegierek-Ciuk, A.; Kruszewski, M.; Lisowska, H.; Banasik-Nowak, A.; Rozga-Wijas, K.; Wojewodzka, M.; Slomkowski, S. *Nanotoxicology* **2013**, *7*, 235.

31. Lesniak, A.; Fenaroli, F.; Monopoli, M. P.; Aberg, C.; Dawson, K. A.; Salvati, A. *ACS Nano* **2012**, *6*, 5845.
32. Treuel, L.; Brandholt, S.; Maffre, P.; Wiegele, S.; Shang, L.; Nienhaus, G. U. *ACS Nano* **2014**, *8*, 503.
33. Tenzer, S.; Docter, D.; Rosfa, S.; Wlodarski, A.; Kuharev, J.; Rekik, A.; Knauer, S. K.; Bantz, C.; Nawroth, T.; Bier, C.; Sirirattanapan, J.; Mann, W.; Treuel, L.; Zellner, R.; Maskos, M.; Schild, H.; Stauber, R. H. *ACS Nano* **2011**, *5*, 7155.
34. Salvati, A.; Pitek, A. S.; Monopoli, M. P.; Prapainop, K.; Bombelli, F. B.; Hristov, D. R.; Kelly, P. M.; Aberg, C.; Mahon, E.; Dawson, K. A. *Nat. Nanotechnol.* **2013**, *8*, 137.
35. Tenzer, S.; Docter, D.; Kuharev, J.; Musyanovych, A.; Fetz, V.; Hecht, R.; Schlenk, F.; Fischer, D.; Kiouptsi, K.; Reinhardt, C.; Landfester, K.; Schild, H.; Maskos, M.; Knauer, S. K.; Stauber, R. H. *Nat. Nanotechnol.* **2013**, *8*, 772.
36. Monopoli, M. P.; Bombelli, F. B.; Dawson, K. A. *Nat. Nanotechnol.* **2011**, *6*, 11.
37. Monopoli, M. P.; Aberg, C.; Salvati, A.; Dawson, K. A. *Nat. Nanotechnol.* **2012**, *7*, 779.
38. Lundqvist, M.; Stigler, J.; Elia, G.; Lynch, I.; Cedervall, T.; Dawson, K. A. *Proc. Natl. Acad. Sci. U.S.A.* **2008**, *105*, 14265.
39. El-Sayed, I. H.; Huang, X.; El-Sayed, M. A. *Nano Lett.* **2005**, *5*, 829.
40. Panariti, A.; Miserocchi, G.; Rivolta, I. *Nanotechnol Sci. Appl.* **2012**, *5*, 87.
41. Pozzi, D.; Colapicchioni, V.; Caracciolo, G.; Piovesana, S.; Capriotti, A. L.; Palchetti, S.; De Grossi, S.; Riccioli, A.; Amenitsch, H.; Lagana, A. *Nanoscale* **2014**, *6*, 2782.
42. Moyano, D. F.; Saha, K.; Prakash, G.; Yan, B.; Kong, H.; Yazdani, M.; Rotello, V. M. *ACS Nano* **2014**, *8*, 6748.
43. Fleischer, C. C.; Payne, C. K. *J. Phys. Chem. B.* **2014**, *118* (49): 14017–14026.
44. Neagu, M.; Piperigkou, Z.; Karamanou, K.; Angin, A. B.; Docea, A. O.; Constantin, C.; Negrei, C.; Nikitovic, D.; Tsatsakis, A. *Arch. Toxicol.* **2017**, *91* (3): 1031–1048.
45. Orts-Gil, G.; Natte, K.; Österle, W. *RSC Adv.* **2013**, *3*: 18203–18215.

# Chapter 4

# Gated Porous Materials for Biomedical Applications

Félix Sancenón*,†,‡, Erick Yu§, Elena Aznar*,‡, M. Dolores Marcos*,†,‡ and Ramón Martínez-Máñez*,†,‡

** Instituto Interuniversitario de Investigación de Reconocimiento Molecular y Desarrollo Tecnológico (IDM), Unidad mixta Universitat Politècnica de València — Universitat de València, Spain*
*†Departamento de Química, Universitat Politècnica de València, Camino de Vera s/n, 46022 València, Spain*
*‡CIBER de Bioingeniería, Biomateriales y Nanomedicina (CIBER-BBN), Spain*
*§Department of Chemical Engineering and Materials Science, University of California Davis, 95616 Davis, California, USA*

## Abbreviations

| | |
|---|---|
| AMF | alternating magnetic field |
| APTES | (3-aminopropyl)triethoxysilane |
| ATP | adenosine triphosphate |
| ATRP | atom transfer radical polymerization |
| AuNPs | gold nanoparticles |
| BCPR | breast cancer resistance protein |
| BSA | bovine serum albumin |
| BTA | benzotriazole |
| CBPQT | cyclobis(paraquat-*p*-phenylene) |
| CFU | colony forming units |
| CPT | camptothecin |
| CTAB | n-cetyltrimethylammonium bromide |
| CytC | cytochrome C |
| DAPI | 4′,6-diamidino-2-phenylindole |
| DNA MTases | DNA methyltransferases |
| DNP | 1,5-dioxynaphthalene |
| DOPC | 1,2-dioleoyl-*sn*-glycero-3-phosphocholine |
| DOPE | 1,2-dioleoyl-*sn*-glycero-3-phosphoethanolamine |
| DOTAP | 1,2-dioleoyl-3-trimethylammoniumpropane |
| DOX | doxorubicin |
| DTT | dithiothreitol |
| EDC | 1-ethyl-3-(3-dimethylaminopropyl)carbodiimide |

| | |
|---|---|
| FA | folic acid |
| FITC | fluorescein isothiocyanate |
| GEM | gemcitabine |
| GFP | green fluorescent protein |
| GOx | glucose oxidase |
| GSH | glutathione |
| IBU | ibuprofen |
| LA | lactobionic acid |
| LCST | lower critical solution temperature |
| LOD | limit of detection |
| MB | methylene blue |
| MMP-9 | metalloproteinase-9 |
| MRI | magnetic resonance imaging |
| MSNs | mesoporous silica nanoparticles |
| NHS | *N*-hydroxysuccinimide |
| NIR | near infrared |
| PC | pericyte |
| PDAC | pancreatic ductal adenocarcinoma |
| PEG | polyethylene glycol |
| PEI | polyethyleneimine |
| PI | propidium iodide |
| PLE | porcine liver esterase |
| PNIPAAm | poly(*N*-isopropylacrylamide) |
| poly(I:C) | polyinosinic–polycytidylic acid |
| QDs | quantum dots |
| Rh | rhodamine |
| SBF | simulated body fluid |
| SERS | surface-enhanced Raman spectroscopy |
| SNTs | silica nanotubes |
| SQ | squaraine |

| | |
|---|---|
| sulfo-SMCC | sulfosuccinimidyl 4-[*N*-maleimidomethyl] cyclohexane-1-carboxylate |
| TBA | thrombin aptamer |
| TGF-$\beta$ | transforming growth factor beta |
| TLR3 | toll-like receptor 3 |
| TMB | tetramethylbenzidine |
| TNF-$\alpha$ | tumor necrosis factor $\alpha$ |
| UCNPs | upconverting nanoparticles |
| US | ultrasounds |

## I. Introduction

The rapidly evolving and advancing field of drug delivery has benefited in the past decade from development in nanoparticle engineering and design. In particular, mesoporous silica nanoparticles (MSNs), which were of significant interest for catalytic applications at the time of its original discovery by scientists at Mobile Crystalline Materials,[1] have been used as an agent for drug delivery applications.

MSNs have a wide variety of porous structures with MCM-type, abbreviated from its original discoverer, Mobile Crystalline Materials, being the most prevalent and commonly used nanostructure. However, other mesoporous silica structures such as SBA (Santa Barbara Amorphous),[2] KIT (Korea Advanced Institute of Science and Technology),[3] MSU,[4] and FSM (Folded Sheets Mesoporous)[5] have been extensively studied in different applications, although the latter four are not typically found in nanoparticle form.

Further experimentation with MSNs ultimately led to their surface functionalization with organic groups to introduce various gating and stimuli-responsive mechanisms.[6] In doing so, "smart" nanoparticle agents could be developed to deliver drugs both with target and trigger capabilities. Advantages of this new class of nanomaterials show selectivity and as a result, minimization of premature drug delivery or instability of the species to be delivered. However, while there have been conflicting arguments on the biocompatibility of bare MSNs, the inclusion of additional functional groups complicates the situation further. Literature generally supports the biocompatibility[7–9] of MSNs, but in combination with the diverse selection of organic functional groups, further clinical investigations are needed in this new emergent field.

Gated materials, based in mesoporous supports, are prepared with the aim of finely tuned release of chemical/biochemical species from the pores of the inorganic phase in response to a predefined stimulus.[10] Those hybrid organic-inorganic materials are mainly composed by two subunits: (i) a porous inorganic support in which cargo is loaded and; (ii) certain molecular/supramolecular entities, grafted onto the external surface that can control mass transport from pores. A typical gated material is depicted in Figure 1. One key point in the design of gated nanomaterials is the selection of the molecular/supramolecular entities that control the release of the entrapped payload. In respect to this, the driving force to switch gating mechanisms (between an open state and a closed state) takes advantage of electrostatic or supramolecular interactions, the rupture/formation of covalent bonds, or changes in the physical properties of molecules or macromolecules.[11–13] Gated nanochemistry has demonstrated the possibility of achieving new advanced pre-designed functions by means of mass transport control. The most studied application of such systems is related to controlled release protocols especially devoted to potential biomedical applications.[14] Moreover, such gated materials have also been used to develop new sensing/recognition protocols.[15]

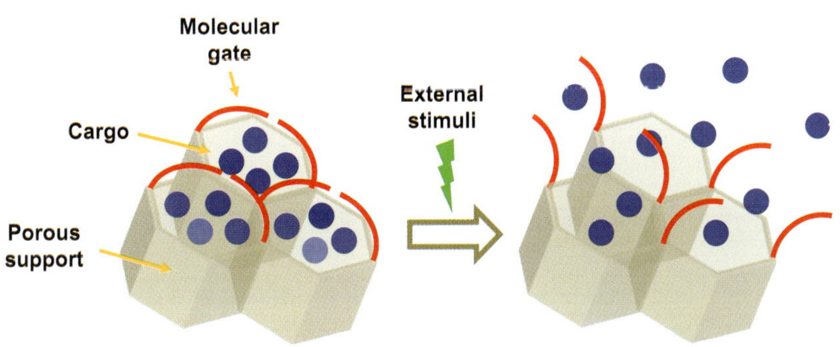


Figure 1. Schematic representation of a gated material.

The application of gated materials in the biomedical field has expanded in the last few years because these systems allow the release of one or several drugs upon the application of external stimuli, and are promising candidates for developing new therapies to improve drug efficacy and stability. Moreover, anchoring biological receptors (e.g. antibodies, peptides, aptamers, carbohydrates) onto an external surface allows targeted delivery of drugs to sites such as cells and tissues that contain over-expressed receptors.[16–18] In recent years, some more complicated systems have been designed with potential applications in the emerging area of "theranostic" devices.[19,20] In these systems, sub-micron, porous inorganic supports are loaded with a therapeutic agent, a diagnostic marker, or with an indicator molecule. Then pores are closed with stimuli-responsive caps and certain biological receptors can be grafted onto the surface for targeting purposes. These nanodevices are capable of targeting selected cells or tissues to avoid the premature degradation of the entrapped cargo, and also facilitate the transfer of the payload across the cell membrane. Moreover, the potential incorporation of signaling units also allows the release process and the accumulation of the gated nanodevice in target tissues to be traced in real time.

This chapter shows selected examples of hybrid mesoporous gated materials based on silica with applications in the biomedical field. Illustrated here first is a brief overview of the structure and synthesis of mesoporous silica (MS). Next, selected examples are then divided in terms of the stimuli used to trigger cargo; i.e. light, changes in redox features, changes in pH, the presence of molecules or biomolecules, enzymes, temperature, ultrasound, and alternating magnetic fields (AMF). In each section we describe the first reported gated material that used the trigger and selected examples of biomedical significance.

## II. Structure of MS

As delivery agents for specific for drug release, MCM, and SBA structures have been the most investigated due to their high surface area, tunable pore size and uniform porous structure. Typically, MCM-type MS have a hexagonal porous structure but depending on the synthesis conditions, cubic crystalline structures are also possible, as found in MCM-48,[21] a variant of the prototypical, hexagonal crystalline MCM-41.[22] Typically the pore size of MCM-41 or MCM-48 is about 2.5–3.5 nm in diameter, whereas other materials such as SBA-type MS have pore sizes around 5–8 nm.[23] This is attributed to the different template used during its synthesis, as will be discussed later.

In addition to uniformly porous MS, other MS nanostructures such as expanded-pore, hollow, and hierarchical-porous MS have been synthesized to further increase their versatility in drug release applications, for example when larger loading species,[24] bimodal porosity,[25] or pre-loading[26] are desirable. Pore-expanded MS are produced during and after the sol–gel synthesis. Strongly small and hydrophobic compounds are typically used for this such as trimethylbenzene (TMB), one of the more commonly used species. TMB has been shown to increase the pore size of MCM-41 and SBA-15 type MS to values up to 23 and 30 nm, respectively.

Hollow MSNs (HMSNs) have gathered significant attention in recent years due to both a larger interior void and potentially less silica content in the nanoparticle itself as well. In the scope of drug delivery, the large void in HMSNs allows for improved loading efficiency as opposed to solid MSNs, where the interior porous network may be more difficult to access. Structurally, the porous shell still retains the original hexagonal or cubic structure.

Hierarchical MSNs share many similarities to HMSNs but often will combine multiple templating steps to produce a diverse distribution of pore sizes or nanostructures within the MSNs. Often this approach is desirable to produce core–shell nanoparticles where the core could be an inorganic metal or complex used for sensing/detection, while the shell is mesoporous silica to retain the storage and release properties. Bi- and trimodal porous MSNs have produced that show the differing pore diameters play an important role on the interaction between the matrix and loading species.[27]

## A. Synthesis of MS

Both MCM and SBA-type MS are synthesized through a sol–gel process, in which a gel-like state is achieved during adsorption of the silica precursor onto a template, for example a surfactant or tri-block copolymer respectively for MCM-41 and SBA-15. The synthesis of MS is strongly dependent on the template selection, through which the pore size can be adjusted. Although many templates are possible and have been investigated in literature,[28] it can be generally defined that the dominant two templates used can be classified into soft and hard templates, of which in the former are quaternary ammonium surfactants and amphiphilic, triblock copolymers; and in the latter, nanoparticles.

Quaternary ammonium surfactants are examples of ionic surfactants that direct MS templating through strong electrostatic interactions between the surfactant micelles and the silica source. For silica, the isoelectric point is approximately 2, where below this value; the surface is positively charged and above, negatively charged. The silicate precursor is typically negatively charged at neutral conditions and as a result, cationic surfactants are much more widespread than anionic surfactants. Quaternary ammonium surfactants such as hexadecyltrimethylammonium bromide (CTAB) are used to produce MCM-41 nanostructures of 2.5–3.5 nm in diameter, but pore size can be adjusted through selection in the alkyl chain length of these surfactants.[29]

Non-ionic and amphiphilic templates are also used to synthesize MS with the most widely used types of these templates being tri-block copolymers, specifically polyethylene oxide-polypropylene oxide-polyethylene oxide (PEO-PPO-PEO). Due to the hydrophilic nature of the PEO tails and the hydrophobic nature of the PPO, these copolymers (also called Pluronics) behave as amphiphilic surfactants, with varying strength in relation with the polymer chain length and ratio of PEO:PPO components. Depending on the composition of the polymer, different pore sizes are achieved.[30]

Controlling the hexagonal and cubic crystallinity of MS is not straightforward but has substantial empirical data concluding that multiple parameters affect the final structure. As noted by Zhao *et al.*,[2] the formation of uniform porous crystallinity is sensitive to all reaction parameters, including the template concentration, pH, and temperature. At non-optimal conditions, amorphous, disordered, or even no silica is formed. The electrostatics between the template and silicate source and the self-assembly of the template itself are crucial parameters that govern the porous crystallinity during the sol–gel formation.

## III. Light Responsive Gated Materials

Light was the first stimulus reported to develop capped MSNs. This powerful tool can be used to modify the conformation of molecules, stimulate the rupture of chemical bonds, or induce significant heating, resulting in the opening of the gated ensemble. Moreover, using light as an uncapping trigger, a spatio-temporal control of application of the stimulus can be achieved. Light has the advantage of being applicable from outside of the patient in a non-invasive manner and it is easily focalized in selected areas, avoiding irradiation of the surrounding tissues. Nevertheless, poor penetration of UV light in deep tissues has refocused the attention in near infrared (NIR) irradiation as a light source, which is more tissue-penetrating.

The first light-driven gated material used photodimerization as a gate controlling mechanism. In this initial example, Fujiwara and coworkers prepared a MCM-41 support functionalized with 7-[(3-triethoxysilyl)propoxy]coumarin groups and loaded with cargos such as phenanthrene, cholestane or progesterone.[31,32] Coumarin moieties are known to undergo a [2+2] photodimerization to give in a cyclobutane dimer. This photoreaction was used to cap the pores by irradiation with >310 nm light. Gated solids only showed cargo leakage when were suspended in *n*-hexane and irradiated with 250 nm UV-light. In contrast, when the solid was not irradiated, cargo release was found to be negligible.

Light can be also used to generate thermal heating (hyperthermia) which induces the destabilization of the capping ensemble in gated materials, thus switching the gate from a closed to an open form and resulting in cargo delivery. In this area, the use of NIR light between 700 and 1100 nm is particularly interesting given the absence of significant absorption from either biological chromophores or water, which allows deep optical penetration into biological samples, such as tissue or blood.

One of the first examples of gated materials whose opening was based in a hyperthermia concept was developed by Martínez-Máñez and coworkers.[33] In this work, a MCM-41 support was loaded with safranin O and functionalized with *N*-(3-triethoxysilylpropyl)gluconamide in the pore outlets. Separately, gold nanoparticles (AuNPs) functionalized with boronic acid were prepared. Then pores were capped by the formation of boroester bonds between boronic acid units in AuNP and the saccharide derivative anchored to MCM-41 external surface. Plasmonic heating of the AuNPs induced by laser light caused the cleavage of boronic ester linkages, which then resulted in payload release. The amount of cargo delivered by simply controlling laser irradiation was finely tuned. The authors also discovered that cargo release was inhibited at pH 5, whereas a rapid release was observed at pH 3. The pH-controlled mechanism was associated with the reversible formation of boroesters (closed-gate), and their quick and facile hydrolysis (open-gate) at an acidic pH.

Following a similar approach, Cheng *et al.* used gold nanorods (AuNRs) covered by a mesoporous silica shell to prepare gated supports able to cargo release triggered by NIR light.[34] The mesoporous solid was loaded with the anticancer drug doxorubicin (DOX) and then functionalized with (3-aminopropyl)triethoxysilane (APTES). Afterward the 4-maleimidobutyric acid *N*-succinimidyl ester was used as a linker to conjugate the amine to the thiol group of a thiolated double stranded DNA. In particular, double stranded DNA was obtained by the hybridization of SH-5′-$(CH_2)_6$-TTT TTC CCG CGC CGG-3′ and 5′-TTT TTC CGG CGC GGG-3′ in phosphate buffer saline (PBS) at 4°C. The final capped material showed that only 10% of DOX leaked after 10 hours. In contrast when NIR light was applied, a pulsatile control of cargo release was observed. Cargo release was due to AuNRs-mediated photothermal conversion, which induced dehybridization of the DNA duplex. The authors also prepared a similar capped system where the non thiolated strand was a green fluorescent protein (GFP)-interfered siRNA. The authors found that upon NIR irradiation, the released siRNAs silenced the GFP expression in GFP expressing HeLa cells.

The same authors implemented the same gating mechanism in MSNs with a $Fe_3O_4$@Au trisoctahedra core.[35] They found that heating the nanocarrier by NIR light irradiation induced the dehybridization of an anchored double stranded DNA (with a denaturing temperature of approximately 47°C), which allowed the cargo to escape. The magnetic $Fe_3O_4$ core also facilitated intracellular drug release by magnetic attraction. For *in vitro* studies, HeLa cells were selected and MTT viability assays were conducted to determine nanomaterial biocompatibility. Cells were incubated with the system at 37°C for 24 hours. Negligible cellular toxicity was detected with at least 90% of cell survival found at very high nanoparticle concentration. Cells were also incubated with the capped nanoparticles for 2 hours, either with or without magnetic attraction, followed by other 24-hour incubation. Then cell cultures were exposed to laser illumination. Interestingly, incubation with the DOX-loaded nanomaterial triggered a 2-fold increase in cell toxicity when compared to that of the empty nanomaterial. Finally, the studies were extended from *in vitro* to *in vivo* to investigate regression efficacy in tumor growth, which was monitored in tumor volume change terms. HeLa cells were transplanted hypodermically in the thighs of nude mice. In the control groups, mice were closely monitored for continuous tumor growth, which grew 7-fold larger on day 14 than it was initially. In the mice intravenously injected with the system and treated with either magnet attraction or laser illumination, minor tumor growth suppression was detected (an increase of 5–6-fold after 14 days). Remarkably in the mice injected with the system and treated with both magnet attraction + laser illumination, tumor growth was only 3-fold larger in 14 days. Most interestingly, a second injection of capped nanoparticles, plus the combination of laser-magnet therapy, fully eliminated the tumor after 14 days. Preliminary toxicological studies performed in mice tissue, such as heart, spleen, kidney, and liver, revealed no acute toxicity of the system. Given its nature, this nanomaterial was also tested for magnetic resonance imaging (MRI) performance. Mice were hypodermically injected with HeLa cells, and MRI experiments were conducted after 1 week of tumor growth with a 9.4 T animal micro MRI system. When magnetic attraction was applied after injecting nanoparticles, an apparent contrast change in signal was found in the MR images, which confirmed that this nanomaterial was an effective MRI T2 contrast agent.

Elbialy *et al.* functionalized MSNs with APTES and loaded the pores with DOX. An Au shell was used to cover nanoparticles and the nanodevice was finally coated with poly(ethylene glycol) to prevent aggregation.[36] Aqueous suspensions of the gold-coated nanoparticles showed negligible dye release at pH 5.8, whereas a clear DOX delivery was observed upon NIR irradiation. The sustained DOX release was ascribed to the gold shell degradation after NIR light irradiation. Confocal laser scanning microscopy (CLSM) was used to study the cellular uptake of the system in human tumor MCF-7 cells. As expected, enhanced cytotoxicity in MCF-7 cells was observed upon the irradiation of internalized DOX-loaded nanoparticles. Flow cytometry studies of MCF-7 treated with the capped system revealed that the nanomaterial induced direct cell cycle arrest and cell death due to DOX delivery. The intravenous application of the system, followed by NIR irradiation of the tumor area, inhibited the growth of subcutaneous Ehrlich carcinoma in mice, and induced a stronger anticancer effect compared to photothermal therapy or chemotherapy alone. Histopathological studies showed significantly higher cell death in mice treated with the DOX-loaded nanomaterial, which is consistent with the data obtained in cellular studies.

Light can be also used to generate free radicals which are able to cause the rupture of chemical bonds and subsequent gate uncapping. For example, Lu and co-workers employed $Mn^{2+}$-doped NaY(Mn)$F_4$:Yb/Er upconverting nanoparticles (UNNPs) coated with a mesoporous silica shell. The silica surface was functionalized with octadecyl hydrophobic chains and with photosensitizer chlorin e6. Moreover, pores were loaded with DOX.[37] The authors also prepared a hydrophobic polymer that contained 9,10-dialkoxyanthracene groups by free radical polymerization of monomers poly(ethylene glycol) methyl ester methacrylate and 3-((10-(3-(methacryloyloxy) propoxy)anthracen-9-yl)oxy)propylstearate. The pores of the functionalized nanoparticles were coated with the polymer through hydrophobic van der Waals interactions with octadecyl chains as depicted in Figure 2. PBS suspensions of the capped nanoparticles showed negligible DOX release without irradiation, whereas a remarkable release

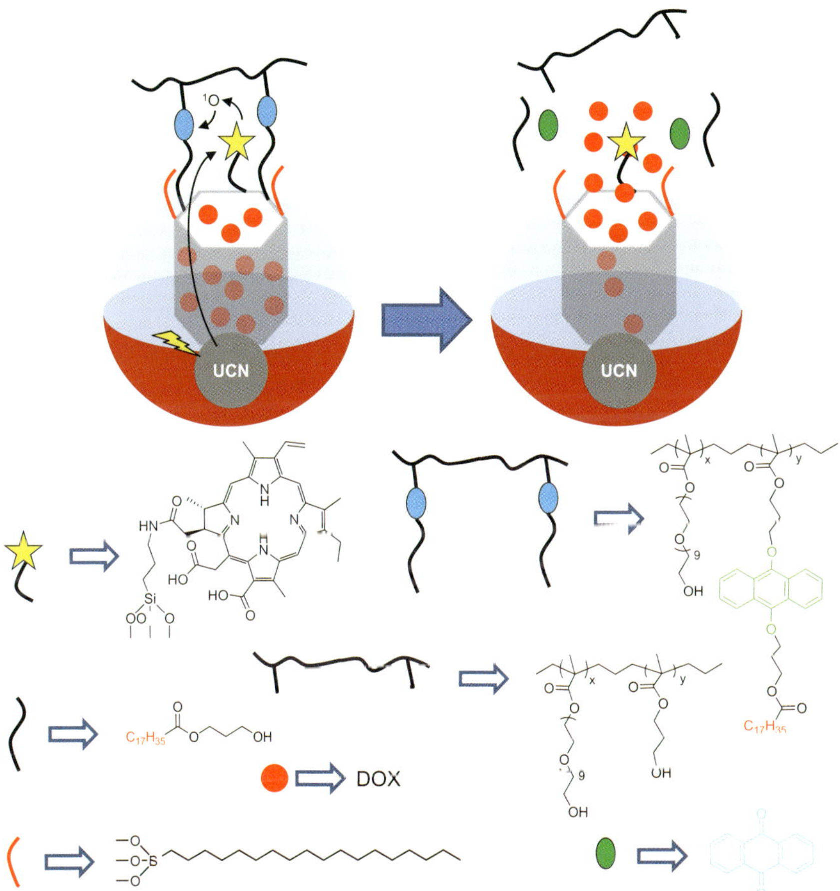


Figure 2. Mesoporous nanoparticle with a UCNP core, loaded with DOX, functionalized with photosensitizer chlorin e6 and capped with a photodegradable copolymer that contained 9,10-dialkoxyanthracene moieties. DOX release was achieved by irradiation at 980 nm.

of the entrapped drug was observed (more than 70% after 8 hours) upon NIR irradiation (with a 980 nm laser). The UCN absorbed the NIR-light converting it into red light (660 nm), which excited the chlorin e6 photosensitizer that produced singlet oxygen. This singlet oxygen oxidized the anthracene moieties of the polymer shell to anthraquinone with the subsequent polymeric cover degradation and drug release. MTT viability studies carried out with KB cells showed that capped nanoparticles were essentially non-toxic, but significant cell death was observed upon NIR irradiation. This cell death was ascribed to a synergic effect due to DOX release and singlet oxygen generation. The nanodevice was also able to reduce tumor growth in nude mice that bore KB cells under NIR irradiation.

AuNRs coated with a mesoporous silica shell were also used as inorganic support for the preparation of NIR light-triggered nanodevices.[38] For this purpose, the inorganic support was prepared and then the pores of the mesoporous silica shell were firstly loaded with DOX and, in a second step, with 1-tetradecanol (phase-changing material with a melting temperature of 39°C). Suspensions of the capped nanodevices in RPMI-1640 medium at room temperature showed nearly zero release of the entrapped drug. However, after 10 min irradiation with a 802 nm laser (1 W $cm^{-2}$) a marked DOX release was observed. DOX delivery was ascribed to the NIR irradiation of the core AuNRs that induced a local heating with the subsequent melting of 1-tetradecanol. In a further step, the external surface of the capped nanodevice was functionalized with APTES and then the amino moieties reacted with folic acid using (1-ethyl-3-(3-dimethylaminopropyl)carbodiimide) EDC/(*N*-hydroxysuccinimide) NHS as coupling agents. Viability studies carried out with KB cells (with overexpressed folate receptors) and the folic acid derivatized and capped nanodevices showed negligible cell death. However, upon NIR irradiation, a marked decrease in KB cell viability was observed, which was ascribed to a synergic effect between the released DOX and the photothermal contribution of the NIR irradiation converted in local heath by AuNRs.

Lin and co-workers used a similar approach for the preparation of NIR light-triggered capped nanodevices but used $Cu_{1.8}S$ nanoparticles instead of AuNRs as core.[39] In a first step, $Cu_{1.8}S$ nanoparticles were prepared and then coated with a mesoporous silica shell. In order to attach gating functionalities the outer surface of the silica shell was functionalized with APTES. Then, the pores of the silica shell were loaded with curcumin and the amino moieties reacted with sulfosuccinimidyl4-[*N*-maleimidomethyl]cyclohexane-1-carboxylate (sulfo-SMCC) linker. At that point, a thiol functionalized anchored single DNA strand (5′- CGA CGA CGA CGA-SH-3′) was covalently linked onto the nanoparticles through a Michael addition reaction. Finally, in order to seal the pores, AS1411 single DNA strand (5′-TCG TCG TCG TCG GG TGG TGG TGG TTG TGG TGG TGG TGG-3′) was conjugated with the anchored strand through base pairing interactions. AS1411 strand acted as capping agent and also as targeting ligand (shows high binding affinity and specificity for the cancer cell membrane-bound overexpressed protein nucleolin). Besides, DOX was intercalated into the grafted double stranded DNA. The NIR controlled release behavior of the final capped nanoparticles was tested in PBS buffered at pH 7.4. Suspensions of the capped nanoparticles in the absence of NIR irradiation showed zero cargo release. However, irradiation with a NIR laser (980 nm, 1 W $cm^{-2}$) induced a marked release of both curcumin and DOX. The combined release of both cargos was ascribed to a NIR-induced uncapping of the pores due to denaturation of the double strand DNA fragment with the subsequent detachment of the larger DNA sequence than act as gate. MTT viability assays carried out with the DNA-capped nanoparticles in MCF-7 cells and without NIR irradiation showed that the capped nanodevice was not toxic. However, upon NIR irradiation a marked cell death was observed due to the synergy between curcumin and DOX release. The authors demonstrated that cell death, induced by the two drugs released upon NIR irradiation, was associated with disruption of mitochondrial function and activation of caspases.

NIR-triggered thermally-responsive nanodevices able to deliver DOX were prepared by Zhang and co-workers.[40] MSNs were selected as inorganic scaffolding and its external surface functionalized with APTES. Then, the single stranded DNA fragment 5′-HS-$C_6$-TTATCGCTGATTCAA was coupled using succinimidyl-4-[*N*-maleimidomethyl]cyclobutane-1-carboxylate as cross-linker. On the other hand, thioglycolic acid-coated CuS nanoparticles were prepared and its surface functionalized with a single stranded DNA sequence (5′-$H_2N$-$C_6$-TTGAATCAGCGATAA complementary to that anchored onto the MSNs surface) using NHS/EDC coupling. The final material was prepared after loading DNA-functionalized MSNs with DOX and sealing the pores upon addition of DNA-coated CuS (by single stranded DNAs hybridization). Aqueous suspensions of the CuS-capped nanoparticles at 37°C showed negligible release whereas, upon heating at 47°C (melting point of the double stranded DNA sequence used to seal the pores) a marked DOX release was observed. The release was also produced upon excitation CuS with a 980 nm laser due to the fact that the generated local heat de-hybridizes the DNA duplex. Confocal microscopy studies showed CuS-capped MSNs internalization in HeLa cells and DOX release upon irradiation with a 980 nm laser. Viability assays carried out with CuS-capped MSNs in HeLa and MCF-7 cells without laser irradiation showed the non-toxicity of the nanocarrier. However, after 30 s irradiation a marked death (90%) was found for both cell lines due to the NIR-induced DOX delivery.

## IV. Redox Responsive Gated Materials

The use of redox processes as a mechanism to control gated supports has been intensively explored. The existence of endogenous reducing agents at the intracellular level makes these gated materials very appealing for biological applications. Moreover, for some diseases (such as cancer) an increase in the concentration of such redox-active species has been reported. Inside of the realm of the redox-triggered capped materials two groups are clearly envisioned: (i) those based on redox-driven supramolecular interactions between macrocyclic compounds (cyclophanes, cyclodextrins or cucurbiturils) and wire-like molecules anchored onto the surface of the support; and (ii) those based on the rupture of disulfide linkages that bind capping agents (biomolecules, inorganic nanoparticles, polymers) onto the porous scaffolds. Most of the reported examples display gating responses in pure water and some of them are also reversible. Finally, some redox-triggered gated materials are frequently decorated with targeting ligands or hydrophilic molecules to increase their stability in biological environments. Some encouraging examples in *in vivo* models have demonstrated that redox-responsive gated nanoparticles loaded with cytotoxics can selectively reach a specific cell type and are able to reduce tumor size in mice models.

The first example of a redox-triggered capped material was reported by Stoddart, Zink and coworkers. These authors used rotaxanes as supramolecular stoppers and redox reactions as triggers to modulate the state of the gate and control cargo release.[41] The prepared material was based on the use of MSNs loaded with $Ir(ppy)_3$ fluorophore and with the external surface functionalized with a 1,5-dioxynaphthalene (DNP) derivative. Finally, the loaded pores were capped by the formation of an inclusion complex between DNP and cyclobis(paraquat-*p*-phenylene) ($CBPQT^{4+}$). Addition of $NaCNBH_3$ induced the rupture of the supramolecular complex with subsequent delivery of the entrapped $Ir(ppy)_3$ fluorophore.

Zhao and coworkers also used [2]rotaxanes functionalized with disulfide bonds as caps for the development of redox-responsive drug delivery supports.[42] The capping system was formed by a tetraethylene glycol encircled by α-CDs and ended with folic acid, which acted as both a stopper and a tumor-targeting agent. The [2]rotaxanes were attached onto the external surface of the MSNs using a disulfide-containing linker and the pores were previously loaded with DOX. No cargo release was observed in TRIS suspensions of the capped nanoparticles. However, in the presence of glutathione (GSH), a clear DOX release was observed at physiological and endosome mimetic conditions (pH 5). Once assessed the *in vitro* behavior the authors tested the controlled release performance of the nanoparticles with HeLa cells which were treated with the nanoparticles either uncapped or capped with FA, and biocompatibility and biodistribution was analyzed by TEM. Cell experiments confirmed the biocompatibility of both materials, as judged by cell and nuclear membrane intactness. Remarkably, the FA-capped nanoparticles were endocytosed more efficiently by HeLa cells than the uncapped ones, suggesting that internalization was enhanced by the targeting ligand FA. Additionally, a 2-fold increase uptake in tumor cells was observed when the internalization of the FA-capped MSNs was quantitated in HeLa cells and normal endothelial cells. This last result can be clearly explained by the fact that the amount of folate receptors was larger on tumor cells when compared with normal cells (favoring receptor-mediated endocytosis of the capped nanoparticles). DOX-loaded FA-capped nanoparticles selectively induced HeLa tumor cell death (through apoptosis). The effects of the DOX-loaded nanoparticles were tested *in vivo* by generating tumor models on nude mice via subcutaneous xenotransplants of HeLa cells. Tumor size was inhibited when treated with DOX-loaded nanoparticles.

Lee *et al.* also prepared redox-triggered capped materials but using, as inorganic supports, $Fe_3O_4$ nanoparticles coated with a mesoporous silica shell.[43] Pores of the support were loaded with DOX and the external surface functionalized with (3-mercaptopropyl)trimethoxysilane. Then, the loaded and thiol-functionalized nanoparticles were reacted with *S*-(2-aminoethylthio)-2-thiopyridine to obtain amino moieties linked by a disulfide bond. Finally, the amino groups were reacted with propargyl bromide and the stalk was further reacted with a mono-6-azido-$\beta$-CD via click chemistry. PBS suspensions of the capped nanoparticles in the absence of GSH were unable to induce DOX release whereas in the presence of the tripeptide a marked delivery of the drug was observed. CLSM studies showed that DOX was released in A549 cells previously incubated with the capped nanoparticles as a result of the cleavage of the gatekeeper mediated by intracellular GSH. The authors also performed *in vivo* MRI experiments to investigate the targeted delivery and intra-tumoral distribution of the system in mice that bore A549 cell xenografts. These studies showed the accumulation of the material in the tumor and the suppression of tumor growth *in vivo* by intravenously injected nanoparticles.

Gaberscek and coworkers synthesized several redox-responsive delivery systems, all of which were based on the use of $\beta$-CDs as stoppers, which were anchored to the silica surface using different linkers.[44] For this purpose, MSNs were functionalized with APTES and then reacted with (*N*-succinimidyl-3-(2-pyridyldithio)propionate, 4-(succinimidyloxycarbonyl)-$\alpha$-(2-pyridyldithio)toluene or 4-(succinimidyloxycarbonyl)-$\alpha$-methyl-$\alpha$-(2-pyridyldithio)toluene to provide stalks with different hindrance levels around disulfide bonds. Finally, pores were loaded with Rhodamine B (Rh B) and capped with monothio-$\beta$-CD or perthio-$\beta$-CD (through a reaction with the pyridyldithio groups of the thiol-containing stalks, which resulted in the formation of disulfide bonds). Capped materials were tested in a non-reducing environment and showed negligible Rh B release, whereas addition of dithiothreitol (DTT) allowed clear dye delivery. The rate of release of the entrapped Rh B was clearly related with the steric hindrance induced by the substituents located around the disulfide bond (the more the hindrance the lower the release rate). Teratogenicity of the prepared materials was evaluated with *in vivo* experiments using zebra fish (Daniorerio). In these experiments, and to avoid excessive animal testing, only the nanoparticles that presented the largest amount of Rh B release were evaluated (material containing an ethylene linker). No tested sample affected hatching of larva on day 4 post-fertilization in all the samples. Also, no effects on fish length were detected at the same time point. The system neither significantly affected embryo development and survival, nor were signs of cell death detected. These data confirmed the non-teratogenicity of the nanomaterial.

Zhao and coworkers used different sized MSNs (48–100 nm) for the preparation of redox-triggered gated materials.[45] The authors functionalized the external surface of the MSNs with (3-mercaptopropyl)trimethoxysilane and then the thiol groups were reacted with 2-carboxyethyl-2-pyridyl disulfide to incorporate carboxylic acid moieties in the MSNs. In a second step, the pores of the MSNs were loaded with DOX and capped with per-6-diamino-$\beta$-CDs (through the formation of amide bonds). Finally, capped MSNs were decorated with PEG chains functionalized with adamantane (formation of inclusion complexes with CDs) and FA (to stimulate the internalization on tumoral cells). The presence of GSH induced the rupture of the disulfide bonds of the gate with subsequent pore opening and DOX release. Also, a drop in pH (from 7.4 to 5.5) induced certain DOX delivery which was attributed to the electrostatic repulsions between the protonated amino moieties in $\beta$-CDs. The active folate targeting of the nanoparticles was tested in the MDA-MB-231 breast cancer cell line, which is well-known to overexpress folate receptors. As expected, incubation of capped nanoparticles with MDA-MB-231 cells led to improved selective uptake of nanoparticles into cells. Also, after 12 hours of incubation, apoptotic cell bodies were observed in cultures. Pre-incubation with free FA prevented capped nanoparticles internalization, and also subsequent DOX cytotoxicity. The reduced toxicity and selective endocytosis of the capped nanoparticles, when compared with free DOX, suggested that this solid can effective be used for controlling tumor growth and reducing systemic toxicity. To test this possibility, *in vivo* research of the therapeutic efficacy of the nanoparticles was carried out in mammalian cancer-induced female nude mice (by implanting MDA-MB-231 breast carcinoma cells). Mice treated with the capped nanoparticles exhibited significantly higher tumor shrinkage rate than those treated with free DOX. No statistically significant body weight loss was observed in any treated group, which indicated that treatment was well tolerated. An *in situ* histological analysis also supported the system's excellent therapeutic effect. Few tumor cells/foci were detected in mice treated with nanoparticles. In addition, examination of selected organs (heart, lung, liver, spleen, and kidney) indicated no apparent abnormality or lesions after 21 days post-treatment. These results indicated that capped MSNs can deliver a sufficient amount of DOX in tumors, which resulted in a remarkable tumor-inhibiting effect when compared with free DOX and non-targeted nanoparticles.

Martínez-Máñez and coworkers reported the design of MSNs capped with oligo(ethyleneglycol) chains containing disulfide bonds.[46] MSNs were selected as inorganic scaffold, the pores were loaded with safranin O and the external surface was functionalized with (3-mercaptopropyl)trimethoxysilane. In a second step, the external thiol moieties were further reacted with 2,2′-dipyridyldisulfide and with *O*-(2-mercaptoethyl)-*O*′-methyl-hexa(ethyleneglycol). Water suspensions of oligo(ethyleneglycol)-capped MSNs at pH 7.5 showed negligible dye release, whereas in the presence of GSH a marked safranin O release was observed. The opening mechanism was based on a selective GSH-induced reduction of the disulfide bond, which resulted in pore opening and entrapped dye release (see Figure 3). CLSM studies, carried out with HeLa cells incubated with the oligo(ethyleneglycol)-capped MSNs, showed safranin O-associated fluorescence in the cytosolic compartment indicating the internalization of nanoparticles, the rupture of the disulfide linkage by the reductive environment and the subsequent release of the entrapped dye. Besides, WST-1 assays showed the non-toxicity of oligo(ethylene

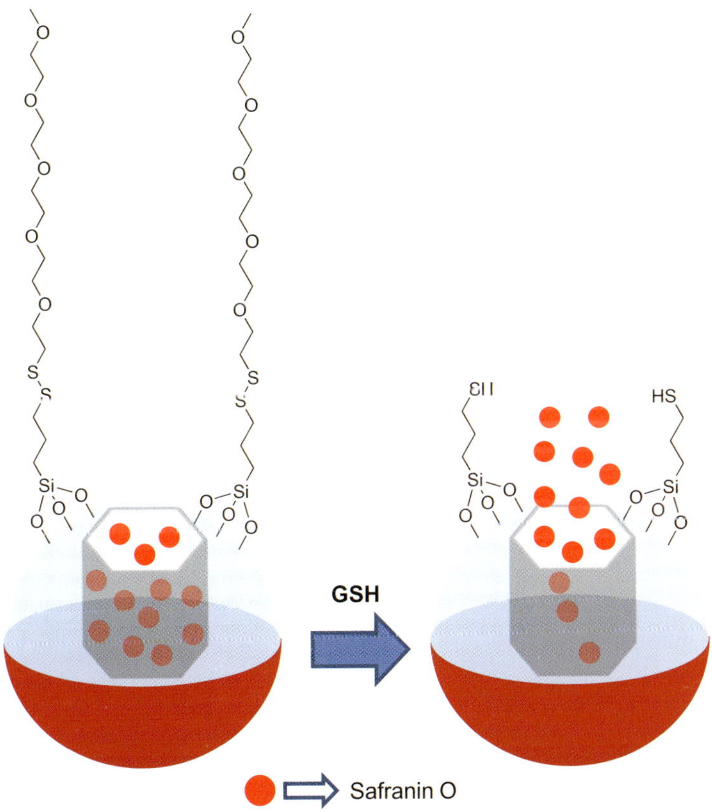


Figure 3. MSNs loaded with safranin O and functionalized with disulfide-containing poly(ethylene glycol) chains. Dye was released in the presence of GSH.

glycol)-capped MSNs for HeLa cells. Finally, MSNs with the same gating mechanism but loaded with DOX instead of safranin O were prepared and characterized. WST-1 assays carried out with HeLa cells and the DOX-loaded MSNs showed a marked decrease in viability (60% of cell killing at a concentration of 150 $\mu$g mL$^{-1}$) due to the internalization and the redox-triggered release of the entrapped DOX. Furthermore, the cytotoxic effect was clearly noted through changes in cell morphology.

Cai and coworkers prepared redox-responsive MSNs loaded with FITC or DOX and grafted with *N*-deacetylated heparin on the outer surface using a disulfide-containing stalk.[47] Also, *N*-deacetylated heparin provided free amino groups to couple lactobionic acid (LA) as a targeting moiety. PBS suspensions of the capped nanoparticles showed nearly zero release in the absence of GSH. However, in the presence of GSH a marked FITC release was observed (due to rupture of disulfide bonds and subsequent detachment of *N*-deacetylated heparin from the MSNs surface). Once corroborated the *in vitro* proper work of the nanoparticles the authors studied the interaction between capped MSNs and Hep G2 hepatoma cells. Unloaded materials did not induce any changes in cellular morphology or in cell viability. Besides, the uptake of capped nanoparticles by Hep G2 cells was much greater than that of unfunctionalized MSNs or capped MSNs without the LA directing agent. The endocytosis of capped MSNs containing FITC was further analyzed by incubating Hep G2 cells. The green fluorescence of endocytosed nanoparticles was mainly located in the cytoplasm. Once again, the cells treated with the capped MSNs that contained LA moieties exhibited better internalization efficiency when compared with the control nanomaterials. This fact was ascribed to galactose receptor-mediated endocytosis as the pre-incubation of cells with free galactose significantly decreased intracellular FITC fluorescence. Next the authors studied, by CLSM, the inhibitory effect on tumor cell viability and apoptosis when using capped nanoparticles loaded with DOX. A significantly higher apoptosis induction was observed for the capped MSNs functionalized with LA targeting subunits. *In vivo* experiments, to evaluate the release and distribution of the nanoparticles in mice that bore Hep

G2 cell xenografts, were also carried out. These experiments showed that tumor growth *in vivo* was effectively suppressed by subcutaneously injected DOX-containing capped nanoparticles. A histological examination to evaluate the damage of major mice organs was carried out. While free DOX induced typical myocardial injury in mice, those injected with the DOX-loaded nanoparticles presented a significant remission of cardiac damage.

Wang and co-workers prepared hybrid lipid-capped MSNs for redox-triggered drug release with the objective to overcoming multidrug resistance.[48] MSNs were prepared and their external surface functionalized with (3-mercaptopropyl)trimethoxysilane. Then, in a second step, the thiol moieties reacted with 2,2′-dithiodipyridine and the obtained solid with octadecanethiol. These two consecutive reactions yielded MSNs coated with $C_{18}$ chains linked to the surface by disulfide moieties. In the next step, the pores of $C_{18}$-functionalized nanoparticles were loaded with DOX and capped with a lipid layer formed by soybean lecithin and D-$\alpha$-tocopherol polyethylene glycol 1000 succinate in 5:1 molar ratio. The *in vitro* controlled release features of the lipid layer-coated MSNs were tested in PBS at pH 5.0 in the absence and in the presence of GSH. In the absence of GSH a tight pore closure was observed and only 8% of the entrapped DOX was released after 48 hours. However, in the presence of GSH (10 mM), 60% of the entrapped drug was released after 48 hours. GSH-induced reduction of disulfide bond induced lipid layer detachment from the surface of the carrier which favors the diffusion of DOX. CLSM studies carried out with MCF-7 (DOX sensitive) and MCF-7/Adr (DOX resistant) cells incubated with the lipid-capped nanoparticles were carried out. These studies showed that lipid-capped nanoparticles were uptake in a similar way by both cell lines and the fluorescence of the released DOX was observed in the cytoplasm and nuclei. These results strongly supports that D-$\alpha$-tocopherol, inserted in the lipid layer, was able to inhibit drug efflux by suppressing the activity P-glycoprotein overexpressed in MCF-7/Adr DOX resistant cells. Finally, $IC_{50}$ values for free DOX and lipid-capped nanoparticles against MCF-7 and MCF-7/Adr cells were calculated. As expected, in MCF-7/Adr cells (with overexpressed P-glycoprotein) the $IC_{50}$ for DOX was 183 $\mu$g/mL, 310-fold higher than that of MCF-7 cells (0.59 $\mu$g/mL). However, stronger cytotoxicity was obtained with the lipid-capped nanoparticles ($IC_{50}$ of 12.31 and 0.22 $\mu$g/mL for MCF-5/Adr and MCF-7 cells respectively). According to these results lipid-capped nanoparticles demonstrated the ability to overcome multi drug resistance.

Amphiphilic peptides were also used as redox-triggered caps in MSNs for cancer targeted drug delivery.[49] For the preparation of the capped nanodevice, the external surface of the MSNs was functionalized with (3-mercaptopropyl)trimethoxysilane and then the pores loaded with DOX. In a second step, the thiol moieties on the surface were reacted with cysteine modified stearic acid using 2,2′-dithiodipyridine as coupling agent. This procedure yielded MSNs decorated with hydrophobic $C_{18}$ chains grafted onto the surface through disulfide linkages. The pores of the loaded solid were then capped upon addition of the amphiphilic $C_{18}$-DSDSDSDSRGDS peptide by the formation of a self-assembled monolayer onto the external surface of the nanoparticles (hydrophobic interactions between $C_{18}$ chains in the peptide and in the surface). The controlled release features of the peptide-capped nanoparticles were tested in buffered suspensions at pH 7.4 in the absence and in the presence of GSH and DTT. In the absence of reducing agents a nearly zero release was observed whereas, when GSH and DTT were present a marked DOX release was found. The redox-triggered DOX release was ascribed to the reduction of the disulfide bond with subsequent detachment of the amphiphilic peptide from the nanoparticles surface. CLSM carried out with UG-87 MG cells (human glioblastoma) and COS7 cells (non-cancerous kidney fibroblasts) incubated with peptide-capped nanoparticles indicated a preferential uptake in cancerous cells. This targeted uptake was due to the fact that integrin $\alpha\beta_3$ (overexpressed in UG-87 MG cells) could bind with the RGD motif of the peptide sequence used as cap. Flow cytometry studies confirmed nanoparticle internalization in UG-87 MG cells via receptor-mediated endocytosis. Finally, cell viability assays carried out with UG-87 MG and peptide-capped nanoparticles showed a marked cell death as a consequence of internalization and GSH-induced DOX release.

Lin and coworkers prepared several capped MSNs and studied their use for DNA and drug co-delivery into plants.[50] For the synthesis of a material, the external surface of MSNs was functionalized with mercaptopropyl groups. Afterward, the surface thiol groups were reacted with 2-(pyridyldisulfanyl)ethylamine and then FITC was used for labeling MSNs through the formation of urea bonds. In another step, a layer of triethyleneglycol was anchored in order to improve the biocompatibility and suspendability of nanoparticles. Finally, a DNA plasmid (containing a GFP gene) was used for coating the solid on the triethyleneglycol layer. The capability of this material for gene transfection was successfully tested in protoplasts. In spite of the remarkable results obtained, the described nanoparticles are not useful for gene transfection in plant tissues by a gene gun system. For this purpose, a second material was prepared using MSNs functionalized with mercaptopropyl subunits and reacted with 2-(pyridyldisulfanyl) ethylamine. Instead of labeling with FITC, the pores of the MSNs were loaded with fluorescein and capped with AuNPs coated with 11-mercaptoundecanoic acid through the formation of amide bonds. Finally, the grafted AuNPs were coated with the DNA plasmid. These new nanoparticles were successfully tested in tissue plants by gene gun techniques. Bombarding tobacco cotyledons with the capped system induced DNA plasmid internalization with subsequent GFP/expressing foci. Also, as a consequence of reduction of disulfide linkages by DTT, FITC delivery was also observed. Similar results were obtained with maize immature embryos. A third solid was obtained, in the same way as the previous one, but was loaded with $\beta$-oestradiol. This third solid was tested also in transgenic tobacco plants in which GFP expression could be detected only if $\beta$-oestradiol was present. As expected, the GFP signal was detected in the transgenic seeds germinated in DTT-containing media. DTT opened the gate by reducing the disulfide linkage in the chemical linker thus release the GFP inducer and the required cargo. These results demonstrated the interaction between the cargo inside MSNs and the DNA coated on AuNPs.

Zhu and co-workers used ZnS QDs as lids for the preparation of capped MSNs able to release an entrapped cargo in the presence of hydroxyl radical (generated in highly oxidative stress conditions such as those presented in inflammatory processes).[51] For this purpose, the authors prepared MSNs and then functionalized its external surface with APTES. The aminopropyl-functionalized nanoparticles were loaded with camptothecin (CPT) or with rhodamine 6G and the pores capped, through carbodiimide chemistry, upon addition of small ZnS QDs coated with 3-mercaptopropionic acid. PBS suspensions of the QD-capped nanoparticles showed negligible release of the cargo. However, in the presence of hydroxyl radical (generated from $H_2O_2/Fe^{2+}$) a marked cargo delivery was observed. The release of the cargo was ascribed to the hydroxyl radical-induced oxidation of thiol moiety of the QDs coating to sulfonic acid with subsequent detachment of QDs from the surface of the nanodevice. The authors also indicated that some etching of the ZnS QDs could be produced in the presence of hydroxyl radical. MTT assays carried out with the CPT loaded nanoparticles in BxPC-3 cells (model of pancreatic cancer cells) showed a marked decrease in cell viability. This decrease was ascribed to the release of CPT due to the intracellularly produced hydroxyl radicals. Also, QD-capped nanoparticles internalization in BxPC-3 cells was studied by confocal microscopy. The fluorescence of CPT and of rhodamine 6G was observed in the cytoplasmic region indicating the cellular uptake of the QD-capped nanoparticles.

Green fluorescent carbon quantum dots (CQDs) have been used as caps for the preparation of gated MSNs for cancer theranostics.[52] CQDs were firstly functionalized with cystamine using an EDC/NHS coupling reaction. Then, folic acid was attached to amine-functionalized CQDs using also EDC/NHS for carboxylic acid activation. On the other hand, MSNs were loaded with DOX and the pores were caped with the FA-functionalized CQDs by simple electrostatic interactions between the silanolate moieties onto the surface of the silica nanoparticles with the free amino groups onto the FA-functionalized CQDs. Aqueous suspensions of CQD-capped nanoparticles at pH 7.0 showed negligible DOX release (4.3% after 24 hours). However, at pH 4.0 a marked drug release was observed (63.9% after 24 hours) due to the partial protonation of silanolates that detached CQDs form the surface of the nanodevice. Besides, at pH 4.0 in the presence of GSH (10 mM) a stronger DOX release was observed (88.9% after 24 hours) due to the additional reduction of the disulfide bond in the coated CQDs. Confocal microscopy studies carried out in HeLa cells showed the uptake of the CQD-capped nanoparticles and the release of DOX in the intracellular acidic media also mediated by GSH.

## V. pH Responsive Gated Materials

Of all the external triggers used in the development of gated materials, pH is perhaps the most widely explored. In the published examples, the open/close mechanism of these capped materials can be classified into four well defined groups: (i) changes in size/shape of the (bio)molecules than acted as gates induced by pH changes; (ii) attraction/repulsion interactions between charged molecules due to proton addition or abstraction; (iii) the use, as capping units, of functional groups that can be hydrolyzed with pH and (iV) organic/inorganic coatings that are dissolved upon pH changes. Moreover, pH-triggered systems offers the advantage of being simple and in most cases the closing and opening mechanism is reversible. pH-triggered gated materials are promising nanodevices for the development of delivery systems for biomedicine applications. In particular, pH difference between the extracellular environment (pH 7.0) and endosomes (pH 5.0) is a useful tool to release therapeutic agents in cells in a selective fashion. Besides, tumors have an acidic environment (due to the production of lactic acid and ATP hydrolysis) when compared with normal tissues and this difference can be used for targeted delivery of payloads.

Martínez-Máñez and coworkers described one of the first examples of a pH-triggered gated silica mesoporous support.[53] In this case, the authors selected a micrometric MCM-41 support as inorganic scaffold and grafted mercaptopropyl subunits in the inner of the pores and functionalized the external surface with 3-[2-(2-aminoethylamino)ethylamino]propylmethoxysilane (a linear polyamine). The gating mechanism was assessed using a blue squaraine (SQ) dye which could react with the thiol moieties inside the pores yielding a colorless derivative. The protonation of the nitrogen atoms at acidic pH induced the polyamine to adopt a rigid-like conformation (due to Coulombic repulsions between the positively charged ammonium moieties) that induced pore blockage and inhibited SQ-thiol reaction (solution remains blue). As a contrast, at neutral pH, the amines (that were partly protonated) adopted a flexible conformation that allowed SQ to enter into the pores (the solution turned colorless). Also, a synergic anion-controlled outcome resulted from the interaction between certain anions and protonated amines. For instance, the presence of large anions (ATP) also inhibited SQ access to the inner of the pores due to the formation of strong bulky ATP-polyammonium complexes.

The same authors developed similar pH-responsive polyamine-containing capped materials for the delivery of Tebuconazole and suggested its potential application to treat candidiasis.[54] For this work, MSNs were used as inorganic scaffold and the same polyamine used in the previous example was anchored onto the external surface. At acidic pH (3.7) the pores were capped and the release of tebuconazole highly inhibited. On the contrary, a marked delivery of the drug was observed at pH 5.5. Cell viability and intracellular cargo delivery studies were carried out with the MSNs using *S. cerevisiae* a microorganism that survives in acidic environments, and perfectly suits the aperture mechanism of the nanomaterial. The non-toxicity of the material was assessed by clonogenic cell viability studies. Besides, a polyamine-capped MSNs loaded with fluorescein was efficiently internalized in yeast cells, and brought about subsequent cargo delivery at pH 5.5 as demonstrated by fluorescence microscopy measurements. Interestingly the material loaded with tebuconazole induced a 9-fold increase in cytotoxicity (when compared with the drug alone) improving its efficacy. This capped material could potentially overcome the adverse side effects associated with topical therapies for vulvovaginal infection.

Teng, Lu and co-workers prepared pH-responsive MSNs by using positively charged natural protein cytochrome C (CytC) as pore blocker.[55] The pores of the MSNs were loaded with DOX and capped with CytC by electrostatic interactions with the silanol groups in the surface of the support. Aqueous suspensions of the CytC-capped nanoparticles at pH 7.4 showed negligible DOX release (16% in 72 hours) whereas a larger delivery of the drug was observed at pH 5.5 (54% in 72 hours). CLSM studies indicated that CytC-capped nanoparticles were internalized by human breast cancer MCF-7 cells and that DOX was released from the nanomaterial. Cytotoxicity and histological assays confirmed that the DOX-loaded CytC-capped nanoparticles possessed lower toxicity than free DOX. Besides, the intratumoral administration of nanoparticles proved significantly more efficacious in tumor reduction than free DOX and unsealed drug carriers in xenograft MCF-7 cancer models.

Meng and coworkers developed a MSNs-based system capable of co-delivering DOX and Pgp siRNA.[56] The authors selected MSNs and grafted, onto the external surface, a polyethyleneimine-polyethylene glycol (PEI-PEG) copolymer. Then, the pores were loaded with DOX and Pgp siRNA attached to the external surface by electrostatic interactions. The simultaneous release of siRNA and DOX from the MSNs was studied in saline-supplemented mouse serum at pH 7.4 and 4.5. When the nanoparticles were suspended in the serum at pH 7.4 a negligible release was observed. However, at pH 4.5, a remarkable delivery of DOX and siRNA was observed. A MCF-7/MDR xenograft model in nude mice was used to test whether the capped nanoparticles provided the protected delivery of stably bound DOX and Pgp siRNA to the tumor site. The capped nanoparticles exhibited effective biodistribution and reduced endothelial reticulum uptake, thus providing 8% enhanced permeability and a retention effect at the tumor site. This dual delivery material induced a synergistic inhibition of tumor growth *in vivo*, when compared to free DOX or the carrier loaded with either drug or siRNA alone. Furthermore, significant Pgp knockdown at the heterogeneous tumor sites was observed, which corresponded to the areas in which DOX was released and triggered apoptosis. These data provide proof-of-principle testing of the use of a dual drug/siRNA nanocarrier to overcome drug resistance in a xenograft.

The same authors developed an engineered approach for the pancreatic ductal adenocarcinoma (PDAC) stromal barrier through the combined use of a liposome that contained gemcitabine (GEM) and MSNs coated with PEI-PEG.[57] For the preparation of the system, PEI was electrostatically attached onto the MSNs and the polymer was also modified with PEG. Finally, a small molecule inhibitor of the transforming growth factor beta signaling pathway (TGF-$\beta$), LY364947, was attached to the PEI polymer through H-bonding interactions. PBS suspensions of the nanoparticles at neutral pH (7.4) showed negligible LY364947 delivery. However, at pH 5.5 a marked release was observed. Dealing with the liposomes, an optimal design was achieved using a transmembrane ammonium sulfate gradient to trap GEM inside the organic nanostructure. Nude mice BxPC3 xenografts were used to explore TGF-$\beta$ delivery to the PDAC tumor. These xenografts elicit a dense infiltrating stroma, which surrounds nests of cancer cells and also covers tumor blood vessel fenestrations. Due to the acidic environment of the tumor stroma, the LY364947-coated nanoparticles were firstly used for the pH-dependent LY364947 delivery. This delivery inhibits the transforming growth factor beta receptor signaling in the xenograft. As a result of TGF-$\beta$ signaling blockage, pericyte (PC) differentiation and attachment to endothelial cells (EC) were also prevented. This allowed liposomal GEM carrier access through the open vascular fenestrations, and the ability to increase GEM delivery. This two-wave approach allowed the effective shrinkage of the tumor xenografts beyond 25 days compared to the treatment only with free drug or GEM-loaded liposomes.

pH-responsive multifunctional nanoparticles using poly(2-vinylpyridine) as gating subunit were prepared by Bein and co-workers.[58] Poly(2-vinylpyridine) is a polymer that presents an hydrophilic extended conformation at pH 5.0 (due to pyridine protonation) and an hydrophobic water-insoluble collapsed conformation at neutral pH. The authors functionalized MSNs with APTES (MSN-$NH_2$) which were further reacted with BOC-protected •-amino-•-carboxy terminated poly(2-vinylpyridine) (BOC-NH-PVP-COOH) using EDC/sulfo-NHS coupling and subsequent BOC deprotection with trifluoroacetic acid. This yielded nanoparticles functionalized with PVP polymer containing free amine moieties (MSN-PVP-$NH_2$). In a further modification, the amine moieties were transformed into carboxylates upon EDC-promoted coupling with oxalic acid (MSN-PVP-COOH). Finally, in order to enhance nanoparticles permeability and retention, the carboxylate groups were reacted with $H_2N$-PEG-$NH_2$ using EDC as coupling agent (MSN-PVP-PEG-$NH_2$). The porous network of MSN-PVP-PEG-$NH_2$ were loaded with different cargoes by suspending nanoparticles in an aqueous suspension at pH 2.0 (pores opened because PVP is in its protonated hydrophilic form) containing selected dyes/drugs. After the loading process the pores were closed by suspending nanoparticles in water at pH 7.0 (PVP in its hydrophobic collapsed conformation). Following this experimental procedure the pores of MSN-PVP-PEG-$NH_2$ nanoparticles were loaded with fluorescein, DOX, 4′,6-diamidino-2-phenylindole (DAPI) and colchicine. *In vitro* controlled release experiments carried out with MSN-PVP-PEG-$NH_2$ nanoparticles loaded with fluorescein showed the expected gating mechanism, namely, no release at neutral pH and a marked fluorophore release upon lowering pH to 5.0. DAPI controlled release was studied with HeLa cells and confocal microscopy studies showed the internalization of the loaded nanoparticles, the delivery of the dye into the endosomes (due to its acidic pH) and, finally, the appearance of strong DAPI-fluorescence in cell nucleus. Colchicine release was also observed in HuH7 cells. MTT assays carried out in T24 bladder carcinoma cells with MSN-PVP-PEG-$NH_2$ nanoparticles loaded with DOX showed a marked decrease in viability (40% after 48 hours) due to carrier internalization and drug delivery in the acidic endolysosomal compartments.

Xu, Chu and coworkers used MSNs as inorganic scaffold, loaded the pores with bovine serum albumin (BSA) protein, and attached chitosan, onto the external surface by reaction with (3-glycidyloxipropyl)trimethoxysilane.[59] Neutral PBS suspensions (pH 7.4) of the capped nanoparticles showed negligible BSA release, whereas pH lowering (4.0) resulted in a marked delivery of the protein. The same nanoparticles, but loaded with tumor necrosis factor $\alpha$ (TNF-$\alpha$) instead of BSA, were prepared for the *in vitro* cellular studies. Besides, antibody ab2428 (that can bind to antigen ErbB 2) was conjugated with the chitosan polymer. After conjugating the antibody onto capped MSNs, nanocarriers were able to deliver TNF-$\alpha$ to MCF-7 breast cancer cells in both *in vitro* and *in vivo* (using nine 3-week-old athymic nude mice). The high affinity between the antibody and the antigen directed the system to achieve targeted delivery to cancer cells. The authors found that loaded TNF-$\alpha$ was gradually released from nanocarriers (due to the acidic micro-environment inside solid tumors) and triggered apoptosis in tumor cells.

Ren, Qu and co-workers synthesized magnetic $Fe_3O_4$ nanoparticles coated with a mesoporous silica shell that was functionalized with aminopropyl groups. Then, fluorescein isothiocyanate (FITC) and TAT peptide were anchored onto the amino-containing surface (the dye through the formation of thiourea bonds and the peptide using an amidation reaction).[60] Then, pores were loaded with CPT and the external surface was coated, via electrostatic interactions, with a polymer prepared by the reaction of a chitosan-folic acid conjugate with citraconic anhydride. Polymer-capped nanoparticles efficiently accumulated in target tissues (H22 tumor-bearing mice) guided by a local strong magnetic field generated by a magnet. Nanocarriers were efficiently internalized in cancer cells via floater acceptor-mediated endocytosis. Once endocytosed, the acidic pH of lysosomes induced the hydrolysis of the amide bonds that linked the chitosan to the nanoparticles. As a result chitosan was detached form the surface of the nanoparticles and exposed the nuclear-targeted TAT peptide groups. The regenerated chitosan broke lysosomes and helped nanoparticles to escape from lysosomes into the cytosol. The TAT peptide on the outer surface then guided nanoparticles to transfer from the cytosol into cell nuclei. Thanks to this continuous 3-stage targeting process, CPT successfully entered the nuclei of tumor cells and induced apoptosis in HeLa, but not in A549, cells (which did not express folate receptors).

Dengler *et al.* prepared two types of MSNs loaded with DNA and caped with two different lipid bilayer (cholesterol-(1,2-dioleoyl-3-trimethylammoniumpropane) DOTAP and 1,2-dioleoyl-*sn*-glycero-3-phosphocholine (DOPC)) for transgene brain and spinal delivery.[61] Suspensions of both nanoparticles at pH 6.0 showed negligible payload release whereas a marked delivery was observed at lower pH values. Also, viability assays showed the non-toxicity of both materials. Besides, the peri-spinal application (subarachnoid, intrathecal) of both bilayer-coated nanoparticles in separate groups of rats resulted in minimal sensory changes, which indicates that the systems presented good biocompatibility. However, the biodistribution of DOTAP-cholesterol and DOPC capped MSNs slightly differed. Only the DOPC coated nanoparticles revealed increased spread to the brain, as observed after 8 weeks. As clear contrast, DOTAP-cholesterol coated nanoparticles revealed superior transfection efficiency to naked pDNA-IL-10. The DOTAP-cholesterol coated nanoparticles loaded with the pDNA-IL-10 were tested for their *in vivo* therapeutic potential as a novel non-viral gene transfer vector delivered to spinal cord to treat peripheral neuropathic pain. Remarkably, this treatment resulted in the functional suppression of pain-related behavior in rats for periods that extended up to 8 weeks.

Li, Gan and coworkers loaded MSNs with irinotecan (an antitumoral drug) and capped the pores with a polymer-lipid bilayer prepared by the reaction of the hydroxyl moieties of Pluronic 123 and amine groups of DOPE or DSPE using *N,N'*-disuccinimidyl carbonate (as linker) and trimethylamine as depicted in Figure 4.[62] Simulated body fluid (SBF) suspensions of the capped nanoparticles at pH 7.4 showed negligible irinotecan release. Upon lowering the pH to 5.0 a marked irinotecan release was observed. Interestingly, the system was able to inhibit breast cancer resistance protein (BCPR)-mediated irinotecan efflux in drug-resistant MCF-7/BCRP breast cancer cells. Incubation with nanoparticles resulted in increased intracellular irinotecan levels, greater cytotoxicity, and stronger cell cycle arrest in MCF-7/BCRP cells, when compared with the results obtained with the free drug. The irinotecan-loaded nanoparticles showed high therapeutic performance and low toxicity in xenografts of drug resistant breast tumors in nude mice.

Zink and co-workers prepared MSNs capped with pH-sensitive nanovalves and functionalized with transferrin protein as targeting element.[63] For the preparation of the nanocarrier the external surface of MSNs was functionalized, in a first step, with (3-iodopropyl) trimethoxysilane and subsequently coupled to *p*-anisidine. In a

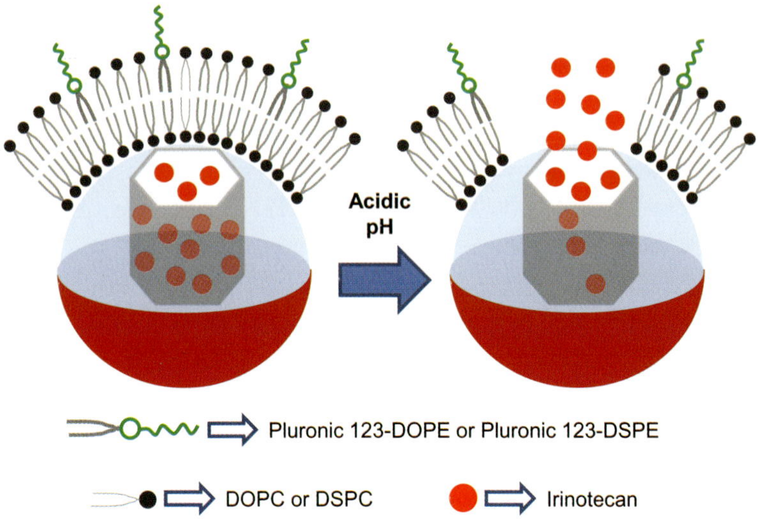


Figure 4. MSNs loaded with irinotecan and capped with a polymer-lipid bilayer. Acidic pH induced irinotecan release.

second step, the external surface was further functionalized with (3-mercaptopropyl)trimethoxysilane. Then, the pores of the functionalized nanoparticles were loaded with DOX and sealed upon addition of α-cyclodextrin (through the formation of inclusion complexes with the grafted *p*-anisidine moieties). Finally, transferrin was covalently attached onto the external surface of the nanoparticles by thiol-disulfide exchange. Aqueous suspensions of the nanovalve-capped nanoparticles at pH 7.0 showed nearly zero release of the entrapped drug whereas, upon lowering pH to 5.5, a marked DOX delivery was observed. Delivery was due because of the detachment of α-cyclodextrin from the nanoparticle surface upon *p*-anisidine protonation. The prepared nanoparticles were uptaken by MiaPaCa-2 cancer cells (that over-expressed transferrin receptor) and were able to release DOX (due to the acidic environment of the lysosomes) as indicated by fluorescence microscopy measurements. This DOX nanocarrier also induced a marked decrease in viability in MiaPaCa-2 cancer cells. Finally, *in vivo* studies carried out with SCID mice xenografts (with MiaPaCa-2 cells) showed successful DOX release in tumor.

MSNs with pH-triggered nanovalves were used to deliver moxifloxacin for the efficient killing of *Francisella tularensis* bacterial pathogen.[64] For this purpose, MCM-41 MSNs containing phosphonate subunits in the inner of the pores were prepared by a co-condensation method using tetraethyl orthosilicate and 3-(trihydroxysilyl)propyl methylphosphonate. Then, the external surface of the MSNs was functionalized with chloromethyltrimethoxysilane and the chloromethyl moieties were further reacted with benzimidazole. The benzimidazole-functionalized nanoparticles were suspended in a PBS (pH 7.4) solution of moxifloxacin and stirred for 24 hours. Then, $\beta$-cyclodextrin was added in order to cap the pores by the formation of inclusion complexes between the cyclic oligosaccharides and the grafted benzimidazole moieties. PBS suspensions of the nanovalve capped nanoparticles at pH 7.4 showed negligible release of moxifloxacin whereas, at pH 6.0, a marked delivery of the drug was observed (due to protonation of the benzimidazole stalk with subsequent dethreading of $\beta$-cyclodextrin from the nanoparticles surface). The prepared nanovalve-capped nanoparticles were uptake by differentiated human THP-1 macrophages infected with *F. tularensis* live vaccine strain (LVS) with subsequent pore opening, release of moxifloxacin and bacterial killing. The release of the drug induced a reduction in bacterial colony forming units (CFU) by 3.4 logs compared with the untreated macrophages at 1 day. The killing efficiency of the nanoformulation was almost twice than that of the same amount of the free drug. *In vivo* experiments, carried out with mice infected with 4000–8000 CFU of *F. tularensis* showed that capped nanoparticles helped to control severe bacterial infection and reduced bacterial burden in lung, liver, and spleen with higher efficacy when compared to that obtained using moxifloxacin alone.

## VI. Molecule/Biomolecule Responsive Gated Materials

Molecular and supramolecular interactions between binding units, anchored on the pore outlets of inorganic mesoporous support, and species present in the solution have also been used as mechanism for the development of gated materials. The chemical species that can be used as triggers for the controlled release of entrapped payloads comprises anions, cations, neutral molecules and biomolecules. One remarkable advantage of these molecule or biomolecule-driven systems is related with their intrinsic specificity and in these systems only when selected triggering molecule or biomolecule is present, the encapsulated cargo is released. This opens the possibility to design precise gated systems for very particular applications in the biomedical field.

The first organic-inorganic capped hybrid material that responded selectively to the presence of certain a chemical species was described in 2006 by Martínez-Máñez and coworkers.[65] The authors selected MCM-41 mesoporous silica support as inorganic scaffold for the preparation of a hybrid material able to selectively detect ATP. For this purpose, the pores were loaded with the $Ru(bipy)_3^{2+}$ complex (indicator dye) and functionalized the external surface with 3-[2-(2-aminoethylamino)ethylamino]propyltrimethoxysilane (the gate ensemble). The gate was opened at neutral pH when the amines were partly protonated, and the $Ru(bipy)_3^{2+}$ complex free to diffuse from the inner of the pores to the bulk solution. The dye release could be inhibited upon addition of ATP because this anion formed strong complexes with the anchored polyamines through hydrogen bonding and electrostatic interactions. Additionally, ADP anion inhibited receptor release but in a lesser extent than that observed with ATP. The material displayed good selectivity and it was found that other anions tested (chloride, sulfate, or GMP) did not effectively cap the pores.

Capped MSNs that respond to sulfide and pH for applications in self-healing coatings have also been reported.[66] For this purpose, the external surface of MSNs was first functionalized with diamine *N*-(3-trimethoxysilylpropyl)ethylenediamine (DiA) and then the pores loaded with a corrosion inhibitor (BTA) and a biocide (benzalkonium chloride, TC). The final capped material was obtained upon addition of $CuSO_4$ because an insoluble Cu-BTA complex formed at the pore openings. At neutral pH the gate remained closed whereas in acidic environments (pH lower than 5) the Cu-BTA complex dissolved allowing cargo release. Sulfide anions (at concentrations higher than

0.02 mM) were able to extract Cu ions from the capping complex inducing again cargo delivery. For anticorrosion and antifouling experiments, steel substrates were coated with a polyester layer incorporating the capped nanocontainer. Corrosion protection was assessed by current density measurements carried out in 0.1 M NaCl aqueous solution using the scanning vibrating electrode technique (SVET). SVET current density measurements showed that the hybrid coating exhibited anticorrosion features in response to pH decrease and the presence of $S^{2-}$ ions. Furthermore, the antifouling ability of the hybrid coating was demonstrated by the suppression of *E. coli* activity due to the release of the biocide from the capped MSNs.

Son, Lee and coworkers used silica nanotubes (SNTs) as inorganic scaffold which were functionalized with a gold nano-ring at the tube opening using a seed-mediated Au growth reaction.[67] In a second step, the nanotubes were loaded with DOX and capped upon addition of 1-octadecanethiol (that was grafted into the surface of gold nano-ring forming a hydrophobic layer around nanotube entrances). A marked DOX release from the nanotubes was observed upon addition of GSH due to the fact that substitution of 1-octadecanethiol by GSH changed the hydrophobic nature of the nanotube entrances allowing delivery of highly polar drug. Real-time live-cell fluorescence spectroscopy in the multi-drug resistant human breast carcinoma MCF-7/ADR cell line was used to study the DOX release from the nanotubes. For this purpose, the cells were pre-incubated with GSH ethyl ester (to enhance intracellular GSH concentration). Under these conditions, an increase in the intracellular emission intensity was observed after 24 hours due to internalization of nanotubes and release of DOX. Similar results were independently obtained in A549 cells. Besides, DOX-loaded capped SNTs were administered via injection into MCF-7/ADR xenografted mice. Total fluorescence intensities increased after treating mice with GSH, showing that the release of DOX appear to increase by two orders of magnitude. Furthermore, an *in vivo* tumor growth test indicated an increased reduction using DOX-loaded capped SNTs.

AuNPs of 5 nm diameter have also been used as caps for the development of GSH-triggered hybrid nanomaterials.[68] The authors functionalized the external surface of MSNs with APTES and then the pores were loaded with carboxyfluorescein. The final capped material was obtained upon addition of AuNPs (through the formation of weak N-Au bonds). PBS suspensions of the AuNPs-capped nanoparticles showed negligible dye release (ca. 20% after 36 hours) whereas in the presence of GSH (10 mM) a marked carboxyfluorescein delivery was observed (77% after 36 hours). The release of the entrapped dye was ascribed to the detachment of the AuNPs from the surface of the silica nanoparticles due to the replacement of Au-N bonds for Au-SH linkages. CPT loaded nanoparticles were also prepared and its therapeutic effect in A549 lung cancer cells tested. Confocal microscopy studies showed the lysosomal internalization of the AuNPs-capped material in A549 cells and the escape (after 4 hours) to the cytosolic compartment. MTT assays showed a marked reduction in A549 cell viability (35%). The authors suggested that the marked reduction in cell viability was due to CPT release and to an oxidative stress exerted by the AuNPs detached from the nanoparticle surface. CPT loaded AuNPs-capped nanoparticles were also used to treat human lung carcinoma xenografted nude mice. Treatment with AuNPs-capped nanoparticles limited growth and induced shrinkage of tumors. Histological and immuno-histochemical analysis showed a low cell density in the tumors and a significant expression of caspase-3 due to CPT-induced apoptosis.

Martínez-Máñez and coworkers used Janus-type nanoparticles (having opposing Au and mesoporous silica faces) as inorganic support for the development of gated materials that were opened in the presence of ethyl butyrate and glucose.[69] For the preparation of the final materials the pores of the mesoporous silica face were loaded with $Ru(bipy)_3^{2+}$ dye, the external surface functionalized with benzimidazole subunits and the pores capped with $\beta$-CD (that formed inclusion complexes with the grafted benzimidazole moieties). Besides, the Au face was covalently functionalized with thiol-modified esterase and glucose oxidase (GOx) enzymes. The presence of either D-glucose or ethyl butyrate in aqueous suspensions of the capped solids induced a local pH decrease, through an enzyme-catalyzed substrate conversion to gluconic or butyric acids, which induced benzimidazole protonation, de-threading of inclusion complex with $\beta$-CD, and subsequent pore opening and dye release. Moreover, Janus ensemble was also loaded with DOX and was tested in HeLa cancer cells. A marked DOX release inside the cells upon exposure to ethyl butyrate or glucose was observed. In addition, a synergistic enhanced DOX delivery was found when HeLa cells were treated with both D-glucose and ethyl butyrate substrates simultaneously.

Li and coworkers prepared different materials with a copolymer shell of dextran-maleic acid bearing 3-amidophenylboronic acid moieties for the controlled delivery of insulin.[70] For the preparation of the capped materials, the external surface of MSNs was functionalized with 3-(methacryloxy)propyltrimethoxysilane and then covered with a polymeric shell of dextran-maleic acid. Finally, the carboxylic acid moieties were reacted with 3-aminophenylboronic acid (using EDC as coupling agent) and the pores loaded with insulin by suspending the final nanoparticles in PBS at pH 9.0. Three capped materials (with different amounts of grafted copolymer) were prepared and their potential *in vivo* application for oral insulin delivery was studied. The cellular uptake of the prepared solids was assessed using a similar solid loaded with insulin and capped with the same copolymer which was additionally marked with FITC. Absorption and bioavailability of protein drugs in intestine is commonly low due to the weak mucosal membrane permeability and enzymatic degradation. Therefore, the absorption properties of capped solids were investigated in epithelial intestinal villi. These studies indicated good nanomaterial adsorption as strong green dots in villi were observed after incubation with copolymer capped nanoparticles. Finally *in vivo* studies, via oral administration of diverse insulin-loaded capped solids having a different polymer shell thickness, in diabetic rats were carried out and compared to free insulin oral and injected administration. Oral administration of insulin-loaded nanoparticles decreased blood glucose levels at 1.5 hours, and remained low even at 7 hours after administration. Also, studies showed that a more sustained reduction of glucose was achieved when using the capped material having the thickest polymer shell. These assays confirmed that the capped materials were able to reduce more efficiently blood glucose levels than oral administration of insulin.

Martínez-Máñez and coworkers prepared aptamer-capped MSNs which where selectively opened in the presence of thrombin.[71] The external surface of the MSNs was functionalized with APTES, the pores were loaded with Rh B and capped upon addition of a thrombin aptamer (TBA) which was adsorbed on the surface through electrostatic interactions with the positively charged amines grafted onto the support. In the absence of thrombin the solid was tightly capped and nearly a zero release of Rh B was observed. However, in the presence of thrombin, a marked delivery of the entrapped dye was found due to the formation of TBA-protein complexes. The authors reported a limit of detection (LOD) for thrombin of 2 nM in simulated human blood plasma. Moreover, other non-exclusive binding proteins, such as ovalbumin and BSA, were unable to induce uncapping of the pores.

Biomolecules, in particular oligonucleotides, were also used as triggers in uncapping protocols. At this respect, in a pioneering work published in 2010, Climent *et al.* prepared MSNs loaded with fluorescein and functionalized with APTES in the outer surface.[72] Then, the pores were capped upon addition of oligonucleotide 5′-AAT GCT AGC TAA TCA ATC GGG-3′ (through electrostatic interactions with the amines at the surface that were partially protonated at neutral pH). Fluorescein release was achieved upon addition of complementary single strand of the capping oligonucleotide as a consequence of the hybridization of both single oligonucleotides and detachment from the nanoparticles' surface. Dye delivery was also studied in the presence of other oligonucleotides and, in all cases, a negligible fluorescein release was found.

The same authors, using a similar design, prepared capped materials for the detection of genomic DNA.[73] Again, MSNs were selected as inorganic scaffold, its pores loaded with Rh B and the external surface functionalized with APTES. In this example, as capping oligonucleotide, a sequence highly conserved in the *Mycoplasma* genome that corresponds to a fragment of the 16S ribosomal RNA subunit was selected. The capped material showed no Rh B release until the addition of *Mycoplasma fermentans* genomic DNA. Pore opening was highly selective because other genomic DNA, from other bacteria such as *Candida albicans* or *Legionella pneumophila*, were unable to induce dye release. A LOD as low as ca. 70 DNA copies $\mu L^{-1}$ was determined. The capped nanoparticles were used to detect *Mycoplasma* in real contaminated cell-culture media without the use of PCR techniques.

In a more recent work, Pascual *et al.* developed a similar system to detect genomic DNA. In this case, the authors employed MSNs loaded with Rh B and capped with covalently attached DNA.[74] For the gating mechanism two single-stranded oligonucleotides were selected: (i) a short DNA sequence ($NH_2$-$(CH_2)_6$-5′-GAC TAC GAC GGT ATC-3′), which was covalently anchored to the surface of MSNs through an amidation reaction; and (ii) a single-stranded oligonucleotide selective for *Mycoplasma* (5′-AAG CGT GGG GAG CAA ACA GGA TTA GAT ACC CTG GTA GTC-3′) as depicted in Figure 5. The capped material was able to detect *Mycoplasma fermentas* genomic DNA at a concentration as low as 70 DNA copies $\mu L^{-1}$. Moreover, *Candida albicans* and *Legionella pneumophila* genomic DNAs were unable to induce pore opening and subsequent Rh B delivery.

Abnormal DNA methylation can lead to the development of cancer, Huntington's disease and psychiatric disorders. Therefore, precise detection of DNA methyltransferases (DNA MTases) could provide valuable information for an early clinical cancer diagnosis. Taking these facts into account Zhang and co-workers reported DNA-capped MSNs for surface-enhanced Raman spectroscopy (SERS) detection of DNA methyltransferases.[75] For this purpose, MSNs were selected as inorganic scaffold and, in a first step, its outer surface functionalized with APTES. Then, the pores of the MSNs were loaded with a selected single stranded DNA sequence (loading DNA: 5′-ATT CAG GGG TGT ATA CAG TG -3′) and capped with a double DNA strand (formed by the hybridization of DNA1: 5′-CTA AGG ATG CAG CTG AGG GAG TCC TGC TGA GGG A5-COOH-3′ and DNA2: 5′- CTC CCT CAG CTG CAT CCT TAG TCC TGC TGA GGG A5-COOH-3′) through the formation of

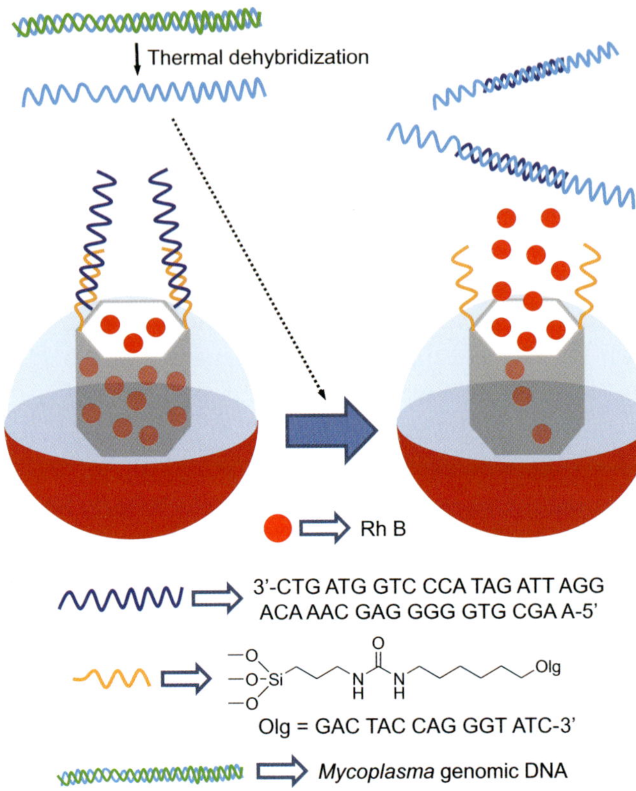


Figure 5. MSNs loaded with Rh B and functionalized with a specific oligonucleotide. Rh B release was achieved only in the presence of Mycoplasma genomic DNA.

amide bonds using EDC/NHS coupling agents. DNA1 and DNA2 were designed to leave a short single strand at the 3′ end. The opening mechanism of the DNA1/DNA2-capped MSNs was related with the presence of a selected hairpin DNA (5′-ACT TAT CAG CTT AAG GAT CCC TCA GCA GGA TCC TTA AGC TGA TAA GT-3′), DNA MTases and Dpn I (endonuclease enzyme). In these conditions, hairpin DNA was methylated (by DNA MTases) and cleaved (by Dpn I) generating a methylated trigger DNA. This trigger DNA hybridized with the short single strand of the DNA1/DNA2 gate thus offering splice site for Nb.BbvCI enzyme. Action of Nb.BbvCI enzyme induced the hydrolysis of DNA1/DNA2 biogate yielding again methylated trigger DNA (that could open other nanoparticles with an amplification effect) and opening the pores with subsequent release of loading DNA. The SERS signaling was accomplished by using AuNPs functionalized with a single strand DNA bearing 5-carboxy-X-rhodamine (DNA-Rox: 5′- Rox-GAC GAC GAC T6-SH -3′) and magnetic beads functionalized with another single strand DNA (DNA-MB: 5′-$NH_2$-T15-CAC TGT ATA C-3′). Both single stranded DNA (grafted into the AuNps and the MB) were complementary to half of the loading DNA. By using this protocol DNA MTases could be detected in the range from 0 to 10 U with a limit of detection of 0.02 U $mL^{-1}$.

Tang and co-workers used AuNPs as caps for the development of a gated material able to detect and MTH1 mRNA and suppress its activity.[76] In a first step, the authors prepared MSNs loaded with Rh B and with aminopropyl moieties linked onto the external surface. On the other hand, a thiol modified single stranded DNA (5′-GAA ATT CCA CGG GTA CTT CAA AAA A-$(CH_2)_6$-SH-3′) was designed as recognition sequence to target MTH1 and hybridized with a small sequence bearing a carboxylate group and Cy5 fluorophore (5′-HOOC-(Cy5)-TGA AGT ACC CG-3′). In a second step, the double stranded DNA was linked to AuNPs through its thiol moieties. Finally, the double stranded DNA-functionalized AuNPs were linked onto the external surface of the loaded MSNs through the formation of amide bonds using EDC as coupling agent. PBS suspensions of the AuNP-capped nanoparticles showed negligible dye release (less than 8% after 24 hours) in the absence of MTH1 mRNA. However, in the presence of MTH1 a marked RH B delivery was observed (ca. 80% after 24 hours). Dye release was ascribed to the preferential hybridization of MTH1 mRNA with the recognition DNA sequence grafted onto the external surface of the AuNPs. This preferential binding detached AuNPs from the surface of silica nanoparticles with subsequent pore opening. For intracellular applications, Rh B was substituted with S-crizotinib, a MTH1 inhibitor. Human cancer cells (HepG2 and HeLa, MTH1-positive) and normal liver cells (HL7702, MTH1-negative) were treated with

AuNP-capped nanovehicles. MTT viability assays showed a significant cytotoxic effect in the two cancer cells but non on HL7702 normal cells. Finally, the nanoplatform was able to decrease volume tumor in mice xenografted with HepG2 cancer cells and also was used as imaging agent due to the presence of the linked Cy5.

Capped MSNs able to deliver antagomir-122 (Ant122) and small molecules able to inhibit miR-122 in hepatocellular carcinoma cells were prepared by Yao and co-workers.[77] In order to prepare the capped nanovehicles the authors selected MSNs as inorganic scaffold and grafted, onto its outer surface, APTES and a triethoxysilane derivatized tetrazine. Then, the grafted tetrazine heterocycles were reacted with the linker *t*-cyclooctene-$PEG_{12}$-dibenzylcyclooctyne (TCO-$PEG_{12}$-DBCO) by biorthogonal tetrazine-TCO ligation. Then, the pores were loaded with Rh B and capped upon addition of Ant122 (through electrostatic interactions with the grafted protonated amines). Finally, azide-functionalized RGD peptide (targeting ligand) was covalently anchored onto the external surface of the nanoparticles by using copper-free azide-DBCO click chemistry. PBS suspensions (pH 7.4) of the Ant122-capped nanoparticles showed negligible Rh B release (less than 5% after 24 hours). However, upon addition of synthetic miR-122 a marked dye release was observed due to the uncapping of the nanoparticles as a consequence of hybridization of the complementary Ant122 and miR-122 sequences. Confocal microscopy studies carried out with Huh7 cells indicated internalization of the Ant122-capped nanoparticles (by endocytosis) and its escape from the endosomes/lysosomes via proton sponge effect. Also, Rh B was released and localized in the mitochondria due to the presence of endogenous miR-122. In a further step, the same Ant122-capped nanoparticles were prepared but loaded with DOX instead of Rh B. These nanoparticles were also internalized by Huh7 cells, were able to release DOX and induced apoptosis and cell death. Finally, the Ant122-capped nanoparticles were loaded with sm122 (a small molecule able to inhibit endogenous miR-122 function by effectively blocking the synthesis of pri-miRNA). Again, these nanoparticles were internalized with Huh7 cells and were able to inhibit endogenous miR-122 function due to the release of Ant122 (block endogenous miR-122) and of encapsulated sm122 (inhibition of pri-miRNA).

Martínez-Máñez and coworkers, employed capped MSNs to enhance the efficacy of antimicrobial drugs ε-poly-L-lysine (ε-PL) and to extend the antimicrobial spectrum of antibiotic vancomycin.[78] For these purposes, MSNs were prepared, loaded with vancomycin or Rh B, and its external surface functionalized with *n*-[(3-trimethoxysilyl) propyl]ethylenediaminetetraacetic acid trisodium salt. The pores of the final material were capped upon addition of cationic polymer ε-PL through electrostatic interactions with the negatively charged nanoparticles. Aqueous suspensions of the capped nanoparticles in the absence of *E. coli* bacteria showed nearly zero payload release. However, cargo delivery occurred in the presence of *E. coli*. This release was ascribed to the adhesion of the positively ε-PL-capped material with the negatively charged bacterial wall that weakened the interaction between ε-PL and the carboxylate-functionalized MSNs. *In vitro* dye delivery studies carried out in water at pH 7 in the absence and presence of *E. coli* DH5α at different cell concentrations confirmed the proposed gating mechanism. The interaction of ε-PL-capped nanoparticles with different Gram-negative bacteria (i.e. *E. coli* DH5α, *E. coli* 100, *E. coli* 405, *S. typhi* and *E. carotovora*) were also carried out. The Rh B-loaded material showed significantly enhanced toxicity of ε-PL, when compared with that of free ε-PL in solution. Also, a potent synergistic effect was observed when ε-PL and vancomycin were used in the nanoformulation. This synergic effect did not occur when using the free drugs alone.

Ozalp and co-workers prepared aptamer-capped MSNs able to release an entrapped antibiotic in the presence of *Staphylococcus aureus*.[79] The external surface of the MSNs was functionalized with mercaptopropyl moieties and then the pores loaded with vancomycin. Finally, the pores of the loaded nanoparticles were capped upon addition of an amino terminated aptamer ($H_2N$-GCG CCC TCT CAC GTG GCA CTC AGA GTG CCG GAA GTT CTG CGT TAT AGG GCG C) using sulfo-GMBS as linker. The selected aptamer formed a hairpin structure that is able to close the pores of the nanoparticles with the subsequent inhibition of vancomycin release. To test the *in vitro* performance of the aptamer-capped nanoparticles serum was selected to mimic endogenous background. Serum suspensions of the aptamer-capped nanoparticles showed nearly zero vancomycin release (5% after 24 hours incubation) whereas in the presence of $10^4$ cells of *S. aureus* a marked release was observed. Vancomycin delivery was ascribed to a change in the hairpin conformation of the aptamer to an open form upon interaction with *S. aureus* cells. In these nanoparticles, the grafted aptamer acted as gate and also as targeting ligand because is selective to *S. aureus*. In fact, *S. epidermis* presented a very low affinity for the aptamer and was able to induce a poor vancomycin release (minimum inhibitory concentration of 0.42 and 6.29 $\mu$g mL$^{-1}$ for *S. aureus* and *S. epidermis*, respectively).

## VII. Enzyme Responsive Gated Materials

In 2008, researchers started to explore the use of enzymes to uncap gated materials. Since then, this is one of the most popular stimuli as the appeal of taking into account that the use of tailor-made derivatives and specific enzymes provides exquisite selectivity in the design of advanced gated devices for on-command delivery of drugs and biomolecules in realistic biological environments. For instance, enzyme overexpression in a certain disease can be used to design personalized capped systems able to be open in a specific cell type, which means a significant reduction of undesirable side effects and a large increase in the effectivity of the drug delivered. Synthetic capping molecules can be designed to contain different functional fragments such as moieties to be recognized by target-enzymes, groups for targeting certain cells, or for cell entry-control. In the design of the systems, it is important to consider the orientation of the gating structure so that cleavable groups must be accessible to the enzyme active site. The huge potential of this kind of systems can be clearly envisioned.

The first enzyme-driven gated MSNs were described by Zink, Stoddart and coworkers. In their work, they used an esterase as a trigger, which is an enzyme capable of catalyzing the hydrolysis of esters into an acid and an alcohol.[80] The authors functionalized MSNs with PEG stalks and encircled on it $\alpha$-CDs. Finally, they used an adamantyl bulky stopper to retain α-CDs and to effectively block pores, which was attached to stalks via an ester linkage. The capped system was loaded with Rh B. The authors demonstrated that cargo release was only accomplished in the presence of porcine liver esterase, which was able to hydrolyze the ester group linking PEG stalk and the adamantyl moiety as depicted in Figure 6. The dethreading of α-CDs was then induced, and subsequent cargo delivery was registered. In a control experiment, the authors synthesized a similar system which contained

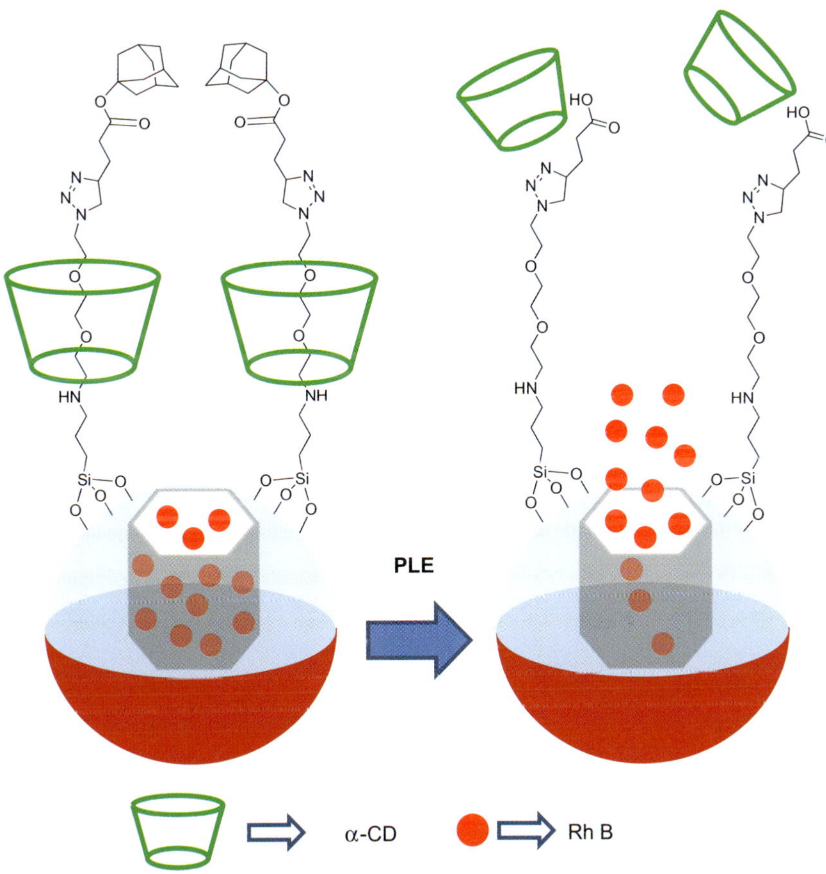


Figure 6. MSNs loaded with Rh B, functionalized with poly(ethylene glycol) chains, encircled by an α-CD and capped with an ester-linked adamantyl stopper. Rh B release was triggered in the presence of PLE.

an amide-linked adamantyl stopper that was unable to be cleaved by the used esterase.

In a different design, esterase responsive MSNs capped with poly($\beta$-amino ester) were prepared.[81] In a first step, the external surface of MSNs was functionalized with APTES. Then, amino groups were reacted with 4-carboxy-3-fluorophenylboronic acid and subsequently the nanoparticles were loaded with DOX or propidium iodide (PI). Finally, the pores were capped upon addition of poly($\beta$-amino ester), which was prepared by a Michael addition between triethylene glycol diacrylate and dopamine, through the formation of boronate esters. When nanoparticles were suspended in PBS (pH 7.4) a negligible release of the cargo was detected. In contrast, in the presence of esterase a marked release (77% and 71% for DOX and PI, respectively) was observed after ca. 100 minutes due to the hydrolysis of the ester linkages in the capping poly($\beta$-amino ester) by esterase. Besides, a marked release at pH 4.0 was observed due the hydrolysis of the boronic ester bonds that linked poly($\beta$-amino ester) onto the surface of the nanoparticles. Finally, viability tests were carried out on MDA-MB-231 breast cancer cells. In this case, DOX loaded nanoparticles capped with the poly($\beta$-amino ester) showed a marked cell killing efficiency. Reduction in viability was ascribed to nanoparticle internalization and DOX release due to the cleavage of the polymer cap as a consequence of acidic pH and esterase presence in cancer cells.

Acid phosphatase was used as trigger in a design of a gated macroporous scaffold by Vallet-Regí, Martínez-Máñez and coworkers.[82] In their system, these authors loaded MSNs with $Ru(bipy)_3^{2+}$ and functionalized the external surface with 3-[2-(2-aminoethylamino)ethylamino]propyl-trimethoxysilane. Pores were capped with ATP via the formation of supramolecular complexes between the negatively charged ATP anions and the positively charged ammonium groups on MSNs. The authors confirmed that $Ru(bipy)_3^{2+}$ was only released in the presence of acid phosphatase in aqueous buffered solutions. Following the work, the authors prepared a gated scaffold. First, a paste was obtained by mixing capped MSNs with gelatin and using glutaraldehyde as a cross-linker. Then, a scaffold was produced by rapid prototyping 3D printing techniques. In this scaffold, the acid phosphatase-driven release mechanism was also assessed. Finally, the biocompatibility of the scaffold was confirmed. In particular, the authors performed cell proliferation, cytotoxicity and spreading assays with a human osteoblast-like cell line. It was found that cells did not elicit any cytotoxic effect and adhered, proliferated and spread well in the scaffold.

Glycosidases which catalyze the hydrolysis of glycosidic linkages were also used to control the release from gated materials. In a first example, several MSNs loaded with $Ru(bipy)_3^{2+}$ and capped with a trialkoxysilane derivative of the hydrolyzed starch Glucidex 47, Gludicex 39 or Glucidex 29 were prepared.[83] Capped solids showed zero delivery, whereas cargo release was observed in the presence of pancreatin, which contains amylases that induced the hydrolysis of the glycosidic bonds in anchored saccharides. It was found that the lesser the hydrolysis of the starch anchored on nanoparticles, the lower the delivery rate in the presence of pancreatin. Furthermore, internalization and biocompatibility studies of gated material capped with Glucidex 47 were conducted in HeLa and LLC-PK1 cell lines. Confocal microscopy detected a clear intracellular dotted signal in both cell lines, which indicated that nanoparticles were internalized and co-localized with endosomes and lysosomes. In experiments at different temperatures, the researchers demonstrated that cellular uptake of capped MSNs was achieved by endocytosis and following an energy-dependent path. In a further experiment, cell viability studies were performed in both cell lines and the studies assessed that MSNs were biocompatible and no cytotoxic. Finally, a solid loaded with DOX and capped with Glucidex 47 was prepared. This new solid showed an enhancement of DOX cytotoxicity in HeLa cells.

Giving a step forward, the same authors prepared MSNs loaded with Rh B and functionalized the external surface with an alkylgluconamine derivative of a galacto-oligosaccharide polymer (GOS).[84] The capped MSNs showed no cargo delivery, but opened in the presence of $\beta$-galactosidase due to the selective hydrolysis of the GOS cap. In additional experiments, the performance of the GOS-capped nanoparticles *in vitro* was also studied for the potential targeting of senescent cells, in which $\beta$-galactosidase is known to be overexpressed. In a first step, GOS nanoparticles were incubated with yeast cells and Rh B staining was only observed in the β-galactosidase overexpressing strains. GOS-capped MSNs were also selectively opened in human senescent fibroblasts from X-linked Dyskeratosis Congenita (X-DC) patients, in which β-galactosidase was overexpressed. Remarkably, no emission was detected in lung carcinoma H460 cells (in which β-galactosidase was not present), even after 48 hours. In all cases, cell viability remained unaffected.

Proteases were also widely used as driving force to open selectively capped MSNs. For example, Wang, Zhao and coworkers used gelatin as a coating to prepare enzyme-responsive gated nanoparticles, which were uncapped using metalloproteinase-9 (MMP-9).[85] In their work, MSNs were functionalized with aminopropyl moieties and further reacted with glutaraldehyde to obtain –CHO groups. Then, pores were loaded with DOX and capped upon gelatin addition through the formation of imine bonds between the —$NH_2$ groups of the biopolymer and the –CHO groups on the silica surface. When capped nanoparticles were suspended in PBS buffer at pH 7.4, a zero DOX release was found. In contrast, a moderate drug release (8% after 24 hours) was found in the presence of MMP-9. Release of DOX was ascribed to the hydrolysis of gelatin coating by the MMP-9 enzyme. In following experiments, the intracellular uptake and *in vitro* cytotoxicity of the prepared gelatin-capped nanoparticles were studied using the LO2 (MMP-9 negative) and cancer HT-29 (MMP-9 positive) cell lines. Interestingly, gelatin-capped nanoparticles were only uptaken by HT-29 cells. One, inside of the cells, the presence of MMP-9 enzyme induced pore opening, DOX release and a subsequent reduction in cell viability. Finally, *in vivo* studies were performed in a xenograft mice model. When DOX-loaded gelatin-capped nanoparticles were injected into the xenograft, a significant delay in tumor growth was found, which confirmed the efficiency of the system.

Specific peptidic designs have also been used as caps for the selective and controlled release of drugs from MSNs.[86] At this respect, Bein and coworkers prepared MSNs functionalized with thiol moieties in the inner of the pores and with amino groups onto the external surface. Then, the external surface of the nanoparticle was functionalized with a biotin-heptapeptide (biotin-PLGMWSR) using EDC/sulfoNHS coupling reaction. The heptapeptide sequence used is selectively recognized and hydrolyzed by MMP-9. The pores were loaded with fluorescein, cisplatin and with a combination of cisplatin + bortezomib and, finally, capped upon addition of avidin. HBSS (Hank's balanced salt solution) buffered suspensions of the avidin-capped nanoparticles in the absence of MMP-9 showed nearly zero fluorescein release. However, in the presence of MMP-9 a marked dye release was observed (reaching a plateau after ca. 16 hours). Fluorescein release was ascribed to MMP-9-induced hydrolysis of the heptapeptide linker with subsequent detachment of the avidin protein from the surface of the nanoparticles. Viability studies carried out with the nanoparticles loaded with cisplatin in A549 and H1299 lung cancer cell lines showed the non-toxicity of the nanovehicles. However, a marked reduction in the viability in both cell lines was observed when cells were transfected with MMP9 cDNA (induced the overexpression of MMP9 enzyme). Moreover, an increased cytotoxicity was observed in both cell lines (with MMP9 overexpressed) when treated with the nanoparticles loaded with cisplatin + bortezomib. Additionally, both nanoparticles were able to deliver the entrapped drugs in lung tumor region of Kras mutant mice (MMP9 overexpressed), without causing toxicity in tumor-free areas. The same results were observed in human lung tumor tissues.

Royo, Martínez-Máñez and coworkers also prepared MSNs loaded with the cytotoxic DOX, functionalized with triple bonds obtained by the nucleophilic addition of propargylamine to 3-(triethoxysilyl)propylisocyanate and capped with a CXCR4 receptor targeting peptide via a "click chemistry" reaction.[87] The capping aminoacid sequence had a double role as gating mechanism and targeting moiety. In a first step, DOX release from pore voids was evaluated. In the absence of the proteolytic enzyme, a remarkable flat baseline (less of 2% of DOX release after 24 hours) was obtained. In contrast, cargo release in the presence of protease was increased within time, due to the protease-induced hydrolysis of the peptidic bonds of the anchored peptide. In a further step, the efficiency of the targeted and capped system was studied in a cell line of B-cell non-Hodgkin's lymphoma. In their work, the authors found a rapid internalization in CXCR4 cells when were exposed to the capped MSN. In contrast, bare MSNs did not affect the receptor expression. Also, antitumoral effect of the nanoparticles was demonstrated. Moreover, conducted experiments showed that interaction between the gating and targeting peptide in the nanoparticles and the CXCR4 chemokine receptor was partially reversed by the pretreatment with the CXCR4 antagonist AMD3100, which demonstrates that nanoparticles act as a targeted nanotherapeutic system.

The same authors prepared a thrombin-driven nanosystem taking advantage of the ability of this serine protease to highly and specifically recognize the Leu-Val-Pro-Arg-Gly-Ser peptide sequence and selectively cleave the Arg-Gly bonds.[88] Their design consisted of MSN, loaded with the anticoagulant drug acenocoumarol and externally functionalized with 3-(azidopropyl)triethoxysilane. In a further step, peptide LVPRGSGGLVPRGSGGLVPRGSK-pentynoic acid, which contains three repeats of the thrombin-specific cleavage site, was anchored to the solid via the copper(I)-catalyzed Huisgen azide/alkyne 1,3-dipolar cycloaddition "click" reaction. In the presence of thrombin, peptidic chains of the capped solid were hydrolyzed and acenocoumarol was freely released from pore voids. Also, the authors performed anticoagulation assays using rabbit blood plasma. In their experiments, they found that thrombin time was delayed from $2.6 \pm 0.3$ to $5.0 \pm 0.3$ minutes when plasma was incubated with the capped solid. In contrast, calcined MSN or a solid capped but loaded with a dye were unable to change thrombin time form normal values.

Ultimo *et al.* prepared a polyinosinic–polycytidylic acid (poly(I:C)) capped MSN as a new therapeutic tool to target Toll-like receptor 3 (TLR3), which is overexpressed in different cancer types.[89] To prepare their system, MSN were loaded with sulforhodamine B and functionalized with APTES. In a further step, the synthetic double stranded RNA poly(I:C) was coupled with the amine groups of the MSN surface via the formation of a phosphoramidite bond. The capped material showed negligible release in PBS buffer whereas sulforhodamine freely released from pore voids in the presence of lysosomal extract.

Following their work, breast cancer SK-BR-3 cells, which express high levels of the membrane TLR3 were incubated with the prepared (poly(I:C)) capped MSN. Interaction of TLR3 receptor with poly(I:C) induced rapid internalization of the nanoparticles, triggering caspase-3 activation and subsequent cell death. After confirm that capped MSNs were internalized and were able to release sulforhodamine B, cytotoxicity studies were conducted. The authors found that poly(I:C) was much more effective in killing SK-BR-3 cells when incorporated in the nanoparticles, compared with free poly(I:C), probably due to an "enhanced concentration effect" associated with the presence of a large amount of poly(I:C) on the surface of the nanoparticles. Moreover, the specific targeting of the gated nanoparticles in SK-BR-3 cells, was confirmed by competition assays with a TLR3–dsRNA complex inhibitor and by Western blot assays after treatment of SK-BR-3 cells with the capped solid, where activation of caspase-3 was detected. Finally, the researchers prepared a new capped solid but loaded with the chemotherapeutic DOX. Treatment of SK-BR-3 cells with this second solid resulted in a significant decrease in cell viability when compared with free poly(I:C) and with the poly(I:C)-capped solid loaded with sulforhodamine B. These results were confirmed by confocal microscopy, where cells showing apoptotic degradation and co-localization in the nuclei of DOX and DAPI dye were observed.

## VIII. Temperature Responsive Gated Materials

Temperature, as exogenous stimulus, has the advantage of a fine external control for tunable dose delivery. In this field, systems capped with thermosensitive polymers that were able to deliver cargo after a temperature-dependent phase transition represent the larger body of examples. Also, the melting of double stranded DNA sequences or other organic coatings such as paraffins has also been explored. It is important to note that, changes in temperature are a fingerprint in some diseases such as inflammation, infection or tumoral tissues, which can be used to deliver selected cargoes in a specific manner. However, the increase in temperature in biological environments is no more than 4 or 5°C, which needs of highly sensitive systems working on a very narrow range of temperatures. Besides, temperatures beyond 45°C must be carefully used to avoid undesired cellular death.

In a groundbreaking paper, Lopez and coworkers described poly(*N*-isopropylacrylamide) (PNIPAAm) as temperature-responsive cap.[90] In their work, the authors grew the thermosensitive polymer on the inner pores of the mesoporous support by atom transfer radical polymerization (ATRP) of *N*-isopropylacrylamide. At temperatures below 32°C, PNIPAAm polymer was hydrated, while at higher temperatures adopted its collapsed form. Taking advantage of this behavior, cargo was loaded in the pore voids at high temperatures (polymer collapsed form). A decrease of temperature below 32°C induced the polymer hydration which blocked the pores inhibiting the release of dye (fluorescein and Rh 6G). In contrast, when temperature was increased (50°C, above the lower critical solution temperature (LCST), the polymer adopted a hydrophobic collapsed form and permitted dye delivery.

In a different approach, Stroeve, Martínez-Máñez, and coworkers prepared magnetic pore-expanded MSNs capped with PNIPAAm as a thermosensitive coating.[91] Specifically, the mesoporous support was functionalized with APTES and then the amine groups were further reacted with $\alpha$-bromoisobutyryl bromide. Finally PNIPAAm was generated by ATRP. Two different solids were prepared. One loaded with the dye $Ru(bipy)_3^{2+}$ and other with the antibacterial enzyme lysozyme. Both solids showed negligible release at 25°C whereas cargo delivery is easily detected at 37°C due to the change in conformation of the PNIPAAm layer from an expanded (25°C) to a collapsed form at 37°C. In further experiments, the authors explored the possibility of using the capped solids as innovative antibacterial agent taking advantage of the lysozyme ability to break down Gram positive bacteria cell walls. *Bacillus cereus* and *Micrococcus luteus* bacteria were used to test the effectivity of the lysozyme loaded solid. In the studies, when bacteria were incubated at 25°C in the presence of the capped MSNs no effect was found. In contrast, a reduction in the colony count of near 60% for *B. cereus* and 45% for *M. luteus* was found when temperature was set at 37°C confirming the temperature-dependent antibacterial activity of the capped solid.

Song and coworkers also followed this approach to develop a drug delivery system using a co-polymer composed of *N*-isopropylacrylamide and methacrylic acid as capping agent.[92] They functionalized the external surface of MSNs with methacryloxypropyl subunits. Then *N*-isopropylacrylamide and methacrylic acid were copolymerized on the external surface by seeded precipitation polymerization. The mix of polymers conferred thermo- and pH-sensitive properties to the gating agent due to the *N*-isopropylacrylamide and methacrylic acid moieties, respectively. Fluorescein was loaded in mesopores. Also, a second capped solid further derivatized with FA was prepared to improve cellular uptake. When capped solids were suspended in aqueous solutions, cargo delivery was only registered at temperatures over LCST and in an acidic environment as depicted in Figure 7. Following their experiments, cellular studies with both solids were performed in Hep2 cells. Internalization was assessed by CLSM and by flow cytometry. Also, a targeting effect to Hep2 cells was observed when the FA-functionalized capped MSNs were tested, when compared to A549 cells. Finally, the authors prepared two similar capped materials, but loaded with the anticancer drug cisplatin. The capped solid without FA did not significantly affect to cell viability. In contrast, evident cell death enhancement was achieved when the FA-functionalized cisplatin-loaded nanoparticles were used. Finally, the effects on tumor growth on Hep2-tumor bearing nude mice resulted in the same tendency as that observed in cell viability studies.

In a different configuration, Bhatia and coworkers prepared MSNs loaded with DOX and with *N*-(3-aminopropyl)methacrylamidehydrochloride electrostatically adsorbed on the outer surface.[93] The final nanoparticles were coated with a thermosensitive polymer shell obtained by radical polymerization of *N*-isopropylacrylamide and poly(ethylene glycol)diacrylate. At temperatures below 37°C, DOX delivery from the nanoparticles was negligible, whereas a remarkable drug release was observed above 37°C as a result of the phase transition that the thermosensitive polymer shell underwent.

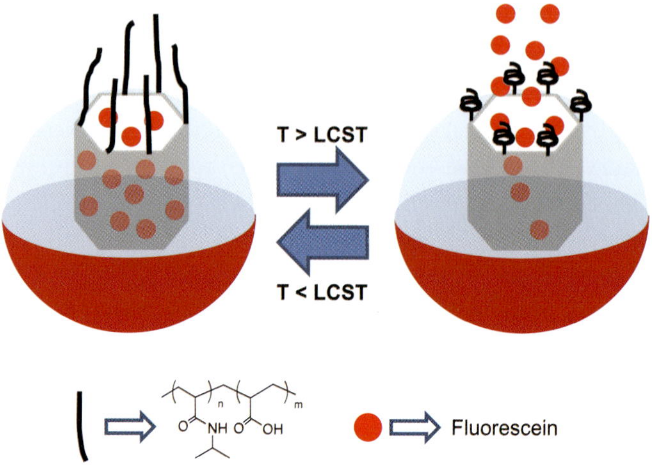


Figure 7. MSNs loaded with fluorescein and capped with a copolymer composed of *N*-isopropylacrylamide and methacrylic acid. Fluorescein was released at temperatures higher than polymer LCST.

Also paraffins have been used to develop temperature-sensitive capped MSNs.[94] In this case, MSNs were loaded with safranin O and then the external surface of the mesoporous support was functionalized with octadecyltrimethoxysilane hydrophobic chains. Finally, different paraffins were used to cap the pores by the formation of a hydrophobic layer around the nanoparticle. Three different materials were prepared using heneicosane, docosane, and tetracosane as capping paraffin. These paraffins have melting points of 39°C, 42°C and 49°C, respectively. When solids were suspended in water at temperatures below the melting point of the capping paraffin, an almost zero release of safranin O was detected. In contrast, when temperature was raised above the paraffin melting point, a clear safranin O release was found. Heneicosane-capped nanoparticles were also tested in HeLa cells. In the experiments, the authors found that safranin O can be released in the cells cytoplasm when incubation was carried out at temperatures above 39°C. Finally, heneicosane-capped nanoparticles loaded with DOX were prepared and used to release this chemotherapeutic agent in HeLa cells after heating.

## IX. Ultrasound and Magnetic Field Responsive Gated Materials

In the field of physical stimuli, ultrasound and magnetic fields have promising roles to control cargo release from gated materials. Both techniques allow a pulsable delivery of the cargo, are non-invasive, can penetrate depth in tissues, and can be carefully controlled by changing frequency, power, cycles and time of application. Ultrasound (US) is one of the most appealing external stimuli because is an excellent method to achieve spatial and temporal control of drug release at desired site. US presented several features (non-invasiveness, absence of ionizing radiations, cost effectiveness and easy regulation of tissue penetration depth) than will most likely boost its use as external trigger in the next years. Magnetic stimuli provide extra features such as the possibility of guidance of the gated support, for example to accumulate in a selected tumor area, tracking of the nanoparticles by MRI or in applications where separation and recovery of the gated particles could be of interest.

For example, Honma and coworkers developed a controlled delivery system using MSNs embedded in a poly(dimethylsiloxane) thin film.[95] MSNs were loaded with ibuprofen (IBU) and then mixed with a poly(dimethylsiloxane) solution. After drying, a solid thin film was obtained. IBU release experiments were performed using a poly(dimethylsiloxane) film loaded with IBU and with a film with embedded gated nanoparticles. The films alone did not present significant cargo delivery but in contrast, both films exhibited IBU delivery when were irradiated with US. Moreover, in the film with embedded nanoparticles, the amount of released drug was remarkably larger. The authors attributed this performance to the generation of cavitation inside the polymeric matrix. Moreover, a pulsable delivery capability was also demonstrated.

Vallet-Regí and coworkers prepared MSNs capped with an US-sensitive polymer for the controlled release of fluorophores and drugs.[96] For this purpose, in a first step, 2-(2-methoxyethoxy) ethyl methacrylate (temperature-responsive monomer) and 2-tetrahydropyranyl methacrylate (monomer that contains an US-sensible tetrahydropyranyl moiety) were copolymerized. Then, the copolymer was silylated through a DCC/NHS coupling with APTES. MSNs were selected as inorganic scaffold and its external surface functionalized with the silylated copolymer. The copolymer functionalized MSNs were suspended in a fluorescein PBS solution at 4°C (at this temperature the copolymer is its extended conformation and the pores were opened) in order to achieve maximum loading of the porous network. Then, the temperature was raised to 37°C in order to change the conformation of the copolymer to its collapsed form closing the pores. PBS suspensions of the copolymer-functionalized nanoparticles at 37°C showed negligible fluorescein release (ca. 12% after 7 hours) whereas upon US application (10 minutes, 1.3 MHz, 100 W) a marked release was observed (ca. 75% after 7 hours). Upon US irradiation the tetrahydropyranyl subunit of the copolymer was hydrolyzed yielding hydrophilic methacrylic acid groups that induced a change from a collapsed to an extended conformation with subsequent fluorescein release. MTS viability assays showed that fluorescein loaded nanoparticles were non-toxic to LNCaP cells either under without or with US irradiation. Finally, the same copolymer was used to prepare nanoparticles loaded with DOX. MTS viability assays showed the non-toxicity of the capped nanodevices whereas a marked decrease in cell viability was observed upon US irradiation. Besides, confocal fluorescence microscopy images showed DOX fluorescence in the cytoplasm upon US irradiation indicating drug release from the nanoparticles.

The same authors have also developed a magnetically triggered system using poly-NIPAM-*block*-PEI thermoresponsive copolymer as cap.[97] As support, the authors used mesoporous silica with embedded superparamagnetic iron oxide nanocrystals. The magnetic support was then functionalized with (trimethoxysilyl)propyl methacrylate and in a subsequent step poly-NIPAM-*block*-PEI polymer was grafted around the nanoparticles by radical polymerization. The pores of the silica support were loaded with fluorescein and Soybean Trypsin Inhibitor type II-S (STI) was incorporated into the copolymer shell. The PNIPAM polymer exhibited a phase transition at a

LCST of approximately 32°C in water. At temperatures around 30°C, the copolymer was hydrated and both fluorescein and Soybean Trypsin Inhibitor type II-S were effectively encapsulated. In contrast, at temperatures above LCST (around 40°C), fluorescein and the protein were freely released as depicted in Figure 8. Alternating magnetic field (AMF)-driven capability of the gated system was demonstrated by exposing the solid to an AMF of 24 kA·m$^{-1}$ and 100 kHz for 6 hours. Upon application of the stimulus, both protein and fluorescein were delivered.

Bringas *et al.* also developed a magnetically triggered system.[98] A mesoporous support with superparamagnetic cores of iron oxide was loaded with methylene blue (MB) and capped with a DOPC lipid

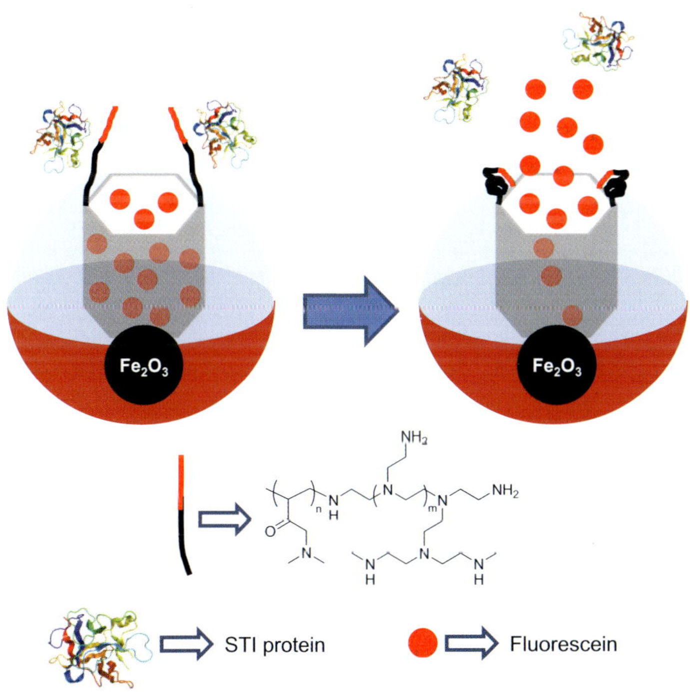


Figure 8. MSNs loaded with fluorescein, functionalized a with poly-NIPAM-block-PEI thermoresponsive copolymer with embedded STI protein. Fluorescein release was achieved upon the application of an AMF.

bilayer. When the gated solid was suspended in water and exposed to an AMF (50 Hz, 1570 G), release of MB from pore voids was triggered, whereas in the absence of the AMF, the dye was effectively retained inside of the porous framework. The observed effect was ascribed to lipid bilayer permeation and its local warning due to the application of the AMF. In continuation of their work, toxicity of the capped nanoparticles was studied in a collection of human cell lines, which included human nerve (A172), liver (Hep G2), heart (HCM), lung (A549), kidney (293T), colon (SW480), brain (BE(2)-C), and skin (A431) models. The capped solid was not toxic at concentrations below 0.5 mg $mL^{-1}$.

## X. Conclusions and Perspectives

Research at the forefront of functional materials has covered the way to the development of stimuli-responsive devices such as hybrid materials, hydrogels, liposomes and nanocapsules. Among them, we have shown here selected examples on the use of porous supports with gating functionalities for the on-command delivery of target cargo for biomedical applications. The design of such gated mesoporous materials, able to control the release of species in the presence of a certain stimulus, have proved very successful and a clever way to apply the versatility of molecular, supramolecular, and biomolecular-based ideas to design smart supports with controlled release features. Using a wide range of interactions and different external stimuli, such as light, changes in redox features, changes in pH, the presence of molecules or biomolecules, enzymes, temperature, ultrasounds, and AMFs, a number of capping examples and uncapping protocols have been described. Although the first reports centered on the design of different gating systems, today gated nanochemistry focuses more on their use on practical applications, especially in the biomedical field. In this sense, a number of reported gated materials have explored the possibility to deliver drug in cells or in mice models. However, despite the exponential development of this field, drawbacks still need to overcome and new advances in the gated nanochemistry area are anticipated in the near future.

## Acknowledgments

We thank the Spanish Government (project MAT2015-64139-C4-1-R (MINECO/FEDER)) and the Generalitat Valenciana (project PROMETEOII/2014/047) for support.

## References

1. Kresge, C. T.; Leonowicz, M. E.; Roth, W. J.; Vartuli, J. C.; Beck, J. S. Ordered Mesoporous Molecular Sieves Synthesized by a Liquid-crystal Template Mechanism. *Nature* **1992**, *359*, 710–712.
2. Zhao, D.; Feng, J.; Huo, Q.; Melosh, N.; Fredrickson, G. H.; Chmelka, B. F.; Stucky, G. D. Triblock Copolymer Syntheses of Mesoporous Silica with Periodic 50 to 300 Armstrong Pores. *Science* **1998**, *279*, 548–552.
3. Ryoo, R.; Kim, J. M.; Ko, C. H.; Shin, C. H. Disordered Molecular Sieve with Branched Mesoporous Channel Network. *J. Phys. Chem.* **1996**, *100*, 17718–17721.
4. Bagshaw, S. A.; Prouzet, E.; Pinnavaia, T. J. Templating of Mesoporous Molecular Sieves by Nonionic Polyethylene Oxide Surfactants. *Science* **1995**, *269*, 1242–1244.
5. Inagaki, S.; Koiwai, A.; Suzuki, N.; Fukushima, Y.; Kuroda, K. Syntheses of Highly Ordered Mesoporous Materials, FSM-16, Derived from Kanemite. *Bulletin of the Chemical Society of Japan* **1996**, *69*, 1449–1457.
6. Slowing, I. I.; Vivero-Escoto, J. L.; Wu, C. –W.; Lin, V. S. –Y. Mesoporous Silica Nanoparticles as Controlled Release Drug Delivery and Gene Transfection Carriers. *Adv. Drug Deliver. Rev.* **2008**, *60*, 1278–1288.
7. Lu, J.; Liong, M.; Li, Z.; Zink, J. I.; Tamanoi, F. Biocompatibility, Biodistribution, and Drug-delivery Efficiency of Mesoporous Silica Nanoparticles for Cancer Therapy in Animals. *Small* **2010**, *6*, 1794–1805.

8. Huang, X.; Li, L.; Liu, T.; Hao, N.; Liu, H.; Chen, D.; Tang, F. The Shape Effect of Mesoporous Silica Nanoparticles on Biodristribution, Clearance, and Biocompatibility *in Vivo*. *ACS Nano* **2011**, *5*, 5390–5399.
9. Fu, C.; Liu, T.; Li, L.; Liu, H.; Chen, D.; Tang, F. The Absorption, Distribution, Excretion and Toxicity of Mesoporous Silica Nanoparticles in Mice Following Different Exposure Routes. *Biomaterials* **2013**, *34*, 2565–2575.
10. Aznar, E.; Oroval, M.; Pascual, Ll.; Murguía, J. R.; Martínez-Máñez, R.; Sancenón, F. Gated Materials for On-command Release of Guest Molecules. *Chem. Rev.* **2016**, *116*, 561–718.
11. Saha, S.; Leung, K. C. –F.; Nguyen, T. D.; Stoddart, J. F.; Zink, J. I. Nanovalves. *Adv. Func. Mater.* **2007**, *17*, 685–693.
12. Wang, F.; Liu, X.; Willner, I. DNA Switches: from Principles to Applications. *Angew. Chem. Int. Ed.* **2015**, *54*, 1098–1129.
13. Trewyn, B. G.; Giri, S.; Slowing, I. I.; Lin, V. S. –Y. Mesoporous Silica Nanoparticle Based Controlled Release, Drug Delivery, and Biosensor Systems. *Chem. Commun.* **2007**, 3236–3245.
14. Aznar, E.; Martínez-Máñez, R.; Sancenón, F. Controlled Release Using Mesoporous Materials Containing Gate-Like Scaffoldings. *Expert Opin. Drug Deliv.* **2009**, *6*, 643–655.
15. Sancenón, F.; Pascual, Ll.; Oroval, M.; Aznar, E.; Martínez-Máñez, R. Gated Silica Mesoporous Materials in Sensing Applications. *ChemistryOpen* **2015**, *4*, 418–437.
16. Koutsopoulos, S. Molecular Fabrications of Smart Nanobiomaterials and Applications in Personalized Medicine. *Adv. Drug Deliver. Rev.* **2012**, *64*, 1459–1476.
17. He, Q.; Shi, J. MSN Anti-Cancer Nanomedicines: Chemotherapy Enhancement, Overcoming of Drug Resistance, and Metastasis Inhibition. *Adv. Mater.* **2014**, *26*, 391–411.
18. Taylor-Pashow, K. M. L.; Della Rocca, J.; Huxford, R. C.; Lin, W. Hybrid Nanomaterials for Biomedical Applications. *Chem. Commun.* **2010**, *46*, 5832–5849.
19. Ambrogio, M. W.; Thomas, C. R.; Zhao, Y. –L.; Zink, J. I.; Stoddart, J. F. Mechanized Silica Nanoparticles: A New Frontier in Theranostic Nanomedicine. *Acc. Chem. Res.* **2011**, *44*, 903–913.
20. Rosenholm, J. M.; Sahlgren C.; Lindén, M. Multifunctional Mesoporous Silica Nanoparticles for Combined Therapeutic, Diagnostic and

Targeted Action in Cancer Treatment. *Curr. Drug Targets* **2011**, *12*, 1166–1186.

21. Schumacher, K.; Ravikovitch, P. I.; Du Chesne, A.; Neimark, A. V.; Unger, K. K. Characterization of MCM-48 Materials. *Langmuir* **2000**, *16*, 4648–4654.
22. Chen, C. –Y.; Li, H. –X.; Davis, M. E. Studies on Mesoporous Materials: I. Synthesis and Characterization of MCM-41. *Micropor. Mater.* **1993**, *2*, 17–26.
23. Wang, S. Ordered Mesoporous Materials for Drug Delivery. *Micropor. Mesopor. Mater.* **2009**, *117*, 1–9.
24 Na, H. –K.; Kim, M. –H.; Park, K.; Ryoo, S. –R.; Lee, K. E.; Jeon, H.; Ryoo, R.; Hyeon, C.; Min, D. –H. Efficient Functional Delivery of siRNA using Mesoporous Silica Nanoparticle with Ultralarge Pores. *Small* **2012**, *8*, 1752–1761.
25. Antonietti, M.; Berton, B.; Göltner, C.; Hentze, H. –P. Synthesis of Mesoporous Silica with Large Pores and Bimodal Pore Size Distribution by Templating of Polymer Latices. *Adv. Mater.* **1998**, *10*, 154–159.
26. Zhao, Y.; Lin, L. –N.; Lu, Y.; Chen, S. –F.; Dong, L.; Yu, S. –Y. Templating Synthesis of Preloaded Doxorubicin in Hollow Mesoporous Silica Nanospheres for Biomedical Applications. *Adv. Mater.* **2010**, *22*, 5255–5259.
27. Chen, Y.; Chen, H.; Ma, M.; Chen, F.; Guo, L.; Zhang, L.; Shi, J. Double Mesoporous Silica Shelled Spherical/Ellipsoidal Nanostructures: Synthesis and Hydrophilic/Hydrophobic Anticancer Drug Delivery. *J. Mater. Chem.* **2011**, *21*, 5290–5298.
28. Wu, S. –H.; Mou, C. –Y.; Lin, H. P. Synthesis of Mesoporous Silica Nanoparticles. *Chem. Soc. Rev.* **2013**, *42*, 3862–3875.
29. Kruk, M.; Jaroniec, M. Application of Large Pore MCM-41 Molecular Sieves to Improve Pore Size Analysis Using Nitrogen Adsorption Measurements. *Langmuir* **1997**, *13*, 6267–6273.
30. Li, Y.; Shi, J. Hollow-structured Mesoporous Materials: Chemical Synthesis, Functionalization and Applications. *Adv. Mater.* **2014**, *26*, 3175–3205.
31. Mal, N. K.; Fujiwara, M.; Tanaka, Y. Photocontrolled Reversible Release of Guest Molecules from Coumarin-Modified Mesoporous Silica. *Nature* **2003**, *421*, 350–353.

32. Mal, N. K.; Fujiwara, M.; Tanaka, Y.; Taguchi, T.; Matsutaka, M. Photo-Switched Storage and Release of Guest Molecules in the Pore Void of Coumarin-Modified MCM-41. *Chem. Mater.* **2003**, *15*, 3385–3394.
33. Aznar, E.; Marcos, M. D.; Martínez-Máñez, R.; Sancenón, F.; Soto, J.; Amorós, P.; Guillem, C. pH- and Photo-Switched Release of Guest Molecules from Mesoporous Silica Supports. *J. Am. Chem. Soc.* **2009**, *131*, 6833–6843.
34. Chang, Y.-T.; Liao, P.-Y.; Sheu, H.-S.; Tseng, Y.-J.; Cheng, F.-Y.; Yeh, C.-S. Near-Infrared Light-Responsive Intracellular Drug and siRNA Release Using Au Nanoensembles with Oligonucleotide-Capped Silica Shell. *Adv. Mater.* **2012**, *24*, 3309–3314.
35. Li, W.-P.; Liao, P.-Y.; Su, C.-H.; Yeh, C.-S. Formation of Oligonucleotide-Gated Silica Shell-Coated Fe3O4-Au Core-Shell Nanotrisoctahedra for Magnetically Targeted and Near-Infrared Light-Responsive Theranostic Platform. *J. Am. Chem. Soc.* **2014**, *136*, 10062–10075.
36. Elbialy, N.; Mohamed, N.; Monem, A. S. Synthesis, Characterization and Application of Gold Nanoshells Using Mesoporous Silica Core. *Micropor. Mesopor. Mater.* **2014**, *190*, 197–207.
37. Yang, S.; Li, N.; Liu, Z.; Sha, W.; Chen, D.; Wu, Q.; Lu, J. Amphiphilic Copolymer Coated Upconversion Nanoparticles for near-Infrared Light-Triggered Dual Anticancer Treatment. *Nanoscale*, **2014**, *6*, 14903–14910.
38. Liu, J.; Detrembleur, C.; De Pauw-Gillet, M. –C.; Mornet, S.; Jérôme, C.; Duguet, E. Gold Nanorods Coated with Mesoporous Silica Shell as Drug Delivery System for Remote Near Infrared Light-Activated Release and Potential Phototherapy. *Small* **2015**, *11*, 2323–2332.
39. Zhang, Y.; Hou, Z.; Ge, Y.; Deng, K.; Liu, B.; Li, X.; Li, Q.; Cheng, Z.; Ma, P.; Li, C.; Lin, J. DNA-Hybrid-Gated Photothermal Mesoporous Silica Nanoparticles for NIR-Responsive and Aptamer-Targeted Drug Delivery. *ACS Appl. Mater. Interfaces* **2015**, *7*, 20696–20706.
40. Zhang, L.; Li, Y.; Jin, Z.; Yu, J. C.; Chan, K. M. A NIR-Triggered and Thermally Responsive Drug Delivery Platform Through DNA/Copper Sulfide Gates. *Nanoscale* **2015**, *7*, 12614–12624.

41. Hernandez, R.; Tseng, H. R.; Wong, J. W.; Stoddart, J. F.; Zink, J. I. An Operational Supramolecular Nanovalve. *J. Am. Chem. Soc.* **2004**, *126*, 3370–3371.
42. Luo, Z.; Ding, X. W.; Hu, Y.; Wu, S. J.; Xiang, Y.; Zeng, Y. F.; Zhang, B. L.; Yan, H.; Zhang, H. C.; Zhu, L. Liu, J.; Li, J.; Cai, K.; Zhao, Y. Engineering a Hollow Nanocontainer Platform with Multifunctional Molecular Machines for Tumor-Targeted Therapy in Vitro and in Vivo. *ACS Nano* **2013**, *7*, 10271–10284.
43. Lee, J.; Kim, H.; Kim, S.; Lee, H.; Kim, J.; Kim, N.; Park, H. J.; Choi, E. K.; Lee, J. S.; Kim, C. A Multifunctional Mesoporous Nanocontainer with an Iron Oxide Core and a Cyclodextrin Gatekeeper for an Efficient Theranostic Platform. *J. Mater. Chem.* **2012**, *22*, 14061–14067.
44. Nadrah, P.; Maver, U.; Jemec, A.; Tisler, T.; Bele, M.; Drazic, G.; Bencina, M.; Pintar, A.; Planinsek, O.; Gaberscek, M. Hindered Disulfide Bonds to Regulate Release Rate of Model Drug from Mesoporous Silica. *ACS Appl. Mater. Inter.* **2013**, *5*, 3908–3915.
45. Zhang, Q.; Wang, X.; Li, P.-Z.; Kim Truc, N.; Wang, X.-J.; Luo, Z.; Zhang, H.; Tan, N. S.; Zhao, Y. Biocompatible, Uniform, and Redispersible Mesoporous Silica Nanoparticles for Cancer-Targeted Drug Delivery In Vivo. *Adv. Funct. Mater.* **2014**, *24*, 2450–2461.
46. Giménez, C.; de la Torre, C.; Gorbe, M.; Aznar, E.; Sancenón, F.; Murguía, J. R.; Martínez-Máñez, R., Marcos, M. D., Amorós, P. Gated Mesoporous Silica Nanoparticles for the Controlled Delivery of Drugs in Cancer Cells. *Langmuir* **2015**, *31*, 3753–3762.
47. Dai, L.; Li, J.; Zhang, B.; Liu, J.; Luo, Z.; Cai, K. Redox-Responsive Nanocarrier Based on Heparin End-Capped Mesoporous Silica Nanoparticles for Targeted Tumor Therapy in Vitro and in Vivo. *Langmuir* **2014**, 30, 7867–7877.
48. Han, N.; Zhao, Q.; Wan, L.; Wang, Y.; Gao, Y.; Wang, P.; Wang, Z.; Zhang, J.; Jiang, T.; Wang, S. Hybrid Lipid-Capped Mesoporous Silica for Stimuli-responsive Drug Release and Overcoming Multidrug Resistance. *ACS Appl. Mater. Interfaces* **2015**, *7*, 3342–3351.
49. Xiao, D.; Jia, H. –Z.; Ma, N.; Zhuo, R. –X.; Zhang, X. –Z. A Redox-responsive Mesoporous Silica Nanoparticle Capped with Amphiphilic

Peptides by Self-assembly for Cancer Targeting Drug Delivery. *Nanoscale* **2015**, *7*, 10071–10077.

50. Torney, F.; Trewyn, B. G.; Lin, V. S. Y.; Wang, K. Mesoporous Silica Nanoparticles Deliver DNA and Chemicals into Plants. *Nat. Nanotechnol.* **2007**, *2*, 295–300.
51. Muhammad, F.; Qi, W.; Wang, A.; Gu, J.; Du, J.; Zhu, G. Using Oxidant Susceptibility of Thiol Stabilized Nanoparticles to Develop an Inflammation Triggered Drug Release System. *J. Mater. Chem. B* **2015**, *3*, 1597–1604.
52. Prasad, R.; Aiyer, S.; Chauhan, D. S.; Srivastava, R.; Selvaraj, K. Bioresponsive Carbon Nano-gated Multifunctional Mesoporous Silica for Cancer Theranostics. *Nanoscale* **2016**, *8*, 4537–4546.
53. Casasús, R.; Marcos, M. D.; Martínez-Máñez, R.; Ros-Lis, J. V.; Soto, J.; Villaescusa, L. A.; Amorós, P.; Beltran, D.; Guillem, C.; Latorre, J. Toward the Development of Ionically Controlled Nanoscopic Molecular Gates. *J. Am. Chem. Soc.* **2004**, *126*, 8612–8613.
54. Mas, N.; Galiana, I.; Hurtado, S.; Mondragón, L.; Bernardos, A.; Sancenón, F.; Marcos, M.D.; Amorós, P.; Abril-Utrillas, N.; Martínez-Máñez, R.; Murguía, J. R. Enhanced Antifungal Efficacy of Tebuconazole Using Gated pH-driven Mesoporous Nanoparticles. *Int. J. Nanomedicine* **2014**, *9*, 2597–2606.
55. Tang, Y.; Teng, Z.; Liu, Y.; Tian, Y.; Sun, J.; Wang, S.; Wang, C.; Wang, J.; Lu, G. Cytochrome C Capped Mesoporous Silica Nanocarriers for pH-Sensitive and Sustained Drug Release. *J. Mater. Chem. B* **2014**, *2*, 4356–4362.
56. Meng, H.; Mai, W. X.; Zhang, H. Y.; Xue, M.; Xia, T.; Lin, S. J.; Wang, X.; Zhao, Y.; Ji, Z. X.; Zink, J. I.; Nel, A. E. Codelivery of an Optimal Drug/siRNA Combination Using Mesoporous Silica Nanoparticles to Overcome Drug Resistance in Breast Cancer in Vitro and in Vivo. *ACS Nano* **2013**, *7*, 994–1005.
57. Meng, H.; Zhao, Y.; Dong, J. Y.; Xue, M.; Lin, Y. S.; Ji, Z. X.; Mai, W. X.; Zhang, H.Y.; Chang, C. H.; Brinker, C. J.; Zink, J. I.; Nel, A. E. Two-Wave Nanotherapy to Target the Stroma and Optimize Gemcitabine Delivery to a Human Pancreatic Cancer Model in Mice. *ACS Nano* **2013**, *7*, 10048–10065.

58. Niedermayer, S.; Weiss, V.; Hermann, A.; Schmidt, A.; Datz, S.; Müller, K.; Wagner, E.; Bein, T.; Bräuchle, C. Multifunctional Polymer-capped Mesoporous Silica Nanoparticles for pH-responsive Targeted Drug Delivery. Nanoscale 2015, 7, 7953–7964.
59. Deng, Z.; Zhen, Z.; Hu, X.; Wu, S.; Xu, Z.; Chu, P. K. Hollow Chitosan silica Nanospheres as pH-Sensitive Targeted Delivery Carriers in Breast Cancer Therapy. *Biomaterials* **2011**, *32*, 4976–4986.
60. Li, Z.; Dong, K.; Huang, S.; Ju, E.; Liu, Z.; Yin, M.; Ren, J.; Qu, X. A Smart Nanoassembly for Multistage Targeted Drug Delivery and Magnetic Resonance Imaging. *Adv. Func. Mater.* **2014**, *24*, 3612–3620.
61. Dengler, E. C.; Liu, J.; Kerwin, A.; Torres, S.; Olcott, C. M.; Bowman, B. N.; Armijo, L.; Gentry, K.; Wilkerson, J.; Wallace, J.; Jiang, X.; Carnes, E. C.; Brinker, C. J.; Milligan, E. D. Mesoporous Silica-Supported Lipid Bilayers (Protocells) for DNA Cargo Delivery to the Spinal Cord. *J. Control. Release* **2013**, *168*, 209–224.
62. Zhang, X.; Li, F.; Guo, S.; Chen, X.; Wang, X.; Li, J.; Gan, Y. Biofunctionalized Polymer-Lipid Supported Mesoporous Silica Nanoparticles for Release of Chemotherapeutics in Multidrug Resistant Cancer Cells. *Biomaterials* **2014**, *35*, 3650–3665.
63. Hwang, A. A.; Lu, J.; Tamanoi, F.; Zink, J. I. Functional Nanovalves on Protein-coated Nanoparticles for in vitro and in vivo Controlled Drug Delivery. *Small* **2015**, *11*, 319–328.
64. Li, Z.; Clemens, D. L.; Lee, B. –Y.; Dillon, B. J.; Horowitz, M. A.; Zink, J. I. Mesoporous Silica Nanoparticles with pH-sensitive Nanovalves for Delivery of Monofloxacin Provide Improved Treatment of Lethal Pneumonic Tularemia. *ACS Nano* **2015**, *9*, 10778–10789.
65. Casasús, R.; Aznar, E.; Marcos, M. D.; Martínez-Máñez, R.; Sancenón, F.; Soto, J.; Amorós, P. New Methods for Anion Recognition and Signaling Using Nanoscopic Gatelike Scaffoldings. *Angew. Chem. Int. Ed.* **2006**, *45*, 6661–6664.
66. Zheng, Z. L.; Huang, X.; Schenderlein, M.; Borisova, D.; Cao, R.; Mohwald, H.; Shchukin, D. Self-Healing and Antifouling Multifunctional Coatings Based on pH and Sulfide Ion Sensitive Nanocontainers. *Adv. Func. Mater.* **2013**, *23*, 3307–3314.

67. Wang, L.; Kim, M.; Fang, Q.; Min, J.; Jeon, W. I.; Lee, S. Y.; Son, S. J.; Joo, S. –W.; Lee, S. B. Hydrophobic End-Gated Silica Nanotubes for Intracellular Glutathione-Stimulated Drug Delivery in Drug-Resistant Cancer Cells. *Chem. Commun.* **2013**, *49*, 3194–3196.
68. Lu, H. –Y.; Chang, Y. –J.; Fan, N. –C.; Wang, L. –S.; Lai, N. –C.; Yang, C. –M.; Wu, L. –C.; Ho, J. A. Synergism Through Combination of Chemotherapy and Oxidative Stress-induced Autophagy in A549 Lung Cancer Cells Using Redox-responsive Nanohybrids: A New Strategy for Cancer Therapy. *Biomaterials* **2015**, *42*, 30–41.
69. Diez, P.; Sanchez, A.; Gamella, M.; Martínez-Ruiz, P.; Aznar, E.; de la Torre, C.; Murguia, J. R.; Martínez-Máñez, R.; Villalonga, R.; Pingarrón, J. M. Toward the Design of Smart Delivery Systems Controlled by Integrated Enzyme-Based Biocomputing Ensembles. *J. Am. Chem. Soc.* **2014**, *136*, 9116–9123.
70. Sun, L.; Zhang, X. G.; Wu, Z. M.; Zheng, C.; Li, C. X. Oral Glucose- and pH-Sensitive Nanocarriers for Simulating Insulin Release in Vivo. *Polym. Chem.* **2014**, *5*, 1999–2009.
71. Oroval, M.; Climent, E.; Coll, C.; Eritja, R.; Avino, A.; Marcos, M. D.; Sancenón, F.; Martínez-Máñez, R.; Amorós, P. An Aptamer-Gated Silica Mesoporous Material for Thrombin Detection. *Chem. Commun.* **2013**, *49*, 5480–5482.
72. Climent, E.; Martínez-Máñez, R.; Sancenón, F.; Marcos, M. D.; Soto, J.; Maquieira, A.; Amorós, P. Controlled Delivery Using Oligonucleotide-Capped Mesoporous Silica Nanoparticles. *Angew. Chem. Int. Ed.* **2010**, *49*, 7281–7283.
73. Climent, E.; Mondragón, L.; Martínez-Máñez, R.; Sancenón, F.; Marcos, M. D.; Murguia, J. R.; Amorós, P.; Rurack, K.; Pérez-Paya, E. Selective, Highly Sensitive, and Rapid Detection of Genomic DNA by Using Gated Materials: *Mycoplasma* Detection. *Angew. Chem. Int. Ed.* **2013**, *52*, 8938–8942.
74. Pascual, L.; Baroja, I.; Aznar, E.; Sancenón, F.; Marcos, M. D.; Murguía, J. R.; Amorós, P.; Rurack, K.; Martínez-Máñez, R. Oligonucleotide-capped Mesoporous Silica Nanoparticles as DNA-responsive Dye Delivery Systems for Genomic DNA Detection. *Chem. Commun.* **2015**, *51*, 1414–1416.

75. Wang, X.; Cui, M.; Zhou, H.; Zhang, S. DNA-hybrid-gated Functional Mesoporous Silica for Sensitive DNA Methyltransferase SERS Detection. *Chem. Commun.* **2015**, *51*, 13983–13985.
76. Gao, W.; Cao, W.; Sun, Y.; Wei, X.; Xu, K.; Zhang, H.; Tang, B. AuNP flares-capped Mesoporous Silica Nanoplatform for MTH1 Detection and Inhibition. *Biomaterials* **2015**, *69*, 212–221.
77. Yu, C.; Qian, L.; Uttamchandani, M.; Li, L.; Yao, S. Q. Single-vehicular Delivery of Antagomir and Small Molecules to Inhibit miR-122 Function in Hepatocellular Carcinoma Cells by Using "Smart" Mesoporous Silica Nanoparticles. *Angew. Chem. Int. Ed.* **2015**, *54*, 10574–10578.
78. Mas, N.; Galiana, I.; Mondragón, L.; Aznar, E.; Climent, E.; Cabedo, N.; Sancenón, F.; Murguia, J. R.; Martínez-Máñez, R.; Marcos, M. D.; Amorós, P. Enhanced Efficacy and Broadening of Antibacterial Action of Drugs via the Use of Capped Mesoporous Nanoparticles. *Chem. Eur. J.* **2013**, *19*, 11167–11171.
79. Kavruk, M.; Celikbickak, O.; Ozalp, V. C.; Borsa, B. A.; Hernandez, F. J.; Bayramoglu, G.; Salih, B.; Arica, M. Y. Antibiotic Loaded Nanocapsules Functionalized with Aptamer Gates for Targeted Destruction of Pathogens. *Chem. Commun.* **2015**, *51*, 8492–8495.
80. Patel, K.; Angelos, S.; Dichtel, W. R.; Coskun, A.; Yang, Y. W.; Zink, J. I.; Stoddart, J. F. Enzyme-Responsive Snap-Top Covered Silica Nanocontainers. *J. Am. Chem. Soc.* **2008**, *130*, 2382–2383.
81. Ferrando I. R.; Ferris, D. P.; Frasconi, M.; Malin, D.; Strekalova, E.; Yilmaz, M. D.; Ambrogio, M. W.; Algaradah, M. M.; Hong, M. P.; Chen, X.; Nassar, M. S.; Botros, Y. Y.; Cryns, V. L.; Stoddart, J. F. Esterase- and pH-Responsive Poly(β-amino ester)-Capped Mesoporous Silica Nanoparticles for Drug Delivery. *Nanoscale* **2015**, *7*, 7178–7183.
82. Mas, N.; Arcos, D.; Polo, L.; Aznar, E.; Sanchez-Salcedo, S.; Sancenón, F.; Garcia, A.; Marcos, M. D.; Baeza, A.; Vallet-Regí, M.; *et al.* Towards the Development of Smart 3D "Gated Scaffolds" for On-Command Delivery. *Small* **2014**, *10*, 4859–4864.
83. Bernardos, A.; Mondragón, L.; Aznar, E.; Marcos, M. D.; Martínez-Máñez, R.; Sancenón, F.; Soto, J.; Barat, J. M.; Pérez-Paya, E.; Guillem, C.; *et al.* Enzyme-Responsive Intracellular Controlled Release Using

Nanometric Silica Mesoporous Supports Capped with "Saccharides." *ACS Nano* **2010**, *4*, 6353–6368.

84. Agostini, A.; Mondragón, L.; Bernardos, A.; Martínez-Máñez, R.; Marcos, M. D.; Sancenón, F.; Soto, J.; Costero, A.; Manguan-García, C.; Perona, R.; *et al.* Targeted Cargo Delivery in Senescent Cells Using Capped Mesoporous Silica Nanoparticles. *Angew. Chem. Int. Ed.* **2012**, *51*, 10556–10560.
85. Xu, J. H.; Gao, F. P.; Li, L. L.; Ma, H. L.; Fan, Y. S.; Liu, W.; Guo, S. S.; Zhao, X. Z.; Wang, H. Gelatin-Mesoporous Silica Nanoparticles as Matrix Metalloproteinases-Degradable Drug Delivery Systems in Vivo. *Micropor. Mesopor. Mater.* **2013**, *182*, 165–172.
86. van Rijt, S. H.; Bölükbas, D. A.; Argyo, C.; Datz, S.; Lindner, M.; Eickelberg, O.; Eickelberg, O.; Köningshoff, M.; Bein, T.; Meiners, S. Protease-Mediated Release of Chemotherapeutics from Mesoporous Silica Nanoparticles to Ex Vivo Human and Mouse Lung Tumors. *ACS Nano* **2015**, *9*, 2377–2389.
87. de la Torre, C.; Casanova, I.; Acosta, G.; Coll, C.; Moreno, M.J.; Albericio, F.; Aznar, E.; Mangues, R.; Royo, M.; Sancenón, F.; Martínez-Máñez, R. Gated Mesoporous Silica Nanoparticles Using a Double-Role Circular Peptide for the Controlled and Target-Preferential Release of Doxorubicin in CXCR4-Expresing Lymphoma Cells. *Adv. Funct. Mater.* **2015**, *25*, 687–695.
88. Bhat, R.; Ribes, A.; Mas, N.; Aznar, E.; Sancenón, F.; Marcos, M.D.; Murguía, J.R.; Venkataraman, A.; Martínez-Máñez, R. Thrombin-Responsive Gated Silica Mesoporous Nanoparticles As Coagulation Regulators. *Langmuir* **2016**, *32*, 1195–1200.
89. Ultimo, A.; Giménez, C.; Pavel Bartovsky, P.; Aznar, E.; Sancenón, F.; Marcos, M.D.; Amorós, P.; Bernardo, A.R.; Martínez-Máñez, R.; Jiménez-Lara, A.M.; Murguía, J.R. Targeting Innate Immunity with dsRNA-Conjugated Mesoporous Silica Nanoparticles Promotes Antitumor Effects on Breast Cancer Cells. *Chem. Eur. J.* **2016**, *22*, 1582–1586.
90. Fu, Q.; Rao, G. V. R.; Ista, L. K.; Wu, Y.; Andrzejewski, B. P.; Sklar, L. A.; Ward, T. L.; Lopez, G. P. Control of Molecular Transport through Stimuli-Responsive Ordered Mesoporous Materials. *Adv. Mater.* 2003, 15, 1262–1266.

91. Yu, E.; Galiana, I.; Martínez-Máñez, R.; Stroeve, P.; Marcos, M.D.; Aznar, E.; Sancenón, F.; Murguía, J.R.; Amorós, P. Poly(N-isopropylacrylamide)-gated Fe3O4/Sio2 Core Shell Nanoparticles with Expanded Mesoporous Structures for the Temperature Triggered Release of Lysozyme. *Colloids Surf. B Biointerfaces* **2015**, *135*, 652–660.
92. Liu, X.; Yu, D.; Jin, C. S.; Song, X. W.; Cheng, J. Z.; Zhao, X.; Qi, X. M.; Zhang, G. X. A Dual Responsive Targeted Drug Delivery System Based on Smart Polymer Coated Mesoporous Silica for Laryngeal Carcinoma Treatment. *New J. Chem.* 2014, *38*, 4830–4836.
93. Singh, N.; Karambelkar, A.; Gu, L.; Lin, K.; Miller, J. S.; Chen, C. S.; Sailor, M. J.; Bhatia, S. N. Bioresponsive Mesoporous Silica Nanoparticles for Triggered Drug Release. *J. Am. Chem. Soc.* **2011**, *133*, 19582–19585.
94. Aznar, E.; Mondragón, L.; Ros-Lis, J. V; Sancenón, F.; Marcos, M. D.; Martínez-Máñez, R.; Soto, J.; Pérez-Paya, E.; Amorós, P. Finely Tuned Temperature-Controlled Cargo Release Using Paraffin-Capped Mesoporous Silica Nanoparticles. *Angew. Chem. Int. Ed.* **2011**, *50*, 11172–11175.
95. Kim, H. J.; Matsuda, H.; Zhou, H.; Honma, I. Ultrasound-Triggered Smart Drug Release from a Poly(dimethylsiloxane)-Mesoporous Silica Composite. *Adv. Mater.* **2006**, *18*, 3083–3088.
96. París, J. L.; Cabañas, M. V.; Manzano, M.; Vallet-Regí, M. Polymer-Grafted Mesoporous Silica Nanoparticles as Ultrasound-Responsive Drug Carriers. *ACS Nano* **2015**, *9*, 11023–11033.
97. Baeza, A.; Guisasola, E.; Ruiz-Hernandez, E.; Vallet-Regí, M. Magnetically Triggered Multidrug Release by Hybrid Mesoporous Silica Nanoparticles. *Chem. Mater.* **2012**, *24*, 517–524.
98. Bringas, E.; Koysuren, O.; Quach, D. V.; Mahmoudi, M.; Aznar, E.; Roehling, J. D.; Marcos, M. D.; Martínez-Máñez, R.; Stroeve, P. Triggered Release in Lipid Bilayer-Capped Mesoporous Silica Nanoparticles Containing SPION Using an Alternating Magnetic Field. *Chem. Commun.* **2012**, *48*, 5647–5649.

# Chapter 5

# Controlled Release from Iron Oxide Nanoparticles

Masoud Rahman

*Department of Chemical Engineering, University of California Davis, Davis, CA, 95616, United States*

## Abbreviations

| | |
|---|---|
| AMF | alternating magnetic field |
| CMC | critical micelle concentration |
| DOX | doxorubicin |
| PEG | polyethylene glycol |
| PVA | polyvinyl alcohol |
| SPIONs | superparamagnetic iron oxide nanoparticles |

## I. Introduction

Superparamagnetism is a form of magnetism that is observed for dispersions of nanoparticles with ferromagnetic and ferromagnetic properties. In the absence of magnetic field, the thermal fluctuation will result in the random flipping of the magnetic field. Therefore, the average magnetization will be zero. With the presence of the magnetic field, the nanoparticles will align with the magnetic field. This phenomenon is shown in Figure 1. The reason that this phenomenon happens in the nanoscale range is that by decreasing the particle size the energy of magnetic anisotropy for each particle will become comparable with the thermal energy. Therefore, the thermal energy can flip the magnetic moment with time. The random magnetic moment of all the nanoparticles in the absence of the external magnetic field will be zero.[1] One of the main advantage of this property is that in the absence of magnetic field, the nanoparticles can be dispersed in a solution easily. However, by applying an external force all the nanoparticles will align with the magnetic field. Two of the major application of superparamagnetic property are: 1) nanoparticle collection and 2) external stimulation of the nanoparticles. These applications will be discussed further in this chapter.

Nanoparticles are famous for their high surface to volume ratio which makes them interesting candidates for catalytic and purification applications. In the catalytic application, the catalyst will be immobilized on the surface of the nanoparticles. The high surface area of the

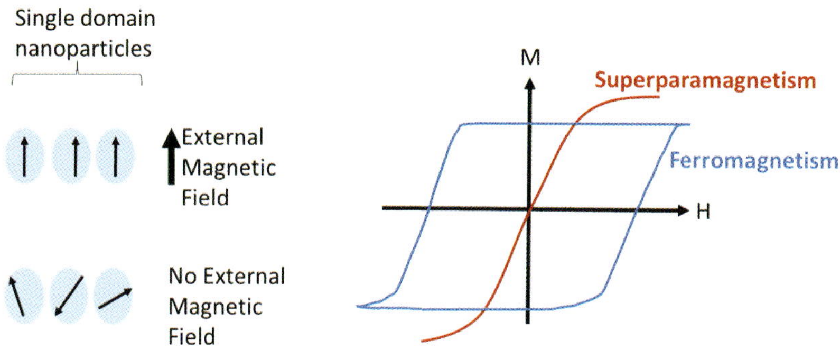


Figure 1. (Left) The superparamagnetism effect is shown for the single domain nanoparticles in the presence of absence of the magnetic field. (Right) The magnetism curve shows that the superparamagnetic nanoparticles will not keep any residual magnetism and therefore, the magnetization curve does not show any hysteresis.

nanoparticles and their dispensability will make the catalytic activity more efficient. However, collection of the catalyst and the nanoparticles after each reaction is a major challenge. The superparamagnetic nanoparticles covered with immobilized catalyst can be easily dispersed and after the reaction, by applying an external magnetic field they are easy to recollect. One example is the immobilization of the lipase on the surface of the iron oxide/mesoporous silica core–shell nanoparticles.[2] The advantage of superparamagnetic nanoparticles compared with the paramagnetic particles is the re-dispersion of the nanoparticles after each collection. In the case of paramagnetic particles, the induced magnetic field, will cause the particles to form internal magnetic fields and hence after the removing the external magnetic field the particles will continue to keep their magnetic field. This means that the paramagnetic particles will agglomerate and separating them will be difficult.

The high surface to volume ratio of the nanoparticles will help them collect pollutants with high efficiency. In this case, the surface of the nanoparticles should be covered with functional materials for efficient pollutant attachment. As mentioned before, the superparamagnetic property will help with collection of the nanoparticles after each process and re-dispersion for future processes. Thiol functionalized superparamagnetic iron oxide nanoparticles (SPIONs) have been studied for the removal of heavy metals such as Hg, Ag, Pb, Cd, and Tl in aqueous systems.[3] The results of this research showed that heavy metals effectively bind to the dimercaptosuccinic acid which is immobilized on the surface of SPIONs. The same concept has been used in the purification of plasmid DNA from bacterial cultures.[4] In this case, after magnetic separation from the bacterial cells, the collected plasmid DNA by the nanoparticles was regenerated and isolated from the nanoparticles by exposure to the buffer solution with adjusted ionic strength. Another strategy for pollutant removal is using the catalytic reaction for decomposition of the pollutants. Phenolic and aniline compounds have been removed from aqueous solution in the presence of SPIONs and hydrogen peroxide.[5] In this case the hydroxyl free radicals in the presence of SPIONS resulted in the degradation of the phenol and aniline.

## II. Superparamagnetic Iron Oxide Nanoparticles

The structure, properties, synthesis, surface engineering, toxicity, and applications of SPIONs have been reviewed in several books and review papers.[6–12] Two of the main crystal structure of iron oxide are maghemite ($\gamma$-$Fe_2O_3$) and magnetite ($Fe_3O_4$). The single magnetic domain iron oxide nanoparticles with the size range of 5–20 nm show superparamagnetic properties. Magnetite has superior biocompatibility which has made it a promising superparamagnetic nanoparticle for biomedical applications.[13] Magnetite has a cubic inverse spinel structure. The oxygen atoms are arranged in FCC crystal structure and iron atoms are located on tetrahedral sites and octahedral sites. SPIONs are mainly based on the magnetite which are covered with proper surface coating for better dispersion. The stable dispersions of SPIONs can be considered as ferrofluids. Ferrofluids can interact with magnetic fields in various applications such as positioning/targeting, MRI imaging, and cancer therapy.

SPIONs have been synthesized by various methods. These methods can be classified as:

(1) Co-precipitation
(2) Reaction in constrained environment (vesicles)
(3) Hydrothermal and high-temperature reactions
(4) Sol–gel reactions
(5) Polyol methods
(6) Flow injection synthesis
(7) Electrochemical methods
(8) Aerosol/vapor methods
(9) Sonolysis

The details of the synthesis methods can be found in recent books and review papers.[6,7,9] The synthesis methods can affect the size distribution, crystal structure, crystallinity, and magnetic properties. One of the challenges of the synthesis of SPIONs is the preparation of uniform nanoparticles with small range of size distribution as well as similar shapes.[14] The size-dependent magnetic properties of iron oxide are shown in Figure 2. In addition to the size, the shape of the nanoparticles could affect the interaction of the nanoparticles with surrounding environment, formation of the protein corona, and cellular uptake.[15]

In order to use SPIONs as colloidal ferrofluids the colloidal stability is a crucial item that should be considered. The colloidal stability will prevent the nanoparticles from aggregation and precipitation. It is important for biological applications that the nanoparticles do not agglomerate. For purification and catalytic applications, it is essentials that after the collection of the nanoparticle by external magnetic field, the nanoparticles can be re-dispersed again without agglomeration and precipitation. In addition to the particle-to-particle forces such as electrostatic, Van der Waals, and steric forces the magnetic dipolar forces should also be considered.

The surface of the iron oxide in the water will be covered with hydroxyl group which can react with both acids and bases. The isoelectric point is around pH 6.8, therefore, the surface charge of the

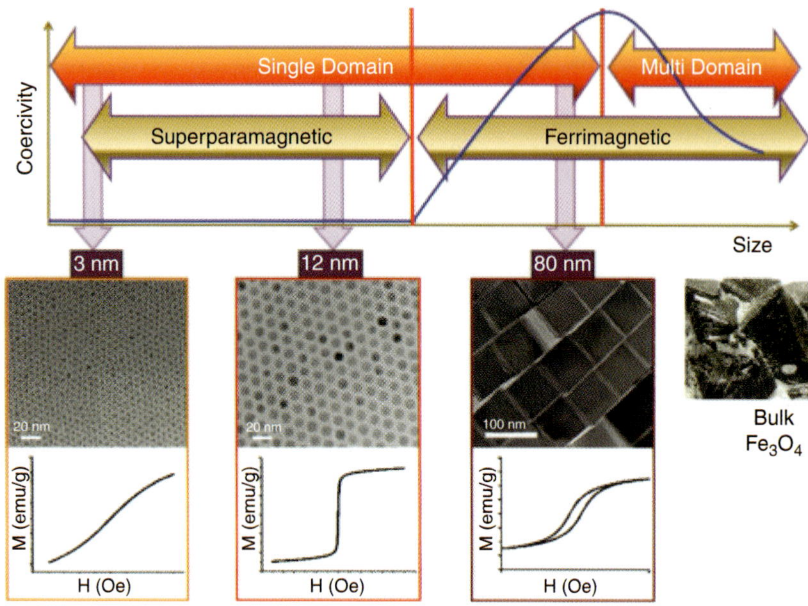


Figure 2. The effect of size on the magnetic properties of iron oxide nanoparticles. Taken from Ref. [14] with permission of America chemical society.

particles can be changed by tuning the pH of the dispersion. In the pH ranges around the point of zero charge, the electrostatic repulsion is negligible and particles have higher tendency to flocculate.[9] Various strategies for stabilization of the SPIONs are listed below:

(1) Monomeric stabilizers
   a. Carboxylates: Adsorption of citric acid will functionalize the surface of SPIONs with carboxylate groups which make the surface negatively charge and hydrophilic.[16]
   b. Phosphates: Alkanesulphonic and alkanephosphonic acid.[17]
(2) Inorganic materials
   a. Silica: Silica coating on the surface of SPIONs shields the magnetic dipole interaction and increase the surface negative charge. Additionally, silica layer will improve biocompatibility and corrosion of SPIONs.[18]

b. Gold: Gold coating will increase the biocompatibility, surface functionalization, and stability in aqueous dispersion.[19]

(3) Polymer stabilizers

a. Dextran: Dextran is a polysaccharide polymer which is famous for its biocompatibility.[20]

b. Polyethylene Glycol (PEG): PEG is a biocompatible, hydrophilic and water-soluble polymer which also enhanced the blood circulation of SPIONs.[21]

c. Polyvinyl Alcohol (PVA): PVA is a biocompatible hydrophilic polymer which has shown to be successful in preventing the SPIONs from agglomeration.[22]

d. Alginate: Alginate has several carboxyl groups which are negatively charged and hence, it increases the electrostatic repulsion of nanoparticles.[23]

e. Chitosan: Due to the biocompatibility and biodegradability of Chitosan, it has become an interesting candidate for encapsulation of SPIONs.[24]

SPIONs have been the center of extensive research for biomedical applications. The biomedical applications of SPIONs can be categorized as below:

(1) Molecular and cellular imaging
(2) *In vitro* bio-separation
(3) Drug delivery
(4) Hyperthermia
(5) Toxin removal

In this chapter, will only focus on the controlled-release for drug delivery application. The detailed review of SPIONs for the above-mentioned application is available in the recent publications.[6–8,10,11]

## III. Controlled-Release Drug Delivery

Various nanoscale drug delivery carriers such as nanoparticles, nanotubes, micelles, vesicles, dendritic polymers, nanocapsules, and nanospheres have been developed and studies. Nanoparticles have presented new opportunities in the field of drug delivery. Some of the unique advantages of the nanoscale drug delivery are:

(1) Targeted drug delivery[25]
(2) Triggered drug release[26]
(3) Time-dependent release of the drug[27]
(4) Delivery of water insoluble drug[28]
(5) Multifunctional targeting, drug delivery and imaging.[29]

Due to the toxicity of some drugs such as tumor and cancer therapy, it is important to prevent the interaction of the healthy cells with the drug. The strategies to prevent damaging healthy cells are: (a) targeted drug release, (b) triggered drug release, and (c) time-dependent release of the drug.

The magnetic property of SPIONs provide an opportunity for magnetic targeting. It has been reported that by applying an external magnetic field the concentration of SPIONs inside tumors increase five times. In addition to the magnetic targeting, the *in vivo* monitoring of SPIONs distribution via MRI gives SPIONs a dual functionality advantage.[30]

The triggered drug release is based on a stimulus such as pH, temperature, magnetic field in order to trigger the drug release. The time-dependent drug release is based on the controlling the releasing rate of the drug. We classify both triggered and time-dependent drug release strategies as controlled-release drug delivery. In this chapter, we will focus on the controlled-release drug delivery methods by SPIONs.

## A. Time-dependent release

The conventional forms of drug delivery systems such as pills, capsules, and injectable drug carriers will result in an immediate increase of the drug concentration followed by time-dependent decay of the drug concentration. The challenges of the conventional forms of the drug delivery are:

(1) The initial high concentration of drug which could be above the therapeutic safe limit.
(2) Fast decay of drug concentration which results in the limited time of drug concentration within the therapeutic limit.
(3) The drug release is not targeted.

Therefore, it is desired to maintain the drug concentration within the therapeutic window for an extended time in the targeted area. Nanocarriers could be potential candidates in order to solve the above-mentioned challenges. There are two approaches for loading the drug on the SPIONs. The first approach is based on the surface functionalization of the SPIONs and direct attachment of the drug to the SPIONs. The second approach is based on synthesis of a porous layer on top of the SPIONs. The drug will be entrapped inside the pores of the top layer.

### 1. *Drug loading via surface functionalization of SPIONs*

With proper surface treatment and functionalization, the pharmaceutical drugs could be loaded directly on the surface of SPIONs. By applying an external localized magnetic field, drug will be released at the targeted site. In this case by administering lower drug concentration the toxicity of drug for healthy cell will be decreased.[31]

Dilnawaz *et al.*[32] coated magnetic nanoparticles with glycerol monooleate (GMO). They loaded hydrophobic anticancer drugs (paclitaxel, rapamycin) by entrapment inside GMO. They showed successful entrapment efficiency of around 95% followed by sustained *in vitro* drug release for more than 2 weeks. They also showed that by proper surface functionalization, HER2 antibody was conjugated to the GMO which resulted in targeting and enhanced cellular uptake in human breast carcinoma cell line. The drug release will be controlled by the diffusion of the drug inside GMO. Additional benefits of GMO are its colloidal stabilization of nanoparticles, ability to incorporate both hydrophilic and hydrophobic drugs as well as its biocompatibility which is approved by food and drug administration (FDA). The compatibility of GMO for loading hydrophobic drugs is especially important for cancer therapy.

Maeng *et al.*[33] developed multifunctional doxorubicin loaded SPIONs for chemotherapy of liver cancer with MRI capability. The SPIONs were decorated with doxorubicin (DOX) and poly (ethylene oxide)-trimellitic anhydride chloride-folate (PEO-TMA-FA). The folate showed successful targeting of the tumors. Therefore, DOX release was more efficient close to the tumors. Folate receptors on the tumors of kidney, brain, lung, and breast provide an opportunity for efficient targeting of these tumors via folate-functionalized nanoparticles. The MRI functionality of SPIONs helped with simultaneous treatment and minoring. Figure 3 shows the structure of these multifunctional SPIONs capable of tumor-targeting, pH-sensitive anticancer doxorubicin drug release, and MRI. The higher release of DOX in acidic environment could be related to the weak binding between doxorubicin and carboxylic groups. Upon cellular

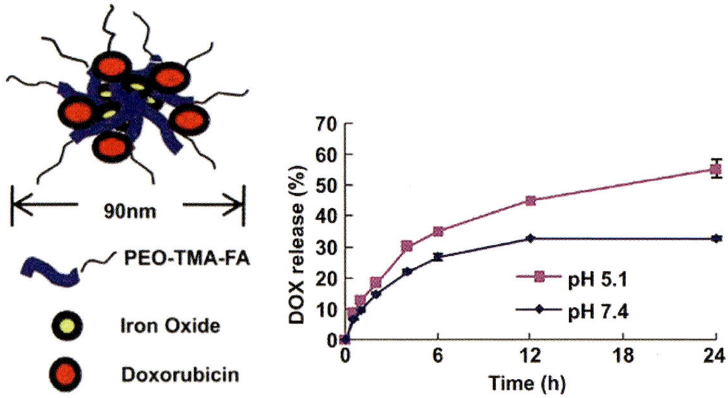


Figure 3. The structure of and multifunctional SPIONs and the doxorubicin drug release at neutral and acidic pHs. Taken from Ref. [33] with permission of Elsevier.

uptake via endocytosis the acidic pH inside endosome and lysosome of the tumor cells will accelerate the DOX release.

Ling *et al.*[34] synthesized dual-functional targeting and MRI-responsive nanoparticles by encapsulating SPIONs and docetaxel anticancer drug inside poly(lactic-co-glycolic) acid polymeric vesicles. The structure of these nanocarriers are shown in Figure 4. The prostate stem cell antigen antibody attached to the surface for targeting the cancer cells. The encapsulation efficiency for SPIONs was 23% and for the docetaxel was around 6%. These multifunctional

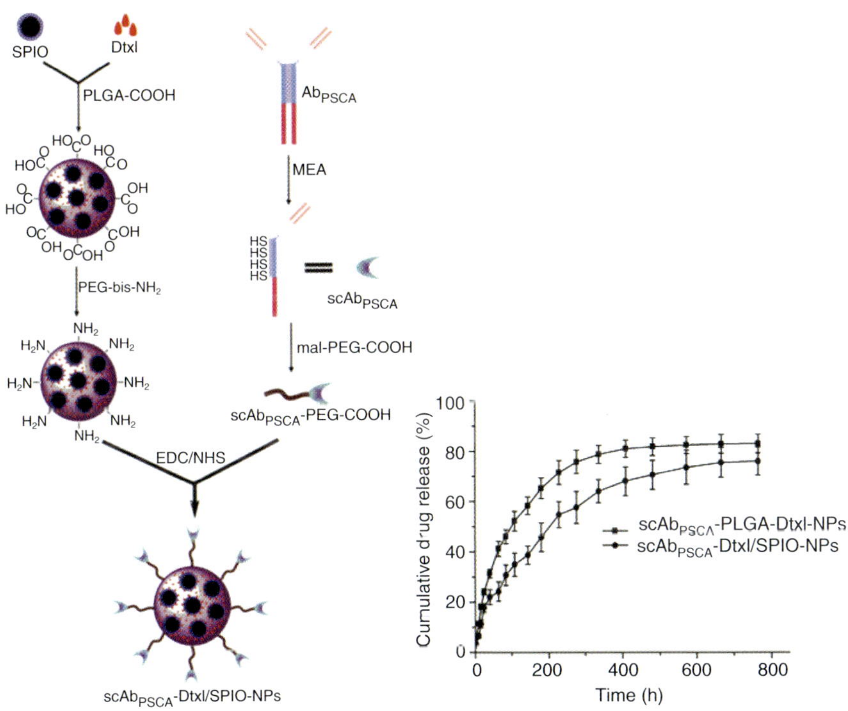


Figure 4. The structure of multifunctional nanoparticles composed of super paramagnetic iron oxide (SPIO) MRI agent, decetaxel (Dtxl) anticancer drug, carboxy-terminated poly (D,L-lactide-co-glycolide) (PLGA-COOH), poly (ethylene glycol)-bis-amine (PEG-bis-$NH_2$), tumor targeting single chain prostate stem cell antigen ($scAb_{PSCA}$)antibody. Taken from Ref. [34] with permission of Elsevier.

nanoparticles showed around 80% drug release for 30 days (Figure 4). The drug release shows three phases. The first phase starts with initial burst effect of around 18% for the first day followed by around 65% release in 15 days. The second 15 days the drug release rate is very slow.

## 2. *Drug loading via entrapping inside porous layer shell around SPIONs*

This strategy is based on entrapping the drug inside the pores of mesoporous silica. The mesoporous silica should be covered with a permeable layer which allows for controlled release of the drug. The MCM-41 is a well know template for the synthesis of mesoporous silica. It is composed of well-ordered arrays of hexagonal close-packed nanopores. One challenge with mesoporous silica is that after loading with the drug, separation of the nanoparticles is not easy. The core-shell structure composed of SPION core covered by mesoporous silica shell solves this problem. The drug can be loaded inside the pores of the mesoporous silica and the magnetic core helps with drug loading, separation, and re-distribution of nanoparticles.

In order to prevent fast release of drug from mesoporous silica, the pores should be coated or gated. Lipid bilayer has been investigated for encapsulating SPION-Silica core–shell nanoparticles. The biocompatibility and ease of synthesis are among advantages of lipid bilayer. The challenge with lipid bilayers is their low permeability, especially for large drug molecules. This challenge has been addressed by adding surfactant molecules such as Pluronics in order to increase the permeability of the lipid bilayer.[35] This strategy is shown in Figure 5. As it is shown in Figure 6, by increasing the concentration of Pluronic in the lipid bilayer the rate of the release will increase. The insertion of Pluronic in the lipid bilayer result in the localized increase of permeability. The release was fitted with a logarithmic function.

The study of pluronics on multiple drug resistant cancer cells showed that they increase the membrane fluidity.[36] The concentration below CMC, total length of pluronic copolymer, and the ration of the hydrophobic propylene oxide and hydrophilic ethylene oxide have strong effect on the pluronic interaction with lipid bilayer. Pluronic

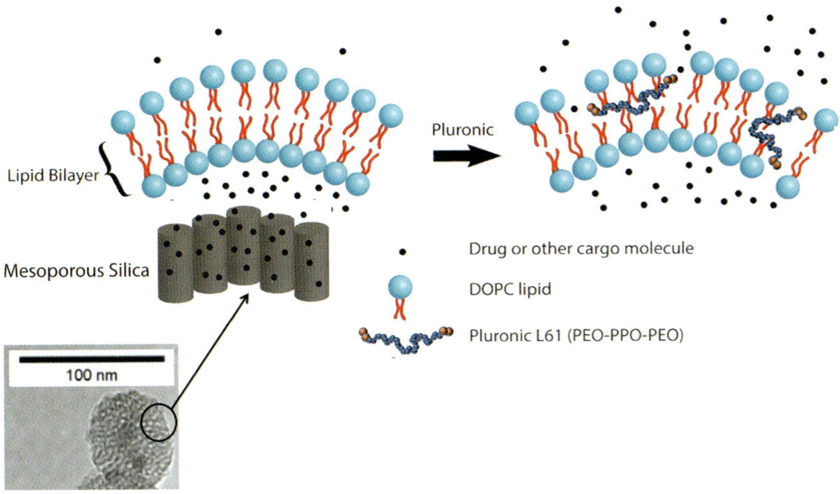


Figure 5. Changing the permeability and drug release rate by adding Pluronic surfactants to the lipid bilayer. Taken from Ref. [35] with permission of Elsevier.

with smaller molecular weight and less hydrophobic content showed higher lipid bilayer membrane permeability.[37] This concept has been shown in Figure 6.

Another strategy to control the drug release is to cover the pores with polymers. In this gating mechanism which is shown in Figure 7, by changing the length and type of the polymers the release rate could be controlled. Aznar *et al.*[38] investigated the effect of various alkyl chains on the release rate from mesoporous silica. The role of the alkyl change length on the release is shown in Figure 8.

## B. Triggered drug release

In this method, drug will be integrated with nanoparticles (drug delivery vehicle). The drug will be released upon the presence of a triggering mechanism such as pH,[39] temperature,[40] enzyme,[41] cation,[42] or magnetic field.[43] Depending on the loading mechanism of the drug various triggering mechanisms could be used. If the drug is attached to the nanoparticle via surface bond the triggering

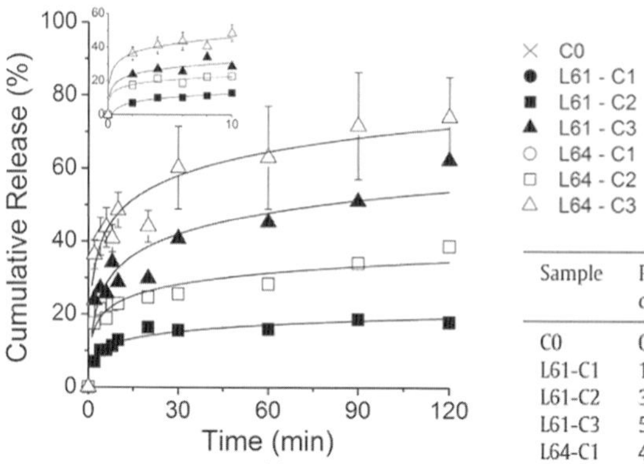


(a)

| Sample | Pluronic concentration (M) | Percentage of Pluronic CMC (PT) |
|---|---|---|
| C0 | 0 | 0 |
| L61-C1 | 1.10E–05 | 10% |
| L61-C2 | 3.30E–05 | 30% |
| L61-C3 | 5.50E–05 | 50% |
| L64-C1 | 4.80E–05 | 10% |
| L64-C2 | 1.44E–04 | 30% |
| L64-C3 | 2.40E–04 | 50% |

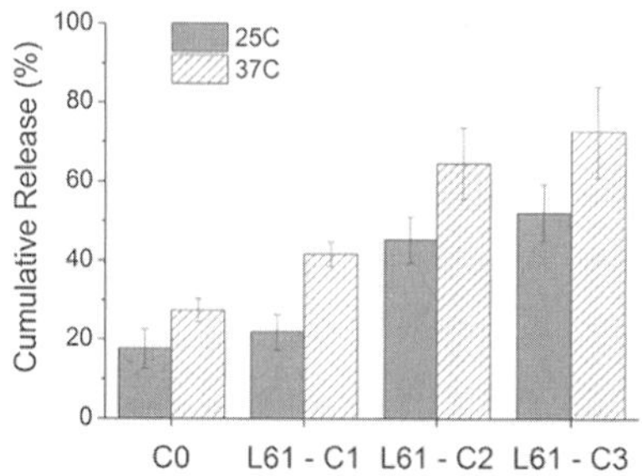


(b)

Figure 6. (a) The cumulative release of methylene blue at various Pluronic concentration for two different Pluronics L61 and L64. The Pluronic concentration in the lipid bilayer increase from C1 to C3. The percentage of Pluronic has been presented versus their critical micelle concentration (CMC). Figure (b) shows the effect of temperature on the release. Taken from Ref. [35] with permission of Elsevier.

mechanism should weaken the surface bond. In the case of entrapped drug inside the gated mesoporous shells, the triggering mechanisms should interact with the gating mechanism. Physical stimuli for triggering the drug release such as light, temperature, and magnetic field provide the flexibility of fine-tuning the spatial and temporal controlled release.

The superparamagnetic property of SPIONs make them ideal candidate for remotely-triggered magnetic field drug release. However,

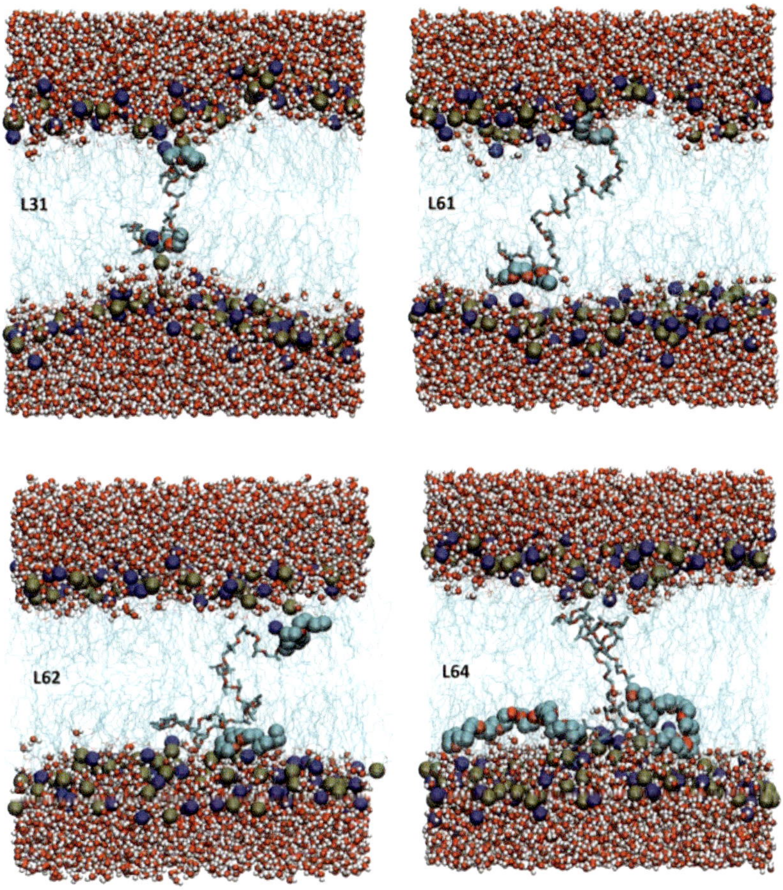


Figure 7. The interaction of various Pluronics (L31 at 50 ns, L61 at 50 ns, L62 at 100 ns, and L64 at 250 ns) with lipid bilayer membrane. Taken from Ref. [37] with permission of Royal society of chemistry.

loading the drug on SPION is a challenge. The drug could be loaded on SPIONs by grafting on the surface of SPIONs. Another approach is based on covering the surface of SPIONs with a mesoporous shell. Mesoporous silica can be easily synthesized on the surface of SPIONs.[44] Drug could be loaded inside the pores of mesoporous silica. Other advantages of SPION-silica core–shell nanoparticles are:

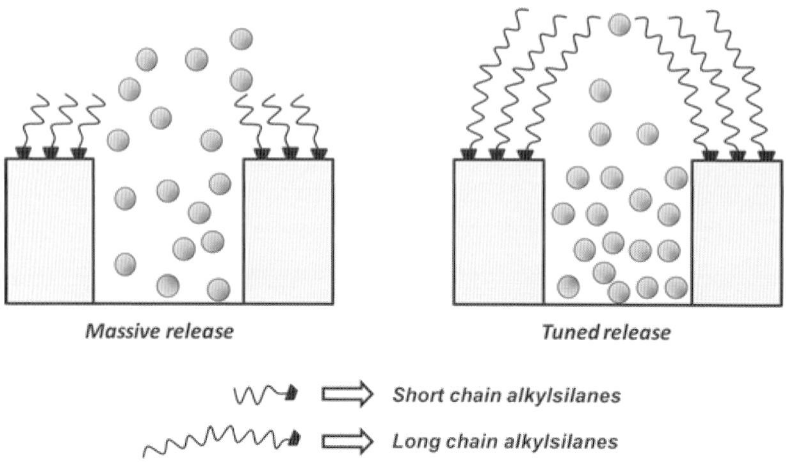


Figure 8. The gating mechanism of alkyl chains on the release rate from mesoporous silica pores. Taken from Ref. [38] with permission of American chemical society.

1) biocompatibility of SPION, 2) enhanced colloid stability, and 3) preventing the corrosion of SPION. The mechanism for the drug release from SPION-silica core–shell nanoparticles will be external alternating magnetic field. The alternating magnetic field (AMF) will result in the release of drug from the mesoporous silica pores (Figure 9). In increased drug release in the presence of AMF has been studied by various groups. Thomas *et al.*[45] concluded that the generated local internal heating is the main reason for the dethreading of the pseudorotaxane gate keepers. The second study concluded that the AMF resulted in the melting of the DNA strands which were covering the pores.[46]

Yang *et al.*[47] prepared multifunctional SPIONs for targeted drug delivery and MRI application (Figure 10). Both the anticancer drug and the tumor targeting ligands were conjugated to the PEGylated SPIONs. The anticancer drug was attached via a pH-sensitive bond. Therefore, the release will be triggered by the pH of the environment. According to Figure 10, the drug release will be triggered at acidic pHs.

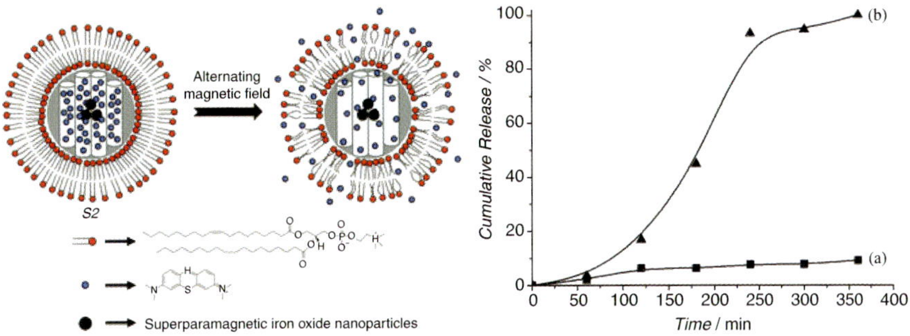


Figure 9. (Left) The magnetically-triggered drug release from SPION-mesoporous silica. (Right) The presence of alternating magnetic field (curve b) has increased the rate methylene blue release from SPION-mesoporous silica core–shell nanoparticles compared to the absence of magnetic field (curve a). Taken from Ref. [44] with permission of Royal society of chemistry.

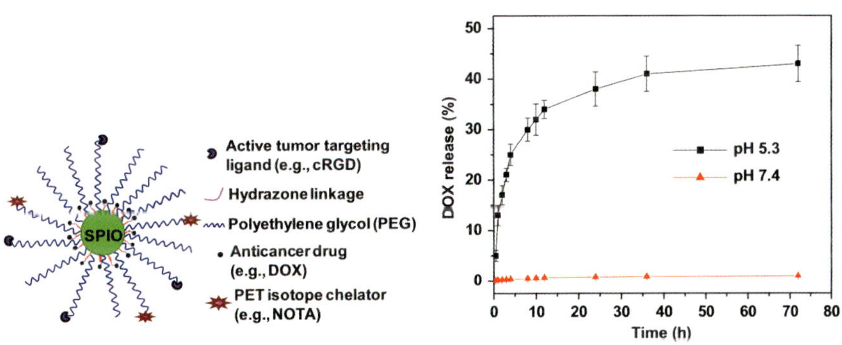


Figure 10. (Left) Multifunctional SPION designed for targeting, drug delivery, and MRI application. (Right) The drug release profile at acidic and neutral pHs. Taken from Ref. [17] with permission of Elsevier.

Chang *et al.*[48] synthesized a pH-sensitive drug release by using hydrazine as the linker via hydrazone bonds. This bond will cleave in acidic environment. The structure of the functional groups on the surface of the SPION and the drug release at acidic pHs versus the neutral pH is shown in Figure 11.

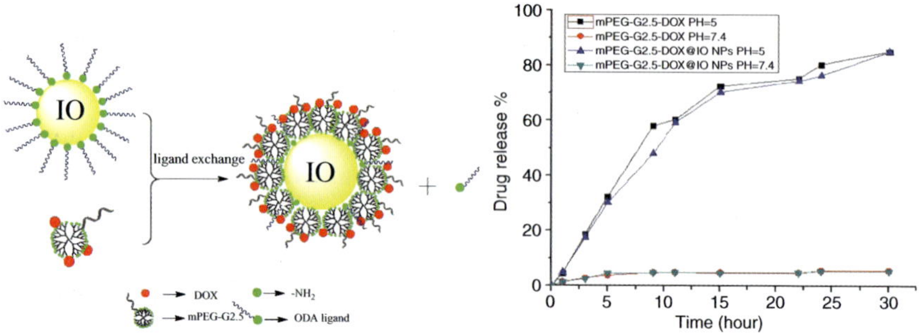


Figure 11. Left: Surface functionalization of SPIONs with mPEG-G2.5 PAMAM dendrimers, doxorubicin (DOX), and monovalent capping ligand ODA. Right: Drug release profile at acidic and neutral pHs. Taken from Ref. [48] with permission of Elsevier.

## IV. Conclusion

Superparamagnetic nanoparticles such as SPIONs show no magnetization in the absence of magnetic field. The thermal fluctuation will result in the random flipping of the magnetic field. With the presence of the magnetic field, the nanoparticles will align with the magnetic field. This behavior provides an interesting opportunity for collection and re-distribution of nanoparticles as well as external stimulation of the nanoparticles. In order to get the required properties of SPIONs such as magnetization, protein corona formation, cellular uptake, and drug release rate controlling the uniform distribution and adjusting the physicochemical properties of the nanoparticles are essential. The magnetic properties of the SPIONs provides the opportunity for the external stimulation of the SPIONs by magnetic field. This property has been used in the triggered drug release. The surface functionalization of the SPIONs is one of the major drug loading methods. However, the surface functionalization needs to be tailored for each drug specifically. Entrapping the drug inside a mesoporous shell around SPIONs is another method which could be used for drug loading. The biocompatibility and superparamagnetic properties of SPIONs provide a unique opportunity for drug delivery and controlled release via SPIONs. However, in order to become a commercial product several challenges such as drug-specific loading, release mechanism, and release-rate should be addressed.

## References

1. Skumryev, V.; Stoyanov, S.; Zhang, Y.; Hadjipanayis, G.; Givord, D.; Nogues, J. *Nature* **2003**, *423*, 850.
2. Karimi, M.; Keyhani, A.; Akram, A.; Rahman, M.; Jenkins, B.; Stroeve, P. *Environ. Technol.* **2013**, *34*, 2201.
3. Yantasee, W.; Warner, C. L.; Sangvanich, T.; Addleman, R. S.; Carter, T. G.; Wiacek, R. J.; Fryxell, G. E.; Timchalk, C.; Warner, M. G. *Environ. Sci. Technol.* **2007**, *41*, 5114.
4. Chiang, C. L.; Sung, C. S.; Wu, T. F.; Chen, C. Y.; Hsu, C. Y. *J. Chromatogr. B Analyt. Technol. Biomed. Life Sci.* **2005**, *822*, 54.
5. Zhang, S.; Zhao, X.; Niu, H.; Shi, Y.; Cai, Y.; Jiang, G. *J. Hazard Mater.* **2009**, *167*, 560.
6. Mahmoudi, M.; Stroeve, P.; Milani, A. S.; Arbab, A. S. *Superparamagnetic Iron Oxide Nanoparticles.* Nova Science Publishers, Inc., 2011.
7. Mahmoudi, M.; Sant, S.; Wang, B.; Laurent, S.; Sen, T. *Adv. Drug. Deliv. Rev.* **2011**, *63*, 24.
8. Gupta, A. K.; Gupta, M. *Biomaterials* **2005**, *26*, 3995.
9. Laurent, S.; Forge, D.; Port, M.; Roch, A.; Robic, C.; Vander Elst, L.; Muller, R. N. *Chem. Rev.* **2008**, *108*, 2064.
10. Laurent, S.; Saei, A. A.; Behzadi, S.; Panahifar, A.; Mahmoudi, M. *Expert Opin. Drug Deliv.* **2014**, *11*, 1449.
11. Santhosh, P. B.; Ulrih, N. P. *Cancer Lett.* **2013**, *336*, 8.
12. Rahman, M.; Saei, A.; Amiri, H.; Mahmoudi, M. *Curr. Organ. Chem.* **2015**, *19*, 982.
13. Konczol, M.; Ebeling, S.; Goldenberg, E.; Treude, F.; Gminski, R.; Giere, R.; Grobety, B.; Rothen-Rutishauser, B.; Merfort, I.; Mersch-Sundermann, V. *Chem. Res. Toxicol.* **2011**, *24*, 1460.
14. Ling, D.; Lee, N.; Hyeon, T. *Acc Chem Res* **2015**, *48*, 1276.
15. Monopoli, M. P.; Walczyk, D.; Campbell, A.; Elia, G.; Lynch, I.; Bombelli, F. B.; Dawson, K. A. *J. Am. Chem. Soc.* **2011**, *133*, 2525.
16. Sahoo, Y.; Goodarzi, A.; Swihart, M. T.; Ohulchanskyy, T. Y.; Kaur, N.; Furlani, E. P.; Prasad, P. N. *J. Phys. Chem. B* **2005**, *109*, 3879.

17. Kreller, D. I.; Gibson, G.; Novak, W.; Van Loon, G. W.; Horton, J. H. *Colloids and Surfaces A: Physicochemical and Engineering Aspects* **2003**, *212*, 249.
18. Ma, D.; Guan, J.; Dénommée, S.; Enright, G.; Veres, T.; Simard, B. *Chem. Mater.* **2006**, *18*, 1920.
19. Lin, J.; Zhou, W.; Kumbhar, A.; Wiemann, J.; Fang, J.; Carpenter, E. E.; O'Connor, C. J. *J. Solid State Chem.* **2001**, *159*, 26.
20. Gamarra, L. F.; Brito, G. E. S.; Pontuschka, W. M.; Amaro, E.; Parma, A. H. C.; Goya, G. F. *J. Magn. Mater.* **2005**, *289*, 439.
21. Moghimi, S. M.; Hunter, A. C.; Murray, J. C. *Pharmacol. L. Rev.* **2001**, *53*, 283.
22. Sairam, M.; Naidu, B. V. K.; Nataraj, S. K.; Sreedhar, B.; Aminabhavi, T. M. *J. Membr. Sci.* **2006**, *283*, 65.
23. Nishio, Y.; Yamada, A.; Ezaki, K.; Miyashita, Y.; Furukawa, H.; Horie, K. *Polymer* **2004**, *45*, 7129.
24. Zhi, J.; Wang, Y.; Lu, Y.; Ma, J.; Luo, G. *Reactive and Functional Polymers* **2006**, *66*, 1552.
25. Mills, J. K.; Needham, D. *Expert Opin. Ther. Pat.* **2005**, *9*, 1499.
26. Brazel, C. S. *Pharm. Res.* **2009**, *26*, 644.
27. Deshpande, A. A.; Rhodes, C. T.; Shah, N. H.; Malick, A. W. *Drug. Dev. Ind. Pharm.* **2008**, *22*, 531.
28. Liu, Z.; Robinson, J. T.; Sun, X.; Dai, H. *J. Am. Chem. Soc.* **2008**, *130*, 10876.
29. Liong, M.; Lu, J.; Kovochich, M.; Xia, T.; Ruehm, S. G.; Nel, A. E.; Tamanoi, F.; Zink, J. I. *ACS Nano* **2008**, *2*, 889.
30. Chertok, B.; Moffat, B. A.; David, A. E.; Yu, F.; Bergemann, C.; Ross, B. D.; Yang, V. C. *Biomaterials* **2008**, *29*, 487.
31. Arruebo, M.; Fernández-Pacheco, R.; Ibarra, M. R.; Santamaría, J. *Nano Today* **2007**, *2*, 22.
32. Dilnawaz, F.; Singh, A.; Mohanty, C.; Sahoo, S. K. *Biomaterials* **2010**, *31*, 3694.
33. Maeng, J. H.; Lee, D. H.; Jung, K. H.; Bae, Y. H.; Park, I. S.; Jeong, S.; Jeon, Y. S.; Shim, C. K.; Kim, W.; Kim, J.; Lee, J.; Lee, Y. M.; Kim, J. H.; Kim, W. H.; Hong, S. S. *Biomaterials* **2010**, *31*, 4995.
34. Ling, Y.; Wei, K.; Luo, Y.; Gao, X.; Zhong, S. *Biomaterials* **2011**, *32*, 7139.

35. Rahman, M.; Yu, E.; Forman, E.; Roberson-Mailloux, C.; Tung, J.; Tringe, J.; Stroeve, P. *Colloids Surf. B Biointerfaces* **2014**, *122*, 818.
36. Batrakova, E.; Lee, S.; Li, S.; Venne, A.; Alakhov, V.; Kabanov, A. *Pharm. Res.* **1999**, *16*, 1373.
37. Nawaz, S.; Redhead, M.; Mantovani, G.; Alexander, C.; Bosquillon, C.; Carbone, P. *Soft Matter* **2012**, *8*, 6744.
38. Aznar, E.; Sancenon, F.; Marcos, M. D.; Martinez-Manez, R.; Stroeve, P.; Cano, J.; Amoros, P. *Langmuir* **2012**, *28*, 2986.
39. Muhammad, F.; Guo, M.; Qi, W.; Sun, F.; Wang, A.; Guo, Y.; Zhu, G. *J. Am. Chem. Soc.* **2011**, *133*, 8778.
40. You, J.; Zhang, G.; Li, C. *ACS Nano* **2010**, *4*, 1033.
41. de la Rica, R.; Aili, D.; Stevens, M. M. *Adv. Drug. Deliv. Rev.* **2012**, *64*, 967.
42. An, J.; Geib, S. J.; Rosi, N. L. *J. Am. Chem. Soc.* **2009**, *131*, 8376.
43. Derfus, A. M.; von Maltzahn, G.; Harris, T. J.; Duza, T.; Vecchio, K. S.; Ruoslahti, E.; Bhatia, S. N. *Adv. Mater* **2007**, *19*, 3932.
44. Bringas, E.; Koysuren, O.; Quach, D. V.; Mahmoudi, M.; Aznar, E.; Roehling, J. D.; Marcos, M. D.; Martinez-Manez, R.; Stroeve, P. *Chem. Commun.* (*Camb*) **2012**, *48*, 5647.
45. Thomas, C. R.; Ferris, D. P.; Lee, J. H.; Choi, E.; Cho, M. H.; Kim, E. S.; Stoddart, J. F.; Shin, J. S.; Cheon, J.; Zink, J. I. *J. Am. Chem. Soc.* **2010**, *132*, 10623.
46. Ruiz-Hernandez, E.; Baeza, A.; Vallet-Regi, M. *ACS Nano* **2011**, *5*, 1259.
47. Yang, X.; Hong, H.; Grailer, J. J.; Rowland, I. J.; Javadi, A.; Hurley, S. A.; Xiao, Y.; Yang, Y.; Zhang, Y.; Nickles, R. J.; Cai, W.; Steeber, D. A.; Gong, S. *Biomaterials* **2011**, *32*, 4151.
48. Chang, Y.; Meng, X.; Zhao, Y.; Li, K.; Zhao, B.; Zhu, M.; Li, Y.; Chen, X.; Wang, J. *J. Colloid Interface Sci.* **2011**, *363*, 403.

# Chapter 6

# The Reverse of Controlled Release: Controlled Sequestration of Species and Biotoxins into Nanoparticles (NPs)

Jenifer Gómez-Pastora*, Eugenio Bringas*, María Lázaro-Díez†, José Ramos-Vivas†, and Inmaculada Ortiz*

** Dept. of Chemical and Biomolecular Engineering, ETSIIT, University of Cantabria, Avda. Los Castros s/n, 39005 Santander, Spain*
*† Dept. of Microbiology, Valdecilla Biomedical Research Institute (IDIVAL), C/ Cardenal Herrera Oria s/n, 39011 Santander, Spain*

## Abbreviations

| | |
|---|---|
| BSA | bovine serum albumin |
| EU | endotoxin units |
| HBS | HEPES-buffered saline |
| LPS | lipopolysaccharide |
| MNPs | magnetic nanoparticles |
| NPs | nanoparticles |
| PBS | phosphate-buffered saline |
| PEG | polyethylene glycol |
| PEI | polyethyleneimine |
| PHMBG | poly-(hexamethylene biguanide) |
| RBCs | red blood cells |

## I. Introduction

Nanoparticles (NPs) are used in a rapidly expanding number of practical applications in medicine, biology, and biotechnology. Recent developments in materials science (synthesis and characterization) have led to significant progress in nanotechnology, where the application of NPs for human health issues has generated the greatest attention and has changed the nature of diagnosis and treatment.[1–4] In these fields, NPs are exploited for different applications such as contrast agents for magnetic resonance imaging, transducers for hyperthermia therapy, carriers for drug and gene delivery, immunoassays, detoxification agents of biological fluids, bio-separation, biological sensing, etc.[1,5,6] Research has demonstrated the potential benefits that nanotechnology can offer and it has been focused on a new and growing discipline known as nanotheranostics, that implements diagnostic and therapeutic functions in one system due to the unique properties of NPs.[7,8]

The use of NPs for therapy against several diseases is extended to both *in vivo* and *ex vivo* applications. The system types divided into 2 categories based on their function: drug delivery and drug sequestration systems. The first category involves particles encapsulated with drugs or other molecules that are usually employed for *in vivo* applications. In this case, the drug coated NPs are introduced into the human body and release the therapeutic agent in the tissue to be treated (e.g. tumor). This method appears very attractive for the treatment of different pathologies but current research focuses mainly in cancer therapy.[2,9]

On the other hand, drug sequestration methods employ NPs as capture agents that recognize a specific target (toxins, microorganisms, etc.). In this regard, the particles are functionalized with a specific antibody tailored to bind to the substances to be removed. Such particles are mainly applied for adsorbing hazardous compounds that circulate in biological fluids, such as blood, in an extracorporeal system. Due to their unique role, NPs used in drug sequestration methods need to satisfy different requirements compared to those used in drug delivery systems. Thus, there exist several important differences in the development of these technologies.

Scientific attention has been focused on the extracorporeal removal of toxic substances from blood. These include biotoxins, which can be defined as poisonous substances produced by biological organisms,[10] microorganisms, or toxic chemicals or drugs that can cause diseases or death. In fact, for a number of clinical conditions such as intoxication, bacteraemia or autoimmune diseases, the removal of the disease-causing agents from blood can be considered as the most direct conceivable treatment.[11,12] Under this scenario, the scientific community has addressed attention to the use of functionalized magnetic NPs (MNPs) to capture a number of toxic substances. Interest in this technology also stems from the limitations of the existing methods (hemodialysis, hemofiltration, plasmapheresis, extracorporeal immunoadsorption, direct injection of chelators and antibodies, etc.) for removing toxic compounds either for patient treatment or diagnosis.[13–15] MNPs-based technology may improve the effectiveness of treatment. Nevertheless, since the long term risks and possible side effects associated with the *in vivo* use of NPs are not yet completely understood, an extracorporeal setting is preferred at this stage of knowledge because it prevents NPs from entering the body.[11]

Extracorporeal blood detoxification through the use of MNPs is a process where the patient's blood is injected with magnetic materials with the ultimate goal of achieving the selective removal of the toxic agent while maintaining the normal functionality of blood constituents. Figure 1 illustrates the stages involved in the extracorporeal blood detoxification process. Once the adsorption is completed, the magnetic separation stage takes place and the toxins are removed along with the material, leading to a clean blood solution that would return to the circulatory system of the patient. Therefore, this technology allows the sequestration of a non-magnetic substance (toxin, microorganism or toxic chemicals) through magnetic means. In the following sections the potential of the MNPs-based processes to carry out the recovery of several toxic substances from blood

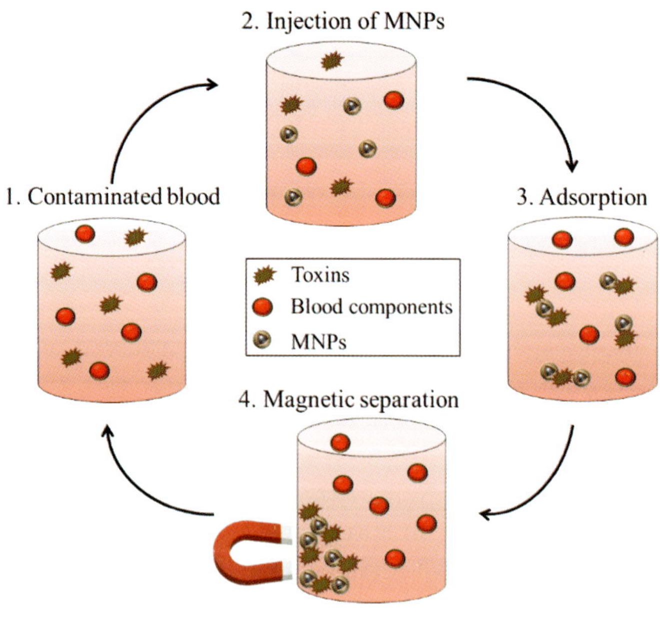


Figure 1. Stages involved in the extracorporeal sequestration of toxins onto MNPs.

solutions was examined. In order to avoid any confusion with the terminology, we refer to all the target compounds as toxins, but we differentiate the biological origin of the microorganisms or their poisonous substance with the term of biotoxins, whereas the toxic substances of no biological origin are termed here as chemical or radioactive toxins.

## II. NPs for Toxin Sequestration from Blood

MNPs need to meet several conditions in order to be effectively used for blood detoxification systems. This section discusses the design factors and functionalization alternatives for the use of MNPs for blood detoxification.

### A. Design factors

MNPs employed for toxin sequestration are generally based on composites, with a diameter in the nanometer scale, in which several materials are combined with the objectives of maximizing the removal capacity of the system and minimizing the unspecific adsorption of blood components, thus maintaining the functionality of the biological fluid that needs to be treated. The design and synthesis of these materials should be carefully planned in order to achieve the best material configuration possible. In this sense, NPs employed as detoxification agents must fulfill several design requirements that are briefly schematized in Figure 2.

Contrary to the requirements of drug delivery systems, where optimized drug encapsulation in the NPs, and therefore a high volume-to-surface-area is preferred,[16] drug sequestration systems require the opposite trend. This is due to the fact that particles with high specific surface areas represent the best configuration for the maximization of surface functionality and capacity for toxin removal, as well as for the reduction of the material dosage. Hence, particles with small diameters are ideal for such applications since surface area inversely scales with particle size. The size of the material can be easily controlled by adjusting the synthesis parameters.[17]

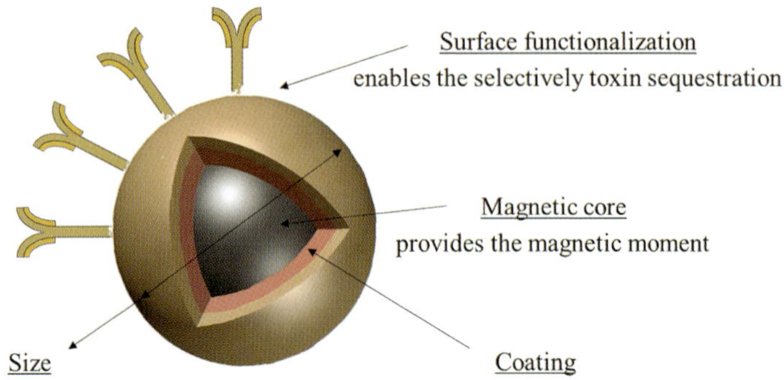


Figure 2. Design factors for the MNPs employed in toxin sequestration systems.

On the other hand, the use of NPs with magnetic properties has gathered great attention in the last few years due to the easy manipulation of the nanocomposite through magnetic fields. In this regard, the inner core of the particle is usually made from a ferromagnetic nanomaterial. The core is further submitted to a surface treatment designed to deposit an external layer of biocompatible polymer chains functionalized with antibodies that capture the target toxins. Overall, the magnetic inner core provides the magnetic aspect of the composite whereas the functionalized surface of the material allocates the capturing antitoxin groups as shown in Figure 2. Certain works found in the scientific literature have reported that despite the apparent importance of the magnetic moment of MNPs, effective separation can also be achieved by adjusting the operating conditions of the magnetic separator.[18] On the other hand, the opposite is true for magnetic drug delivery systems where the magnetic moment of the material must be maximized in order to minimize the external field and gradient required for the separation because of the existing constraints regarding the parameters and operating variables that cannot be easily modified when the particles that have to be separated are inside the body. Therefore, magnetic moments higher than 30 $emu{\cdot}g^{-1}$ are usually required for drug delivery materials, whereas moments of around 10 $emu{\cdot}g^{-1}$ are considered appropriate for toxin sequestration systems because the desired separation efficacy can be reached by tuning the operating variables.[16] Thus, in the design of toxin sequestration systems the volumetric fraction of the magnetic material is not restricted to high values in order to allow for higher magnetic moment, and the flexibility in the MNPs preparation increases.

Finally, close attention has to be paid to the functionalization of the particles to ensure adequate distribution of the active groups through the surface (antitoxin groups) that would increase the adsorption capacity of the material. This requirement has to be met while maintaining the biocompatibility, sufficient dispersion in the medium and the chemical and physical stability of the material. Aggregation effects should be avoided since they decrease the available surface area and therefore the capacity of the system. Due to the aforementioned conditions, surface chemistry is probably the most complex and important aspect in the process design and will be thoroughly discussed in the next subsection.

## B. Functionalized NPs for toxin removal

Several works have shown the potential applicability of MNPs for the selective removal of toxins and microorganisms from human blood samples. These toxin sequestration systems consist mainly of metals, metal oxides or alloys of nanocrystals coated with a functionalized polymeric biocompatible outer shell as mentioned above. Among the ferromagnetic materials that could be employed as inner core compounds, cobalt and nickel have been associated with adverse side effects,[19] while iron cores have reported good magnetic properties (approximately 3–10 times higher saturation magnetization than their oxides), but favor agglomeration owing to their dipole-dipole interactions.[20,21] However, iron oxides (magnetite or maghemite) are non-toxic and have been approved for *in vivo* use as toxin sequestering materials.[13,22] Therefore, they represent excellent candidates for the formulation of the inner core since they also provide sufficient magnetic moment to the particles.

As far as coating materials are concerned, polyethylene glycol (PEG) has shown to enhance the cytocompatibility and dispersibility in fresh human blood solutions.[19] The PEG shell also reduces opsonization (unspecific protein adsorption) and phagocytosis (cell engulfment).[13,23] Other polymeric materials such as poly-(lactic acid) and poly-(lactic-co-glycolic acid) have been approved by the US Federal Drug Administration and can be applied as coatings.[13] Carbon coatings may also represent a viable alternative because of their high stability and biocompatibility.[20]

The selective receptors attached to the nanospherical surface are tailored specifically for each application and care should be taken to avoid the unspecific adsorption of other biomolecules onto the particle surface and thus limiting treatment effectiveness. Therefore, the surface chemistry is strictly dependent on the toxin to be removed. In the following subsections, different surface functionalization options that are available for the selective removal of biological, chemical or radioactive toxins, as well as small molecule drugs that might be present in the bloodstream of intoxicated patients will be explored.

## C. Removal of biological toxins

Microorganisms play a pivotal role in the onset and development of certain pathologies, some of them associated with considerably high morbidity and mortality. For instance, sepsis is a clinical syndrome originating from the presence of pathogenic organisms or their toxins in the body, which produces a complex interplay of exaggerated inflammatory reactions and immune suppression in response to bacteria and bacteria-derived endotoxins circulating in the blood stream.[24] On average, it afflicts 18 million people worldwide every year and accounts for more than 200,000 deaths annually in the US alone.[23,25] Sepsis represents the primary cause of death in intensive care units with hospital mortality rates of approximately 27%, reaching 54% and 80% in the case of septic shock and multiorgan failure, respectively.[15,26–28] The mortality rate can increase as much as 9% for every hour before the administration of correct therapy.[23] Hence, rapid elimination of the toxins and microorganisms is of paramount importance. Although the selective toxin removal from complex biological media, such as blood, remains challenging, the prevention, diagnosis and treatment of sepsis requires the development of novel strategies such as the use of functionalized MNPs as sequestering agents.

Several receptors have been used for the removal of biotoxins and microorganisms from blood solutions in various experimental studies. These receptors can be either terminally attached to the coating material or encapsulated within it.[13] However, the most straightforward method of linking antibodies is to synthetize the external receptor with biotin end groups and bind them to streptavidin coated particles, because the streptavidin-biotin linking is the strongest non-covalent biological interaction known.[12] Since the antitoxin receptor should be selective for the toxin to be removed, its selection depends on the particular intoxication case. In most studies the receptor is designed to target the endotoxins liberated by the bacteria-causing sepsis: the lipopolysaccharides (LPS). The LPS integrate the outer membrane of Gram-negative bacteria and have a key role in the pathogenesis of sepsis.

In order to make the description of the available materials more concise, Table 1[12,21–26,28–30] summarizes the most representative information regarding published studies in which MNPs are employed for the removal of biotoxins from biological solutions. The information reported includes the material used and its characteristics (size and magnetic saturation), the working conditions (MNPs and target toxin concentrations, contact time, etc.) and the most relevant conclusions obtained by their application in both *in vitro* and *in vivo* studies.

As shown in Table 1, cyclic cationic polypeptide antibiotics, such as polymyxin B, are excellent candidates for the neutralization of endotoxins due to their high affinity for lipid A. Lipid A is the toxically active glycolipid component of the bacterial LPS, and its molecular structure is shown in Figure 3. Endotoxins are negatively charged in blood solutions and have strong binding affinity for the positively charged cyclic peptide polymyxin B (Figure 4a), which also has the

**Table 1.** Studies of MNPs Employed for the Removal of Biotoxins from Biological Solutions

| Material | Solution to be treated | Working conditions | Removal efficacy | Considerations | Refs. |
|---|---|---|---|---|---|
| Co/Fe@C-Polymyxin B (21 nm, 207 emu·g$^{-1}$) | Human blood contaminated with LPS | *In vitro* experiments<br>[MNPs] = 1–2 g·L$^{-1}$;<br>[LPS] = 2 mg·L$^{-1}$;<br>Time: 2–15 minutes | ≈80% of removal | Different assays showed a decreased inflammatory response of endothelial cells contacted with the treated blood. The material biocompatibility and its bactericidal properties were confirmed. | [24] |
| $Fe_3O_4$@PEI-PHMBG (332 nm, 56 emu·g$^{-1}$)<br>Co-PHMBG (215 nm, 22 emu·g$^{-1}$) | Tris-HCl buffer contaminated with *E. coli* | *In vitro* experiments<br>[MNPs] = 0.2–2 g·L$^{-1}$;<br>[Pathogens]=$10^6$–$10^7$ CFU·mL$^{-1}$;<br>Time: 3 hours (37°C) | 65–99% depending on the particle concentration | At concentrations of 5 mg·L$^{-1}$, cobalt particles inhibited the cell growth after 10 minutes of contact whereas magnetite particles only were effective at concentrations higher than 10 mg·L$^{-1}$ | [26] |
| $Fe_3O_4$@PEI-PHMBG (50–150 nm, 40–65 emu·g$^{-1}$)<br>$Fe_3O_4$@$SiO_2$-PHMBG (150 nm, 80–89 emu·g$^{-1}$) | HBS buffer contaminated with Gram negative (*P. aeruginosa*, *E. coli* and *S. enterica*) or positive (*S. aureus*, *S. epidermis*, *L. lactis*, *S. phocae* and *L. garvieae*) bacteria | *In vitro* experiments<br>[MNPs] = 0.2–2 g·L$^{-1}$;<br>[Pathogens] = 1.5·$10^6$ CFU·mL$^{-1}$;<br>Time: 3 hours (37°C) | 55–90% and 70–98% for the polymeric and the silica particles, respectively | The antiseptic activity against eight types of bacteria was tested. Biocompatibility tests were carried out showing the silica particles lack of toxicity. Silica coatings also improved the stability of the metallic cores. | [22] |

| | | | | | |
|---|---|---|---|---|---|
| $Fe_3O_4$@PEG-Streptavidin-Genetically engineered mannose-binding lectin (128 nm) | Saline or blood solutions contaminated with *S. albicans*, *S. aureus*, *E. coli*, mixed populations of anaerobic or aerobic cecal microbial flora or LPS | Experiments in a continuous flow device *In vitro*: [MNPs] = 0.5 $g{\cdot}L^{-1}$ (injected at 7.1 $\mu L{\cdot}min^{-1}$); [Toxin] ≈ $10^4$ $CFU{\cdot}mL^{-1}$; [LPS] = 10 $mg{\cdot}L^{-1}$; Blood flow rate: 10 $mL{\cdot}h^{-1}$; *In vivo*: [Pathogens] ≈ $5{\cdot}10^8$ CFU; [LPS] = $13.6{\cdot}10^6$ $EU{\cdot}mL^{-1}$ | >80% for all the toxins from blood and saline solutions; ≈ 90% from the septic rat's blood in 1 hour; 86% of the untreated rats injected with LPS died whereas 89% of the treated animals survived | Flowing the animal blood through the device had no effect on the animal's physiology and did not induce blood coagulation. After the treatment, a decrease in both pathogen load and the level of inflammatory cells was observed. Decreased levels of free-hemoglobin were found in plasma due to the reduced hemolysis. The restoration of normal leukocyte levels was also achieved. | [23] |
| $Fe_3O_4$@$SiO_2$-Peptide-perylene diimide (25 nm) | Tris-buffered saline solution, cell lysates and human serum samples contaminated with LPS from lysate of *E. coli* cells | *In vitro* experiments [MNPs] = 200–2000 $mg{\cdot}L^{-1}$; Time: 15 minutes | 87% from lysates of $10^9$ $CFU{\cdot}mL^{-1}$ (6800 $EU{\cdot}mL^{-1}$) and 100% after 3 or 4 cycles; 89% from human serum samples | The functional group acts as a fluorescent sensor for LPS determination. The fluorescent intensity was LPS concentration dependent. The selectively was confirmed. After the treatment, the sample | [29] |

(Continued)

Table 1. (*Continued*)

| Material | Solution to be treated | Working conditions | Removal efficacy | Considerations | Refs. |
|---|---|---|---|---|---|
| | | | | was incubated with macrophages and the cell stress was decreased. The cytotoxicity was low, with cell viabilities around 87% (at NPs concentrations of 400 $mg \cdot L^{-1}$). | |
| Commercial MyOne particles functionalized with antibodies (Abcam) against *C. albicans* (1000 nm) | Human whole blood contaminated with *C. albicans* | *In vitro* experiments [Pathogen] = $10^6$ $cells \cdot mL^{-1}$; Bead/pathogen ratio of 120; Time: 1 hour (37 °C) | 85–90% | The adsorption followed a Langmuir model, and an average of 28 beads was bound per fungal cell. | [25] |
| $Fe_3O_4$ (9 nm) $Fe_3O_4$-Au (14 nm) and $Fe_3O_4$ @$SiO_2$-$NH_2$ (10 nm) | Saline buffer and body fluids (plasma, serum, abdominal, and cerebrospinal fluids) contaminated with *E. coli*, *S. aureus*, *P. aeruginosa* or *C. albicans* | *In vitro* experiments [Pathogens] = optical density of 600; [MNPs] = 0.3–2.5 $g \cdot L^{-1}$; Time: 3 hours | 45–100% ($Fe_3O_4$ and gold MNPs) and 30–100% (for $NH_2$ MNPs) depending on the pathogen and the MNPs concentration. 100% removal of *C. albicans* from the biological fluids (2.5 $g \cdot L^{-1}$) in 10 minutes. | The MNPs inhibited the pathogen growth. Also, all tested materials did not disrupt the RBCs' membranes, being biocompatible. | [30] |

| | | | | | |
|---|---|---|---|---|---|
| $Fe_3O_4$@PEG-Zn-coordinated bis(dipicolylamine) (100 nm) | PBS and bovine whole blood contaminated with *E. coli* | *In vitro* experiments [*E. Coli*] = $10^7$ CFU·mL$^{-1}$; [MNPs] = 1.4 $10^{10}$-$10^{11}$ particles·mL$^{-1}$; Time: 5 minutes (Room temperature) | 50–100% from PBS depending on the particle concentration. 70% from blood ($10^{11}$ particles·mL$^{-1}$) achieving 100% of removal after 2 cycles | Without the Zn coordination, no adsorption of *E. Coli* was observed. The incubation time required was less than 1 minute. The MNPs were able to capture endotoxins. Hemolysis assay confirmed that no damage was done to the RBCs. Completed separation of *E. Coli* was carried out in a continuous process inside microfluidic channels. | 28 |
| Fe@C-Streptavidin-Anti-interleukin-1$\beta$ (30 nm, 150 emu·g$^{-1}$) | Human blood spiked with interleukin-1$\beta$ (IL-1$\beta$) | *In vitro* experiments in a continuous mode [MNPs] = 0.5 g·L$^{-1}$; Blood flow rate: 15 ml·min$^{-1}$; [IL-1$\beta$] = 1300 ng·L$^{-1}$; Total volume: 0.5 L; Time: 2 minutes | 38% after a single pass through the blood purification device | More removal efficacy could be achieved by recirculating the treated blood through the device more cycles | [21] |

*(Continued)*

**Table 1.** (*Continued*)

| Material | Solution to be treated | Working conditions | Removal efficacy | Considerations | Refs. |
|---|---|---|---|---|---|
| $Fe_3C$@C-Streptavidin-Anti-human interleukin-6 (30 nm, 140 emu·g$^{-1}$) | Human blood spiked with interleukin-6 (IL-6) | *In vitro* experiment [IL-6] = 500–600 ng·L$^{-1}$; Volume: 40–60 mL; Time: 5 minutes | 70% when the particle dosage was 1.5 mg | The purified sample showed increase caspase-3 activity in neutrophils in comparison with the untreated blood, confirming the antiapoptotic effect of IL-6 and the relevance of the separation. Blood clotting and serum potassium, lactate dehydrogenase, bilirubin, and haptoglobin levels in plasma remained in the normal range. The RBCs were unaffected. Iron leaching was not detected. | [12] |

*Note*: PEI: Polyethyleneimine; PHMBG: Poly-(hexamethylene biguanide); HBS: HEPES-buffered saline; PEG: Polyethylene glycol; EU: Endotoxin units; RBCs: Red blood cells; PBS: Phosphate-buffered saline.

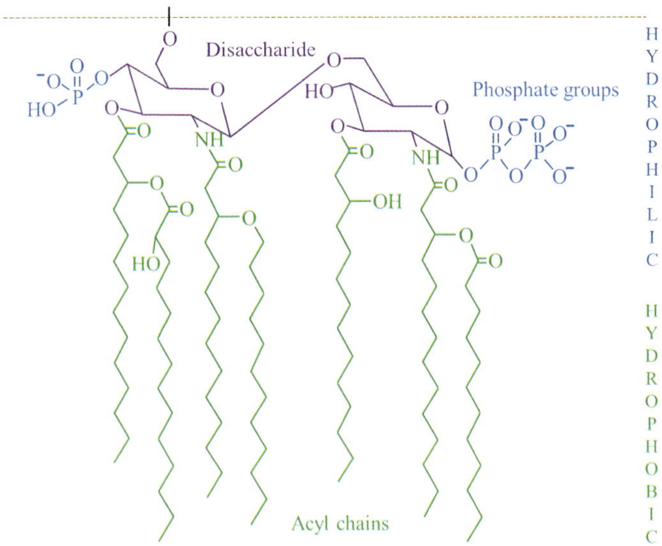


Figure 3. Molecular structure and charge distribution of Lipid A. Taken from Ref. [31] with permission of Elsevier.

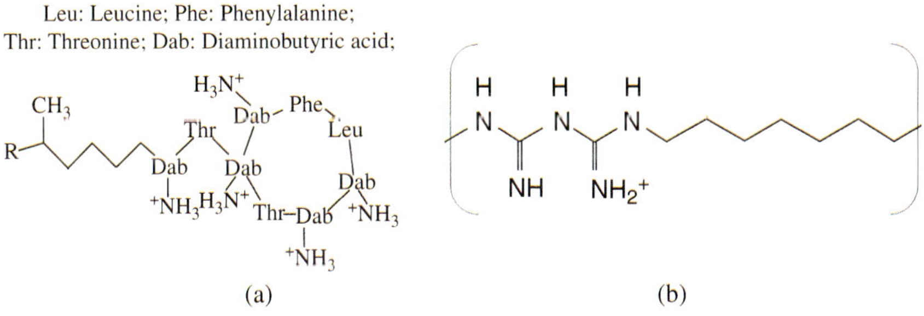


Figure 4. Structure of: (a) Polymyxin B and (b) poly-(hexamethylene biguanide) (PHMBG).

potential to recognize endotoxins of different origins.[31,32] Despite such encouraging preliminary results, the translation of polymyxin B into clinical practice has been limited due to side effects caused by its toxicity (nephrotoxicity, neurotoxicity) and the unwanted stimulation of monocytes to release inflammatory mediators such as interleukin-1.[24,33] Therefore, one of the best strategies to take advantage of the potential of this antibiotic while minimizing its systemic toxic effects is to immobilize it on the surface of adsorbent materials such as MNPs.

For instance, Herrmann *et al.*[24] employed polymyxin B functionalized carbon-coated cobalt/iron alloy NPs (Co/Fe@C-Polymyxin B), for the removal of endotoxins from fresh human whole blood and obtained, after 15 minutes of interaction, a reduced overall endotoxin activity in the purified sample with a surface loading of polymyxin B of 1 wt%. In order to test the selectivity and hence the availability of capture sites, the authors demonstrated that unspecific adsorption of other molecules such as albumin was low. The toxicity of their MNPs was tested with endothelial and renal epithelial cells in order to assure the cytocompatibility. Moreover, functional tests like inflammatory response from endothelial cells or neutrophil chemotaxis were carried out and a significant decrease in the inflammatory response (i.e. reduced expression of inflammatory mediators along with reduced recruitment of neutrophils) was observed when the treated plasma was compared to the non-purified sample. Finally, the authors confirmed that the MNPs affected the viability of *E. coli* bacteria, indicating the material's ability to not only remove endotoxins but also to reduce the viability and culturability of *E. coli*. The bactericidal capabilities of polymyxin B are due to its disturbing effects on the Gram-negative bacteria wall,[31,33] rendering this material as one of the most promising candidates for blood purification.

Another functional receptor that has shown high affinity to bacterial endotoxins is the antiseptic poly-(hexamethylene biguanide) (PHMBG); which exhibits a removal capacity equivalent to polymyxin B.[22,26] MNPs functionalized with this receptor have reported promising bactericidal properties, reducing *E. coli* viability and inhibiting growth, as seen in Table 1. The structure of this endotoxin-selective ligand is given in Figure 4b. One attractive property of PHMBG is its selective interaction with lipid A via steric, electrostatic, and hydrophobic pathways. In fact, one of the studies that demonstrated the bactericidal potential of this material was carried out by Bromberg *et al.*[26] and showed that magnetite and cobalt NPs functionalized with PHMBG completely inhibited *E. coli* cellular growth in 10 minutes, with particle concentrations of 10 $mg{\cdot}L^{-1}$. The bactericidal effect was explained by the affinity of the PHMBG to lipid A. This causes attractive forces between the bacteria membrane and the particles that result in physical contact. At that point, the cellular membrane wraps around the particles, leading to cell death by lysis due to a lack of adequate membrane elasticity.

In a later study,[22] the same group employed PHMBG-functionalized magnetite particles encapsulated with different coating materials (polyethyleneimine or silica) for the removal of eight types of bacteria (see Table 1). These particles were able to capture both Gram-negative and Gram-positive bacteria due to the nonspecific electrostatic and hydrophobic interactions with components of their outer membrane. The tested particles were proven to be effective in capturing these bacteria at concentrations ranging from 0.2 to 2 $g{\cdot}L^{-1}$, showing a broad range of bactericidal capabilities. Particle size was found to be of significant importance, with the smaller particles showing better performance owing to their higher surface area, especially at low concentrations. Finally, in material biocompatibility tests, the silica encapsulated particles were found to be the less toxic to fibroblast cells. Despite the bactericidal potential of PHMBG, more studies with blood solutions are required because of the probability of nonspecific adsorption of blood constituents onto the particle surface. This may result in limitations of their applicability as recoverable antiseptics for treating sepsis.

Engineered receptors immobilized on the surface of MNPs have also been designed for sepsis therapy. More specifically, in a study of particular interest by Kang *et al.*,[23] the authors presented a continuous blood-cleansing device which could remove different pathogens and toxins without first identifying the infectious agent. In their analysis, MNPs coated with a genetically engineered human opsonin (mannose-binding lectin) were shown to capture a broad range of pathogens and toxins including Gram-negative and Gram-positive bacteria, fungi, and LPS from whole human blood with very high efficacy as seen in Table 1. The experiments were carried out both *in vitro* and *in vivo* with rats. When the extracorporeal device was tested with living animals, a 90% reduction in pathogen levels was found along with a decrease in levels of proinflammatory cytokines (granulocyte-macrophage colony-stimulating factor, interferon-ϒ, interleukin-1$\alpha$, interleukin-4 and interleukin-$\gamma$) and toxic cell-free hemoglobin in plasma (due to reduced hemolysis). Whereas 86% of the untreated bacteremic rats died, 89% of the rats treated with their engineered MNPs survived the 5 hours experiment and exhibited a significant improvement in various physiological responses such as respiratory rate, body temperature, and restoration of leukocyte levels.

In addition, the synthesis of functionalized particles with fluorescent activity has attracted significant attention in recent years since such particles can be used as solid sensors for the detection, recovery and removal of toxic molecules from the human body. In this regard, Xing and coworkers have reported the use of peptide-perylene diimide functionalized NPs that selectively interact with LPS from Gram-negative bacteria.[29] Upon the incubation and binding of LPS with this material, high fluorescence intensity was observed, the intensity of which was found to be LPS-concentration dependent. A selectivity study was carried out with other molecules including negatively charged BSA (Bovine serum albumin) and DNA with no significant fluorescence response being detected. Since the ensuing functionalized particles are inherently nonfluorescent, this material holds promise for applications requiring a highly selective fluorescent turn-on sensor for LPS detection and sequestration. Moreover, other functional groups based on zinc,[28] amine groups and gold[30] have indicated abilities to capture, remove and limit the growth of different pathogens in different biological fluids including human blood plasma, serum, abdominal fluids and cerebrospinal fluids as seen in Table 1.

The factors relating to the associated inflammation caused by septic states, namely inflammatory mediators, are also in the crosshairs of scientist involved with sequestration technologies. Particularly, the harmful effects of the host's inflammatory response during sepsis are of significant magnitude because of the cytotoxicity of cytokines and the impaired immunity induced by their prolonged release.[15] Moreover, such inflammatory mediators contribute to multi-organ system failure.[25] In this regard, various studies of MNPs with a surface functionalization based on antibodies have demonstrated their potential for binding to these proinflammaroty proteins in blood solutions, as seen in Table 1.

Despite the wealth of preliminary results demonstrating the potential applicability for a variety of functional groups against microorganisms and endotoxins, more surface receptors should be explored. For instance, ribonucleic acids based-sorbents, such as histidine and histamine, which have been reported to interact with endotoxins, may be as effective as the aforementioned antitoxin groups (e.g. polymyxin B) and equally capable of decontaminating complex protein solutions.[31,33,34] Moreover, cationic polymers, including poly-(ethyleneimine), poly-L-histidine or poly-L-lysine, have shown good endotoxin clearance and could be applied as coating materials. [31,34,35] In conclusion, it seems that more research is required in order to test both the biocompatibility and the removal efficacy of various substances against bacteria and endotoxins.

## D. Removal of radioactive or chemical toxins

Other toxic substances that might be found in the bloodstream of intoxicated patients consist of heavy metal ions (lead, cadmium), radioactive toxins (uranyl) or small molecule drugs (digoxin, diazepam). These substances have been removed from whole blood by using MNPs. Table 2[11,12,21,36–39] shows the most relevant research in the field of radioactive and chemical toxins removal with the use of NPs.

As far as radioactive toxins are concerned, the use of chelators, such as calcium diethylenetriaminepentaacetate or Prussian Blue, is usually the preferred method of treatment after radiological contamination since chelators bind to radionuclides such as plutonium, uranium or cesium.[13,40] After direct administration, these substances bind to the toxins and are excreted from the body, thus reducing their concentration. Many other chelators have been developed in the past with, perhaps, better binding affinities than those used nowadays, but their implementation in clinical trials for detoxification systems has been impaired due to their toxicity.[40] Recent developments in

**Table 2.** Studies of MNPs Employed for Radioactive or Chemical Toxin Removal from Biological Solutions

| Material | Solution to be treated | Working conditions | Removal efficacy | Considerations | Refs. |
|---|---|---|---|---|---|
| $Fe_3O_4$-Dopamine-bisphosphonate (10 nm) | Water and blood contaminated with $UO_2^{2+}$ ions | *In vitro* experiments [MNPs] = 15 g·L$^{-1}$; [$UO_2^{2+}$] = 100 mg·L$^{-1}$ | Elimination of 99 and 69% of uranyl ions from water and blood respectively | The high affinity between the bisphosphonate and uranyl ions is due to the chelating effect. | [36] |
| Ni@$SiO_2$-4,4-difluoro-4-bora-3a,4a-diaza-s-indacene derivative (30–40 nm) | Water and human blood solutions contaminated with $Pb^{2+}$ | *In vitro* experiments [MNPs] = 10 g·L-1; [Pb2+] = 100 μg·L-1; Time: 30 minutes | 96% from blood solutions | When lead was adsorbed onto the particle surface, the functional group showed high fluorescent intensity, proportional to the amount of lead adsorbed. The fluorescent intensity was unchanged in the presence of other metal cations. | [37] |
| $Fe_3O_4$@PEI@PEG-2,2′-(phenylazanediyl) diacetic acid (50 nm, 58 emu·g$^{-1}$) | Water and human blood solutions contaminated with $Cd^{2+}$ | *In vitro* experiments [MNPs] = 0.2 g·L$^{-1}$; [$Cd^{2+}$] = 1 mg·L$^{-1}$; Time: 5–10 minutes (Room temperature) | 98% and 80% from water and blood solutions, respectively | Biocompatibility was confirmed with HeLa cells and blood cells (no hemolysis detected). Competition studies were performed | [38] |

| | | | | | |
|---|---|---|---|---|---|
| | | | | (adsorption of other metals was not significant with the exception of $Cu^{2+}$). | |
| $Fe_3C$@PEI-Diethylenetriaminepentaacetic acid and $Fe_3C$@PEG-Digoxin antibody FAB (30 nm, 140 emu·$g^{-1}$) | Human blood contaminated with $Pb^{2+}$ and digoxin | *In vitro* experiments $[Pb^{2+}] = 1.2$ mg·L-1; [Digoxin] = 5 $\mu$g·L-1; Volume: 40–60 mL; Time: 5 minutes | 56% of $Pb^{2+}$ when the particle dosage was 0.35 mg and 97% of digoxin when the particle dosage was 0.15 mg | Blood clotting and serum potassium, lactate dehydrogenase, bilirubin, and haptoglobin levels in plasma remained in the normal range. RBCs were unaffected by the process. Fe leaching was not detected. | [12] |
| $Fe_3C$@C@PEI-Iminodiacetic acid and Fe3C@C@PEG-Digoxin antibody FAB (50 nm, 123 emu·$g^{-1}$) | Rat's blood contaminated with $Pb^{2+}$ and digoxin | *In vivo* experiments [MNPs] = 0.5 g·$L^{-1}$ (injected at 0.75 mg·$min^{-1}$); Blood flow rate:1.5 ml·$min^{-1}$; $[Pb^{2+}] = 0.16$ mg·$Kg^{-1}$ of rat; [Digoxin] = 0.01 mg·$Kg^{-1}$ of rat; Total volume: 25 mL | Over 40% of the toxin was removed within the first 10 min. and over 75% was achieved after 40 minutes of treatment | Plasma toxin levels in untreated animals remained stable for 40 minutes with a decrease by 10% for $Pb^{2+}$ and 32% for digoxin. When the magnetic separation step was omitted after the adsorption stage for 15 minutes, analysis of lung and liver tissue showed significant accumulation of MNPs. | [11] |

(*Continued*)

Table 2. (*Continued*)

| Material | Solution to be treated | Working conditions | Removal efficacy | Considerations | Refs. |
|---|---|---|---|---|---|
| Fe@C@PEG-Digoxin antibody FAB (30 nm, 150 emu·g$^{-1}$) | Human blood contaminated with digoxin | *In vitro* experiments in a continuous mode [MNPs] = 0.4–0.5 g·L$^{-1}$; Blood flow rate:15 ml·min$^{-1}$; [Digoxin] = 4.45 $\mu$g·L$^{-1}$; Total volume: 0.5 L; Time: 2–2.5 minutes | 66–75% after one single pass through the device; 89% of removal after 1 hour of continuous mode operation | A small amount of MNPs remained in the blood after treatment (1%). | [21] |
| $Fe_3O_4$@3-amino-propyltriethoxysilane-$\beta$-cyclodextrin (196 nm, 63 emu·g$^{-1}$) | Water and plasma solutions contaminated with diazepam | *In vitro* experiments [MNPs] = 10 g·L$^{-1}$; [Diazepam] = 20 mg·L$^{-1}$; Time: 3 hours (37°C) | 74–80% from water and 50% from blood. At high diazepam concentration (40 mg·L$^{-1}$) the removal efficacy was 45% from water. | Biocompatibility was demonstrated with different test (hemolysis ratio, coagulation time and platelet adhesion). Activated carbon showed similar efficacy than the MNPs. At low pH values the adsorption was hindered. | [39] |

*Note*: PEI: Polyethyleneimine; PEG: Polyethylene glycol; RBCs: Red blood cells.

nanotechnology may mean that these chelators can be now utilized in a safe and effective way if permanently immobilized on nanoadsorbents through suitable linking chemistry that prevents unwanted loss of the surface ligands.

The first reported study focusing on the removal of radionuclides from a biological fluid by MNPs was carried out by Xu and coworkers in 2006. In this study,[36] the removal of uranyl ions from blood was conducted by the use of dopamine-bisphosphonate modified magnetite NPs. With this method, 99% and 69% of uranyl was removed from water and blood solutions, respectively, as seen in Table 2. This pivotal study opened the door to the application of functionalized magnetic nanomaterials for the selective and effective removal of radioactive and other toxic metals.

The removal of heavy metal cations such as cadmium or lead by MNPs has been reported in different studies. Due to the long biological half-life of these toxins (approximately 15–20 years),[38] their removal following contamination is of great concern. Although the separation of heavy metals from water matrices can be easily achieved through conventional methods, it is necessary to develop a safe and effective technique for the removal of these compounds from biological fluids. The use of MNPs might be an efficient technology for decorporation of these toxins. As seen in Table 2, Jin *et al.*[38] developed a biocompatible nanocomposite functionalized with 2,2′-(phenylazanediyl) diacetic acid for the removal of cadmium ions. *In vitro* studies showed that the reported material could remove 98% and 80% of $Cd^{2+}$ from water and blood, respectively.

The synthesis of functionalized fluorescent particles can be also used for the detection, recovery and removal of toxic metals from the human body. In this regard, Lee *et al.*[37] reported the use of a fluorescent receptor (4,4-difluoro-4-bora-3a,4a-diaza-s-indacene derivative) immobilized onto the surface of $Ni@SiO_2$ MNPs that selectively interacts with lead. It was observed that with increasing lead concentration in the particle suspension correlates with an increase of fluorescence intensity whereas no fluorescence was detected in the absence of lead in the solution. In this case, the interaction with lead promotes a large chelation-enhanced fluorescence effect, reversed by the desorption of the metal from the particle surface, making this material an ideal chemosensor for $Pb^{2+}$. The MNPs prepared were also tested for the removal of $Pb^{2+}$ from blood solutions. The experiments revealed that more than 96% of $Pb^2$ was removed from blood whereas similar efficacy was achieved when the metal was removed from water. Moreover, the unspecific adsorption of other biologically relevant metal ions was found to be minimal, making this material a highly selective probe for lead cations.

*In vivo* studies have been also developed in recent years showing the extended potential of MNPs in the separation of toxic metals from blood. For instance, in an experimental study by Herrmann *et al.*,[11] rats intoxicated with lead were treated with iminodiacetic acid functionalized MNPs in an extracorporeal system. The blood returned to the animal after the MNPs were separated by a magnetic field generated by a permanent magnet. With this technique, 75% of the toxin was removed in 40 minutes of external blood circulation as seen in Table 2.

The removal of small molecule drugs by MNPs has also been reported in recent years, indicating that these materials might be employed as hemoadsorbents in cases of drug abuse or overdose. For example, Herrmann and coworkers reported the removal of digoxin (cardiac drug) in different studies.[11,12,21] The authors used particles functionalized with digoxin antibody (anti-digoxin FAB), obtaining a removal efficacy of 75% for *in vitro* experiments.[21] In a later study,[11] *in vivo* experiments conducted with rats indicated that similar efficacy could be reached after extracorporeally treating blood for 40 minutes, demonstrating the applicability of this material for its *in vivo* use. As seen in Table 2, other representative studies have also been reported.

In cases of drug overdose, future particle functionalizations could be based on human proteins because of their high affinities to different drug molecules. For instance, digestive drugs have been reported to bind to plasma proteins post-intoxication.[39,41] Since plasma albumin acts as a transport protein that aids in drug distribution within the body, an attractive separation approach could involve its immobilization on the particle surface. Yu *et al.*[41] reported the use of

BSA-functionalized MNPs (as model for human serum albumin protein) that could be used for the removal of warfarin, ibuprofen or diazepam. Finally, it should be noted that the selective adsorption of other drugs like benzodiazepines or antidiabetics with MNPs from biological fluids (human urine or plasma) have been reported in the literature as an effective technology for analysis or diagnosis purposes.[42–46] Thus, the receptors reported in these studies could be the focus of further research in this field.

## III. Design of the Toxin Sequestration Process with MNPs

The design of the extracorporeal detoxification systems must take into account the following stages involved in the process: i) the adsorption of the toxins by the magnetic nanocomposites, ii) the separation of the material before the blood is returned to the circulatory system of the patient, and iii) the integration of the above-mentioned steps into one complete system. In the following subsections, some of the most important design considerations for these stages will be discussed.

### A. Design of the toxin sequestration stage

Although the majority of the studies reported above have been carried out in artificially contaminated blood, the few that involve the implementation of this technology into a feasible *in vivo* process have produced encouraging results. Herrmann and coworkers presented in 2013 the first *in vivo* blood purification system and demonstrated its ability to rapidly clear toxins from rat's blood with high toxin removal rates achieved in less than 1 hour of external circulation.[11] Kang *et al.*[23] have also tested the effectiveness of the pathogen decorporation process *in vivo*, obtaining high cleaning efficacies within 1 hour of external rodent blood adsorption. Despite successful studies in rats, human clinical trials have not been carried out yet, perhaps, due to the underlying uncertainty regarding possible health risks associated with this technology.

In order to improve understanding of the process and facilitate its implementation for human treatment, mathematical models aiming to describe the theoretical sequestration efficacy have been developed. For example, the extracorporeal detoxification process can be considered as a continuous process where blood is extracted from an ideally mixed reservoir (that mimics the patient's circulatory system) and is submitted to treatment. Once the process is concluded, the purified blood is recirculated into the reservoir after leaving the magnetic separation stage, as seen in Figure 5.

If constant removal efficacy for each cycle through the extracorporeal device is assumed, very simple models can be employed for estimating toxin removal. For instance, the change in concentration of chemical toxins (C) remaining over time can be described as follows[21]:

$$C = C_0 \cdot e - \frac{V_R \cdot R \cdot t}{V_{T \cdot \tau}} \quad (1)$$

where $t$ and $\tau$ are the time elapsed from the beginning and the cycle time, respectively, and $C_0$ is the initial concentration of the toxin in the tank. The ratio $t/\tau$ equals the number of cycles through the extracorporeal device. $V_R$ and $V_T$ are the extracorporeal and the total

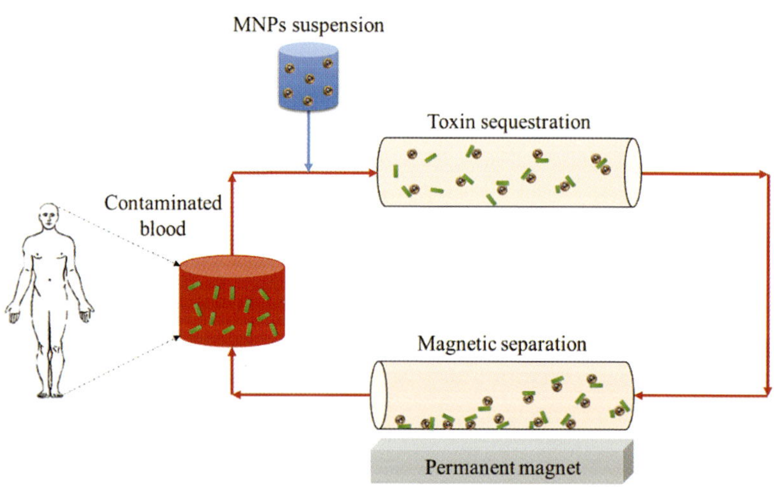


Figure 5. Schematic diagram of the extracorporeal blood purification process.

volume, respectively, and R denotes the cycle removal efficacy (the concentration of the toxin leaving the device relative to the concentration entering the device).

However, in clinically complex situations like sepsis and systemic inflammatory response syndrome, the continuous production of bacteria, endotoxins or cytokines while undergoing treatment hinders the process and increases the difficulty of modeling thereby impairing the design of the detoxification system. In case the target of the process is a biotoxin, the kinetic model that estimates the pathogen removal in a recirculated system can be calculated by the simplified mathematical model of Monod kinetics[23,25]:

$$\frac{dc}{dt} = C\left(\mu - \frac{F \cdot R}{V}\right) \tag{2}$$

where $\mu$ is the specific growth rate of the pathogen in the patient (e.g. 0.7 $h^{-1}$ for *C. albicans*, 0.13 $h^{-1}$ for *S. aureus*, etc.), $F$ is the blood flow rate, $R$ is the removal efficacy per single pass and $V$ denotes the blood volume.

Based on these models, the time required for achieving a desired toxin concentration can be estimated when the geometric characteristics of the device and the per-pass toxin removal are known. Moreover, the process dynamics can be predicted with good accuracy since the theoretical values have shown to be in good agreement with the experimental results.[21,23] Nevertheless, more studies might be required in order to more accurately describe the per-pass toxin removal and the effect of different variables of interest on the process.

Apart from modeling considerations there are other important technical details that should be carefully taken into account in the design of the adsorption stage. For example, particle suspensions that will be mixed with the biological solution during the treatment should be biocompatible and previously decontaminated in order to avoid any possible side effects. In this case, various products based on proteins or biopolymers that have been approved for intravenous injection may be applied as dispersants.[21] It should be noted that the material concentration in the initial dispersion has to be kept as high as possible in order to avoid blood dilution during adsorption (i.e. volume added to the blood solution inside the purification device must be the minimum possible). Moreover, the suspension should be stable and the particles must remain well-dispersed in the solution. This could be achieved through sonication, a method which has been presented as a valid alternative for avoiding particle agglomeration in industrial processes. However, if used for this application, this technique should be applied before the onset of the decontamination process (i.e. prior to the mixing with blood solutions) since it strongly affects blood integrity (hemolysis).[12] Finally, given that the particle suspension is administrated in a continuous mode for this stage of the process, the saturation of the material should not present an issue because fresh material with vacant binding sites is continuously injected into the extracorporeal device.

## B. Design of the magnetic separation stage

The toxicity and the particular long-term effects of most nanomaterials are not yet fully understood.[20] Hence, the nanomaterials have to be reliably separated post-decontamination, limiting exposure time. When toxins are attached to the beads, the addition of a magnetic separation stage is necessary prior to the return of the treated blood to the patient's circulatory system. The toxins bound to particles can be later disposed or submitted for assay or forensics.

Several magnetic separator designs have been proposed in the last years for performing this task. These are commonly batch separators where the particles are separated from flowing blood by a magnet located next to the channel. This magnet generates local gradients that promote the accumulation of the beads on the walls of the channel next to the magnet, leading to a clean solution at the outlet. This process can be repeated several times until the desired separation efficacy is reached. After all the blood to be treated passes through the channel, the permanent magnet is removed and the particles are retrieved from the walls. With this design, particle separation efficacies higher than 99% have been achieved from flowing blood at flow rates comparable to those observed in large veins.[19]

However, recent studies have introduced continuous separators where the particles are deflected from the blood stream and collected into a flowing buffer stream by a magnetic gradient applied perpendicularly to the flow direction. The concept of continuous separators is illustrated in Figure 6.

The use of continuous systems poses several key advantages compared to batch magnetic separators. Firstly, when using batch separators, the accumulation of beads on the walls restricts fluid flow, thus decreasing the efficiency and capacity of the separator.[47] This becomes particularly important when large volumes of blood have to be treated (e.g. 4.5 L for the human body) or when the concentration of the particles is high. Batch separators may also enable the non-specific entrapment of blood components in the capturing regions, degrading the quality of the treated solution. The effect might even be deleterious in certain cases of excessive blood cell entrapment. On the contrary, these issues can be overcome by the use of continuous separators. An additional advantage of continuous separators is the fact that the nature of the process enables a high level of the integration with previous processes (i.e. the adsorption of toxins) leading to a continuous detoxification system. Therefore, their use not only

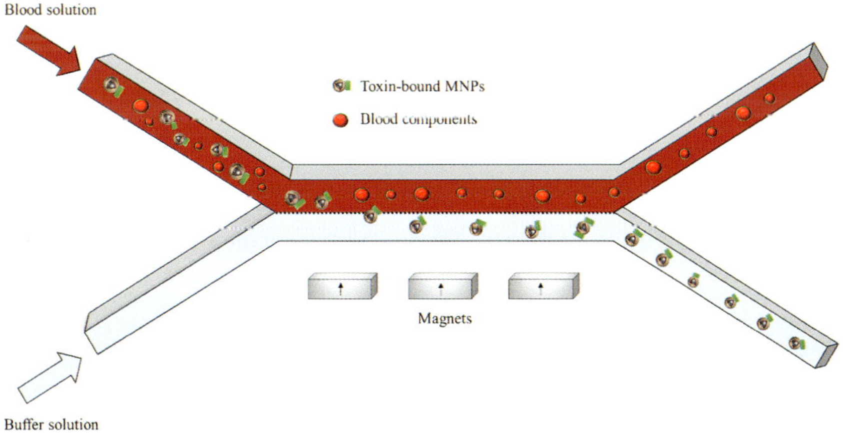


Figure 6. Schematic diagram of the continuous-flow magnetic separation. Taken from Ref. [23] with permission of Nature publication group.

increases separation efficacy and throughput but also allows the integration of both stages into a seamless treatment scheme.

The application of continuous magnetic separators for blood detoxification has been implemented by Kang *et al.*[23] In this study, the adsorption process and the subsequent magnetic separation were carried out in a continuous mode in an *in vivo* animal model. The 128 nm-MNP initial recovery in their magnetic separator found to be approximately 80%. However, 100% particle recovery rates were achieved when 1 $\mu$m magnetic particles were introduced in the continuous separator, acting as local magnets where the nanosized adsorbents (plus the adsorbed toxins) were adsorbed and separated from the clean blood solution.

Various mathematical models describing the magnetic separation of MNPs in different devices have been developed in the last years. These models are based on the sum of forces acting on the particles when the separation takes place.[47–49] In general, the main forces that act on a particle suspended in a fluid under the influence of an external magnetic field are the magnetic (that favors the separation) and the fluidic drag force (that hinders the process). Based on such models, separator design can be carried out in advance of fabrication with the predicted results which accurately emulate experimental results.

## IV. Conclusions

In a variety of clinical situations, including intoxications, severe inflammatory response syndrome or sepsis, the direct removal of the disease-causing agents from biological fluids constitutes the most effective treatment option.[12] Compared to other conventional techniques for toxin sequestration, the use of functionalized MNPs enables higher throughput and requires less specialized requirement while preserving blood integrity.[22] Thus, the adsorption of toxins by affinity adsorbents represents the best technology for direct toxin removal.

One viable option would be the use of porous adsorbents packed in cartridges but it has shown a slow kinetics especially for large biotoxin structures.[31] For these porous materials, adsorption takes place only at the surface and therefore the binding sites inside the porous matrices remain unused. This drawback can be overcome by the use of nanosized non-porous materials where all the active sites are available at the surface, resulting in a higher overall adsorption capacity and faster kinetics than what is observed in macroscopic adsorption columns. Hence, MNPs remain the preferred adsorbent materials since they enjoy a high specific surface area and no pore diffusion hinders the accessibility of the surface binding sites.[3,4]

The ability of MNPs to capture different toxins and restrict pathogen growth from biological fluids depends on the specific toxin and its levels, the surface functionality of the adsorbent, its morphological characteristics (size, antibody loading, etc.) and its concentration. Nevertheless, the separation of these toxins by magnetic forces is a key distinguishing feature of this technology and could present even more advantages in the near future as this technology continues to develop.

Different studies have demonstrated the multifunctional features of these materials. For instance, the particles employed for biotoxin removal can also be applied in a bactericidal role and simultaneously achieve efficient toxin adsorption. Similarly, engineered surface receptors on the particles possess the ability to remove different pathogens and toxins without first identifying the infectious agent. The adsorption capacity and kinetics observed in different studies make this technology a very promising treatment strategy for multiple intoxications and infections since the disease-causing agent is effectively removed within minutes. Nonetheless, some of the developed materials bind the toxins through electrostatic and hydrophobic interactions, thus, the unspecific adsorption of blood constituents might be expected. This issue should be addressed with further studies in order to evaluate the possible side effects of the treatment.

Additionally, compatibility studies have been carried out by several groups, showing, in most cases, that the toxicity of the materials tested was low. Still, the possible long-term effects of the particles in biological systems should be the focus of future study since there seems to be a disequilibrium between the reported synthesis of new materials and their possible toxic effects on biological systems.

Regarding future clinical applications, magnetic separation efficacy remains crucial since the safe implementation of this novel technology relies upon a minimal risk policy which dictates avoiding direct exposure to MNPs and detailed risk evaluation.[11] There exist different separator types for the separation of nanosized magnetic particles from fluid solutions. As mentioned above, continuous flow magnetic separators are preferred over batch separators owing to the advantages that they offer. Regardless of the type employed, the magnetic separator must fulfill two fundamental requirements: i) the complete recovery of the magnetic beads from the blood solution and, ii) the preservation of biological fluid integrity. In addition, it is especially important to evaluate the biological consequences of MNPs entering circulation in order to gain insight and palliate possible unwanted effects.

Finally, it should be noted that blood purification systems based on MNPs can be combined with other conventional treatments. Successful implementation of different therapeutic methodologies could be beneficial for the future treatment of blood infections and intoxications. In this regard, a biomedical extracorporeal device based on the use of MNPs that act as capture agents of pathogens and toxins can be used in combination with conventional therapies such as the administration of antibiotics leading to higher overall treatment efficacy.

## Acknowledgments

Financial support from the Spanish Ministry of Economy and Competitiveness under the project CTQ2012-31639 (FEDER 2007-2013) is gratefully acknowledged. Jenifer Gómez-Pastora also thanks the FPI postgraduate research grant (BES-2013-064415). María Lázaro-Díez holds a contract from Valdecilla Biomedical Research Institute (IDIVAL) and University of Cantabria (PREVAL16/05). José Ramos-Vivas holds a Miguel Servet II contract for Young Researchers from Institute of Health Carlos III (PI13/01310).

## References


1. Gao, Z.; Ma, T.; Zhao, E.; Docter, D.; Yang, W.; Stauber, R. H.; Gao, M. Small is smarter: Nano MRI contrast agents – Advantages and recent achievements. *Small* **2016,** *12* (5), 556–576.
2. Kandasamy, G.; Maity, D. Recent advances in superparamagnetic iron oxide nanoparticles (SPIONs) for *in vitro* and *in vivo* cancer nanotheranostics. *Int. J. Pharm.* **2015**, *496* (2), 191–218.
3. Herrmann, I. K. How nanotechnology-enabled concepts could contribute to the prevention, diagnosis and therapy of bacterial infections. *Crit. Care* **2015**, *19*, 239–243.
4. Herrmann, I. K.; Schlegel, A. A.; Graf, R.; Stark, W. J.; Beck-Schimmer, B. Magnetic separation-based blood purification: A promising new approach for the removal of disease-causing compounds? *J. Nanobiotechnology* **2015**, *13* (1), 49–52.
5. Ruiz, A.; Morais, P. C.; Bentes de Azevedo, R.; Lacava, Z. G. M.; Villanueva, A.; Morales, M.P. Magnetic nanoparticles coated with dimercaptosuccinic acid: Development, characterization, and application in biomedicine. *J. Nanopart. Res.* **2014**, *16*, 2589–2609.
6. Xu, C.; Sun, S. New forms of superparamagnetic nanoparticles for biomedical applications. *Adv. Drug Deliver. Rev.* **2013**, *65*, 732–743.
7. Peng, H.; Liu, X.; Wang, G.; Li, M.; Bratlie, K. M.; Cochran, E.; Wang, Q. Polymeric multifunctional nanomaterials for theranostics. *J. Mater. Chem. B* **2015**, *3*, 6856–6870.
8. Modugno, G.; Ménard-Moyon, C.; Prato, M.; Bianco, A. Carbon nanomaterials combined with metal nanoparticles for theranostic applications. *Brit. J. Pharmacol.* **2015**, *172*, 975–991.
9. Nazir, S.; Hussain, T.; Ayub, A.; Rashid, U.; MacRobert, A. J. Nanomaterials in combating cancer: Therapeutic applications and developments. *Nanomedicine: NBM* **2014**, *10*, 19–34.
10. Zhang, Z.; Yu, L.; Xu, L.; Hu, X.; Li, P.; Zhang, Q.; Ding, X.; Feng, X. Biotoxin sensing in food and environment via microchip. *Electrophoresis* **2014**, *35*, 1547–1559.
11. Herrmann, I. K.; Schlegel, A.; Graf, R.; Schumacher, C. M.; Senn, N.; Hasler, M.; Gschwind, S.; Hirt, A. M.; Gunther, D.; Clavien, P. A.;

Stark, W. J.; Beck-Schimmer, B. Nanomagnet-based removal of lead and digoxin from living rats. *Nanoscale* **2013**, *5*, 8718–8723.
12. Herrmann, I. K.; Urner, M.; Koehler, F. M.; Hasler, M.; Roth-Z'Graggen, B.; Grass, R. N.; Ziegler, U.; Beck-Schimmer, B.; Stark, W. J. Blood purification using functionalized core/shell nanomagnets. *Small* **2010**, *6* (13), 1388–1392.
13. Kaminski, M. D.; Rosengart, A. J. Detoxification of blood using injectable magnetic nanospheres: A conceptual technology description. *J. Magn. Magn. Mater.* **2005**, *293*, 398–403.
14. Mertz, C. J.; Kaminski, M. D.; Xie, Y.; Finck, M. R.; Guy, S.; Rosengart, A. J. *In vitro* studies of functionalized magnetic nanospheres for selective removal of a simulant biotoxins. *J. Magn. Magn. Mater.* **2005**, *293*, 572–577.
15. Rimmelé, T.; Kellum, J. A. Clinical review: Blood purification for sepsis. *Crit. Care* **2001**, *15*, 205–214.
16. Kumar, C. S. S. R.; Hormes, J.; Leuschner, C. (2005). *Nanofabrication Towards Biomedical Applications: Techniques, Tools, Applications and Impact*. Wiley-VCH, Weinheim, Germany.
17. Lu, A. H.; Salabas, E. L.; Schüth, F. Magnetic nanoparticles: synthesis, protection, functionalization and application. *Angew. Chem. Int. Ed.* **2007**, *46*, 1222–1244.
18. Khashan, S. A.; Furlani, E. P. Scability analysis of magnetic bead separation in a microchannel with an array of soft magnetic elements in a uniform magnetic field. *Sep. Purif. Technol.* **2014**, *125*, 311–318.
19. Schumacher, C. M.; Herrmann, I. K.; Bubenhofer, S. B.; Gschwind, S.; Hirt, A. M.; Beck-Schimmer, B.; Günther, D.; Stark, W. J. Quantitative recovery of magnetic nanoparticles from flowing blood: Trace analysis and the role of magnetization. *Adv. Funct. Mater.* **2013**, *23*, 4888–4896.
20. Herrmann, I. K.; Grass, R. N.; Stark, W. J. High-strength metal nanomagnets for diagnostics and medicine: Carbon shells allow long-term stability and reliable linker chemistry. *Nanomedicine* **2009**, *4* (7), 787–798.
21. Herrmann, I. K.; Bernabei, R. E.; Urner, M.; Grass, R. N.; Beck-Schimmer, B.; Stark, W. J. Device for continuous extracorporeal blood purification using target-specific metal nanomagnets. *Nephrol. Dial. Transplant* **2011**, *26*, 2948–2954.

22. Bromberg, L.; Chang, E. P.; Hatton, T. A.; Concheiro, A.; Magariños, B.; Alvarez-Lorenzo, C. Bactericidal core-shell paramagnetic nanoparticles functionalized with poly(hexamethylene biguanide). *Langmuir* **2011**, *27* (1), pp. 420–429.
23. Kang, J. H.; Super, M.; Yung, C. W.; Cooper, R. M.; Domansky, K.; Graveline, A. R.; Mammoto, T.; Berthet, J. B.; Tobin, H.; Cartwright, M. J.; Watters, A. L.; Rottman, M.; Waterhouse, A.; Mammoto, A.; Gamini, N.; Rodas, M. J.; Kole, A.; Jiang, A.; Valentin, T. M.; Diaz, A.; Takahashi, K.; Ingber, D. E. An extracorporeal blood-cleansing device for sepsis therapy. *Nat. Med.* **2014**, *20* (10), 1211–1216.
24. Herrmann, I. K.; Urner, M.; Graf, S.; Schumacher, M.; Roth-Z'graggen, B.; Hasler, M.; Stark, W. J.; Beck-Schimmer, B. Endotoxin removal by magnetic separation-based blood purification. *Adv. Healthc. Mater.* **2013**, *2*, 829–835.
25. Yung, C. W.; Fiering, J.; Mueller, A. J.; Ingber, D. E. Micromagnetic–microfluidic blood cleansing device. *Lab Chip* **2009**, *9* (9), 1153–1308.
26. Bromberg, L.; Chang, E. P.; Alvarez-Lorenzo, C.; Magariños, B.; Concheiro, A.; Hatton, T. A. Binding of functionalized paramagnetic nanoparticles to bactericidal lipopolysaccharides and DNA. *Langmuir* **2010**, *26* (11), 8829–8835.
27. Bosch, T. Recent advances in therapeutic apheresis. *J. Artif. Organs* **2003**, *6*, 1–8.
28. Lee, J. J.; Jeong, K. J.; Hashimoto, M.; Kwon, A. H.; Rwei, A.; Shankarappa, S. A.; Tsui, J. H.; Kohane, D. S. Synthetic ligand–coated magnetic nanoparticles for microfluidic bacterial separation from blood. *Nano Lett.* **2014**, *14*, 1–5.
29. Liu, F.; Mu, J.; Wu, X.; Bhattacharjya, S.; Yeow, E. K. L.; Xing, B. Peptide–perylene diimide functionalized magnetic nano–platforms for fluorescent turn–on detection and clearance of bacterial lipopolysaccharides. *Chem. Commun.* **2014**, *50* (47), 6200–6203.
30. Niemirowicz, K.; Swiecicka, I.; Wilczewska, A. Z.; Markiewicz, K. H.; Surel, U.; Kułakowska, A.; Namiot, Z.; Szynaka, B.; Bucki, R.; Car, H. Growth arrest and rapid capture of select pathogens following magnetic nanoparticle treatment. *Colloid. Surface. B* **2015**, *131*, 29–38.
31. Petsch, D.; Anspach, F. B. Endotoxin removal from protein solutions. *J. Biotechnol.* **2000**, *76*, 97–119.

32. Vetten, M. A.; Yah, C. S.; Singh, T.; Gulumian, M. Challenges facing sterilization and depyrogenation of nanoparticles: Effects on structural stability and biomedical applications. *Nanomedicine: NBM* **2014**, *10*, 1391–1399.
33. Buttenschoen, K.; Radermacher, P.; Bracht, H. Endotoxin elimination in sepsis: Physiology and therapeutic application. *Langenbeck. Arch. Surg.* **2010**, *395*, 597–605.
34. Hirayama, C.; Sakata, M. Chromatographic removal of endotoxin from protein solutions by polymer particles. *J. Chromatogr. B* **2002**, *781*, 419–432.
35. Anspach, F. B. Endotoxin removal by affinity sorbents. *J. Biochem. Bioph. Meth.* **2001**, *49*, 665–681.
36. Wang, L.; Yang, Z.; Gao, J.; Xu, K.; Gu, H.; Zhang, B.; Zhang, X.; Xu, B. A biocompatible method of decorporation: Bisphosphonate-modified magnetite nanoparticles to remove uranyl ions from blood. *J. Am. Chem. Soc.* **2006**, *128*, 13358–13359.
37. Lee, H. Y.; Bae, D. R.; Park, J. C.; Song, H.; Han, W. S.; Jung, J. H. A selective fluoroionophore based on BODIPY-functionalized magnetic silica nanoparticles: Removal of $Pb^{2+}$ from human blood. *Angew. Chem. Int. Ed.* **2009**, *48*, 1239–1243.
38. Jin, J.; Yang, F.; Zhang, F.; Hu, W.; Sun, S. B.; Ma, J. 2, 2'-(phenylazanediyl) diacetic acid modified $Fe_3O_4$@PEI for selective removal of cadmium ions from blood. *Nanoscale* **2012**, *4*, 733–736.
39. Cai, K.; Li, J.; Luo, Z.; Hu, Y.; Hou, Y.; Ding, X. $\beta$-Cyclodextrin conjugated magnetic nanoparticles for diazepam removal from blood. *Chem. Commun.* **2011**, *47*, 7719–7721.
40. Chen, H.; Kaminski, M. D.; Liu, X.; Mertz, C. J.; Xie, Y.; Torno, M. D.; Rosengart, A. J. A novel human detoxification system based on nanoscale bioengineering and magnetic separation techniques. *Med. Hypotheses* **2007**, *68*, 1071–1079.
41. Yu, C. H.; Al-Saadi, A.; Shih, S. J.; Qiu, L.; Tam, K. Y.; Tsang, S. C. Immobilization of BSA on silica-coated magnetic iron oxide nanoparticle. *J. Phys. Chem. C* **2009**, *113*, 537–543.
42. Asgharinezhad, A. A.; Ebrahimzadeh, H.; Mirbabaei, F.; Mollazadeh, N.; Shekari, N. Dispersive micro-solid-phase extraction of benzodiazepines

from biological fluids based on polyaniline/magnetic nanoparticles composite. *Anal. Chim. Acta* **2014**, *844*, 80–89.

43. De Souza, K. C.; Andrade, G. F.; Vasconcelos, I.; Viana, I. M. O.; Fernandes, C.; de Sousa, E. M. B. Magnetic solid-phase extraction based on mesoporous silica-coated magnetic nanoparticles for analysis of oral antidiabetic drugs in human plasma. *Mater. Sci. Eng. C* **2014**, *40*, 275–280.
44. Azodi-Deilami, S.; Abdouss, M.; Asadi, E.; Najafabadi, A. H.; Sadeghi, S.; Farzaneh, S.; Asadi, S. Magnetic molecularly imprinted polymer nanoparticles coupled with high performance liquid chromatography for solid-phase extraction of carvedilol in serum samples. *J. Appl. Polym. Sci.* **2014**, *41209*, 1–10.
45. Azodi-Deilami, S.; Najafabadi, A. H.; Asadi, E.; Abdouss, M.; Kordestani, D. Magnetic molecularly imprinted polymer nanoparticles for the solid-phase extraction of paracetamol from plasma samples, followed its determination by HPLC. *Microchim. Acta* **2014**, *181*, 1823–1832.
46. Morales-Cid, G.; Fekete, A.; Simonet, B. M.; Lehmann, R.; Cardenas, S.; Zhang, X.; Valcarcel, M.; Schmitt-Kopplin, P. In situ synthesis of magnetic multiwalled carbon nanotube composites for the clean-up of (fluoro)quinolones from human plasma prior to ultrahigh pressure liquid chromatography analysis. *Anal. Chem.* **2010**, *82* (7), 2743–2752.
47. Wu, X.; Wu, H.; Hu, Y. Enhancement of the separator efficiency on continuous magnetophoresis by utilizing L/T-shaped microchannels. *Microfluid. Nanofluid.* **2011**, *11*, 11–22.
48. Furlani, E. P.; Sahoo, Y.; Ng, K. C.; Wortman, J. C.; Monk, T. E. A model for predicting magnetic particle capture in a microfluidic bioseparator. *Biomed. Microdevices* **2007**, *9*, 451–463.
49. Furlani E.P. Magnetic biotransport: Analysis and applications. Materials **2010**, *3*, 2412–2446.

# Chapter 7

# Membranes for Controlled Release

Vida Araban*, Neda Aslankoohi†, and Mohammad Raoufi‡

*Faculty of Pharmacy, Tehran University of Medical Science Tehran 1417614411, Iran

†Institute of Biomaterial Science, Helmholtz-Zentrum Geesthacht, Kantstrasse 55, 14513 Teltow, Germany

‡Nanotechnology Research Center, Faculty of Pharmacy, Tehran University of Medical Science, Tehran 1417614411, Iran

## Abbreviations

| | |
|---|---|
| PLA | poly(lactic acid) |
| PEVA | poly(ethylene-co-vinyl acetate) |
| DCA | dichloroacetate |
| PLGA | poly(d, l-lactic-co-glycolide) |
| PDO | poly(dioxanone) |
| CiH | ciprofloxacin hydrochloride |
| TCH | tetracycline hydrochloride |
| PCL | polycaprolactone |
| BSA | bovine serum albumin |
| PTX | paclitaxel |
| Dox | doxorubicin hydrochloride |
| NGF | human-nerve growth factor |
| PC12 | pheochromocytoma cells |
| VEGF | vascular endothelial growth factor |
| PLCL | poly(l-lactic acid-co-$\varepsilon$-caprolactone) |
| PLLA | poly L-lactic acid |
| PEG | poly(ethyleneglycol) |
| PEI | poly(ethyleneimine) |
| PLL | poly(L-lysine) |
| LbL | layer-by-layer |
| CpG | cytosine-phosphate diester-guanine-rich |
| PBS | phosphate buffered saline |

## I. Introduction

The drug delivery research area has received tremendous scientific interest and numerous research works have been carried out in recent years as a result of increasing need for multifunctional systems to deliver therapeutic agents to specific sites and release them in a controlled manner for a desired period of time. Controlled release of drugs can enhance their bioavailability, therapeutic efficacy, and also reduce toxic effects.

In this chapter, membrane-like drug delivery systems are going to be discussed. The term "membrane" is a general concept and covers different areas such as polymeric films, fibrous matrices, vesicles and particle membranes, and also nanoporous inorganic membranes such as alumina, titania, and porous silicon templates. In this chapter, the focus of attention is given to polymeric electrospun fibrous membranes and fabricated membranes prepared by layer-by-layer assembly method.

## II. Electrospun Fibrous Membranes

### A. Introduction to electrospinning

Submicron polymer fibers are different from micro-diameter ones in several aspects like mechanical properties and their high surface area to volume ratio. In this regard, electrospinning of polymers is considered as a promising method for fabricating ultrafine fibers with favorable properties. Electrospun fibers are being used for tissue engineering and controlled drug delivery purposes.

Their potential in delivery of chemical and bioactive molecules such as proteins, growth factors, and nucleic acids has made them an interesting topic for more researches. Most used polymers could be electrospun in solvents, and some of them can form fibers by using melting method.[1]

### B. Electrospinning setup

An electrospinning apparatus consists of three main parts: a syringe pump, a high voltage source, and a collector (Figure 1). High voltage is applied between the reservoir of polymer in solution or in its melted form and the collector. In principle, surface tension prevents polymer solution flowing out of the syringe tip. The electric potential overcomes surface tension and forms a liquid jet. As the liquid jet travels to the collector, its solvent evaporates and fibers are formed. Several parameters can influence electrospinning process and transformation of polymer solution or melt form to ultrafine fibers. There are several contributory factors to electrospinning process that can affect its final results. Polymer reservoir parameters such as surface tension, viscosity and conductivity, ambient parameters such as temperature, humidity,

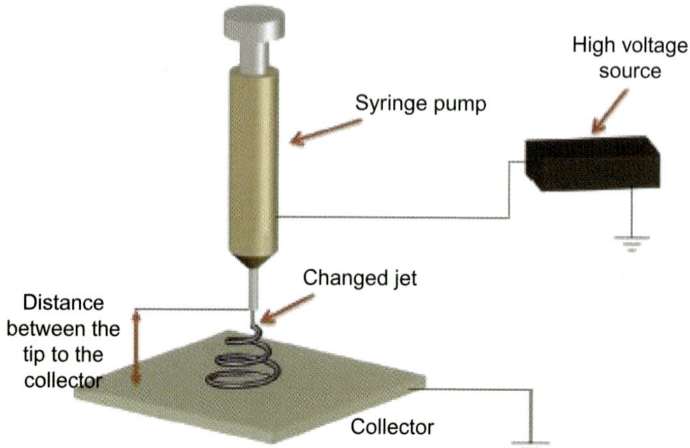


Figure 1. Schematic diagram of electrospinning setup. Taken from Ref. [8] with permission of Dovepress.

air velocity, apparatus-related factors such as applied electrical field strength, and needle and collector properties, distance between the syringe tip and collector are some determining factors in the electrospinning process.[1–5]

Deposition of electrospun fibers on the collector usually results in randomly oriented mats; however, different techniques have been employed to fabricate well-aligned and highly ordered structures. Using rotating disks or wire drums with desired speed, instead of normal flat collectors, can lead to membranes with aligned nanofibers.[1,6,7]

## C. Electrospun fibers in drug delivery

Several properties of electrospun fibers have make them promising candidates in drug delivery field and tissue engineering. High surface to volume ratio, considerable encapsulation and loading capacity, their relative cost-effective fabrication and scaling-up potential for industrial purposes, and also the possibility to design them with similar topological architectures to native ECM components are among the most important characteristics of electrospun fibers. By incorporation of biomolecules and drugs in electrospun fibers, several properties of drugs can be adjusted like their stability and degradation, solubility, bioavailability, and release profile. On the other hand, this incorporation can also affect fiber properties. And this should be taken into consideration in studies.

The first study on using electrospun fiber mats for drug delivery was done by Kenawy *et al.* in 2002. In this study, tetracycline hydrochloride was incorporated in poly(lactic acid) (PLA), poly(ethylene-co-vinyl acetate) (PEVA), and also a 50:50 blend of the two.[9]

Up to now, various bioactive molecules ranging from antibiotics and anticancer agents to proteins, nucleic acids, and growth factors have been incorporated in electrospun fibers.

Different electrospinning methods have been developed to reach desired physicochemical properties of fibers and also to incorporate biomolecules and drugs for biomedical applications.

Blend electrospinning, coaxial electrospinning (co-electrospinning), side-by-side electrospinning, multi-jet electrospinning, emulsion electrospinning, melt electrospinning are examples of some studied methods. Drugs and biomolecules can also be incorporated on electrospun fibers using chemical immobilization or physical attachment to their surface.[10,12,25]

### 1. *Co-electrospinning*

In order to fabricate core–shell fibers, co-electrospinning method has been developed. In this method, core–shell fibers can be reached by either using different polymer reservoirs and a core–shell nozzle or using an emulsion and a single conventional nozzle. Actually, coaxial electrospinning is a developed electrospinning system based on modification of syringe part of a traditional electrospinning setup. By coaxial electrospinning, different polymer solutions can be electrospun simultaneously and coaxially through different capillary channels produced by concentric spinnerets with different diameters, and fibers with a core–shell structure can be produced. Coaxial fibers exhibit controlled release of their encapsulated drugs and prevent initial burst release. Their high encapsulation efficacy, resistant in harsh environment, and low cost of their fabrication make them attractive for drug delivery purpose. Their core–shell structure also brings about the feasibility of multi-drug delivery systems. In addition, the shell protects temporally the core ingredient from the biological environment and helps the drugs remain bioactive.

Dan *et al.* fabricated hollow nanofiber for the first time by using co-electrospinning. Hollow nanofibers are useful for both drug delivery and tissue engineering. Co-electrospinning is preferred for fabricating long nanoscale hollow fibers.[13]

Numerous studies have used co-electrospinning for producing drug-loaded nanofibers.[9,14,15] Both hydrophobic and hydrophilic drugs have been incorporated in electrospun fibrous mats with this method. Studies have shown that dissolution of drug in polymer solution can affect morphology and diameter of fibers. For instance, Kim *et al.*[14] has incorporated an antibiotic (Mefoxin) in PLGA and PLGA/PEG-b-PLA/PLA by co-electrospinning method. They observed that mixing 5 wt% of the drug with polymer solution resulted in continuous fibers, while polymer solution alone produced fibers with beads. According to the authors, this effect is for increased conductivity of the solution that contains drug. Moreover, incorporation of drug resulted in fiber diameter decrease from $360 \pm 220$ nm to $260 \pm 90$ nm.[14] Dissolution of drugs in different solvents and applied electric field for electrospinning might influence the bioactivity of the drug. These probable effects necessitate performing appropriate analyses in order to control the bioactivity of the drug after co-electrospinning process. Electrospinning can be used for incorporating different drugs and bioactive molecules like anticancer agents in polymer fibers. In a study done by Xu *et al.*,[16] it was shown that by incorporating the anticancer drug BCNU in PEG-PLLA fibers, the activity of the drug can be maintained for 72 hours comparable to 48 hours for free BCNU. Using PLLA as carrier material, Liu *et al.*[17] fabricated doxorubicin fibrous mats which have showed inhibitory effect on tumor growth of secondary hepatic carcinoma. In another study done by Xie *et al.*, paclitaxel as an anti-cancer drug was embedded in electrospun PLGA mats with more than 90% encapsulation efficiency and controlled release profile obtained up to 61 days.[18] Liu *et al.*[17] used PLA and dichloroacetate (DCA) to fabricate electrospun mats by co-electrospinning and covered solid tumors of cervical carcinoma in mice with it. Necrosis of tumor cells and significant decrease in the volume and weight of the tumors, up to the point of disappearance of the tumor in half of mice, in 19 days was obtained by DCA-loaded electrospun mats.[19] Xie *et al.* embedded cisplatin in PLA/PLGA (30/70) electrospun fibers and achieved more than 90% encapsulation efficiency and sustained release over more than 75 days without significant initial burst release.[20] These studies exhibit good examples of anticancer drug-loaded electrospun fibrous mats and controlled local release of therapeutic agent from them, especially for the prevention of post-surgery tumor recurrence.[27]

Sustained release of antibiotics in a controlled manner through electrospun fibers could be used in prophylaxis of several infectious problems such as postoperative infections.[21] Coaxial electrospinning can be used for making core–shell nanofibers[22] (Figure 2). Coaxial

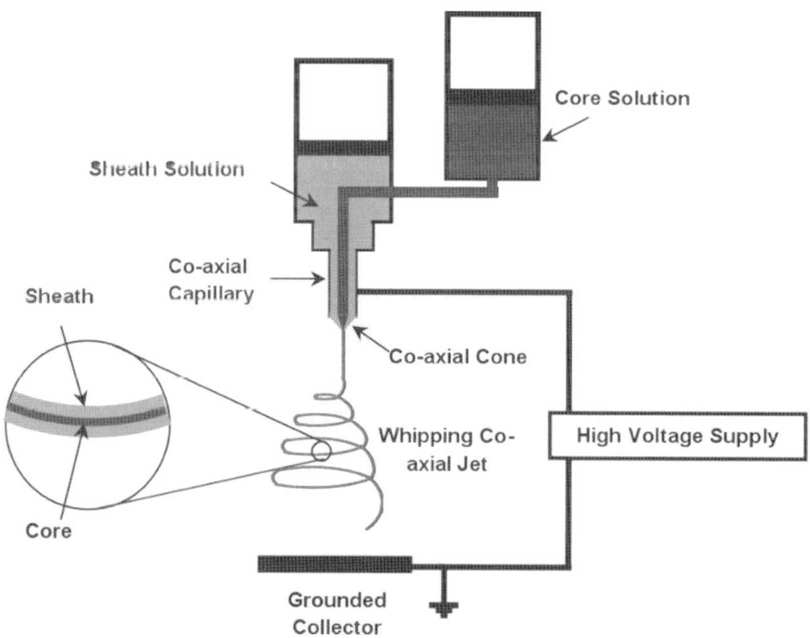


Figure 2. Coaxial electrospinning. Taken from Ref. [47] with permission of Taylor & Francis.

electrospinning is a suitable method for encapsulating fragile molecules and water-soluble biomolecules such as DNA, growth factors, and even living organisms.[25–28] In this method, direct contact of drug or agent to harsh conditions such as organic solvents have been eliminated and the shell layer not only protects the fragile agent but also inhibits premature release of water soluble drug or agent.[23]

## 2. *Emulsion electrospinning*

Core–shell structures can be also produced by emulsion electrospinning.[29] (Figure 3) In this process, the emulsion spheres are elongated with the polymer solution under the applied electric field. Core–shell structure can alleviate initial burst release of drug and shell can protect bioactivity of proteins or drugs in the core.[30]

Xu *et al.*[31] produced PEG-PLLA fibers containing Doxorubicin hydrochloride (Dox), a water-soluble anticancer agent, through emulsion electrospinning. The core–shell structure reduced the initial

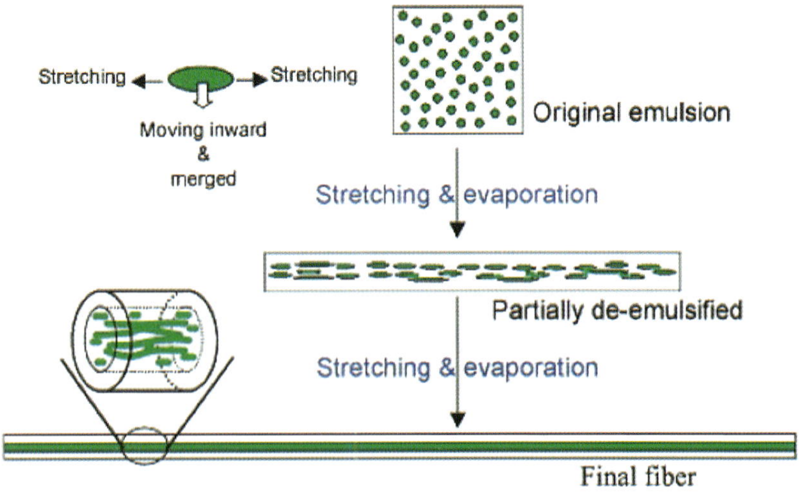


Figure 3. Schematic mechanism for the formation of core–sheath fibers through emulsion electrospinning. Taken from Ref. [29] with permission of Wiley and Sons.

burst release. In another study, a hydrophobic and a hydrophilic drug, paclitaxel (PTX) and doxorubicin hydrochloride (DOX), were successfully incorporated into PEG–PLA nanofiber mats through emulsion electrospinning. High hydrophilicity of DOX resulted in its higher release rate and also accelerated the release rate of PTX.[32] In another study done by Yang *et al.*[33] lysozyme, as a model protein, was incorporated in poly(DL-lactide) fibers. These fibers were prepared by emulsion electrospinning method and had bead-free, porous core–shell structure. It was shown that electrospinning process did not induce significant denaturation of protein. However, protein incorporation resulted in faster degradation rate of polymeric material.[34]

Emulsion electrospinning can also be used for encapsulation of drug reservoir in fibers. Hangxu *et al.* loaded bovine serum albumin (BSA) in Ca-alginate microspheres as a drug reservoir for hydrophilic drugs and encapsulated them in electrospun fibers of PLLA. *In vitro* release results showed that BSA was retained in fibers after 120 hours, while in naked BSA-loaded microspheres complete release occurred after 10 hours.[35] In an other study, (R)-(+)- limonene, a highly volatile fragrance, encapsulated in poly(venylalcohol) matrix by emulsion electrospinning showed 15-days sustained release of fragrance. This result shows the potential of using nanofibers and this method in variety of fields such as cosmetics and cosmeceuticals.[36] Emulsion electrospin ning can also be used in sustained release of growth factors in tissue engineering. In one study, vascular endothelial growth factor (VEGF) was encapsulated into poly(l-lactic acid-co-ε-caprolactone) (PLCL) nanofiber core by emulsion electrospinning. Controlled release of VEGF through this core–shell nano fibers has made this structure promising for future use as tissue regeneration substrates and scaffolds such as for cardiac tissue engineering.[30]

## D. Drug release

Drug release from electrospun nanofibers is mostly through these mechanisms: (1) diffusion from nanofiber channels and pores (2) desorption (3) degradation of the scaffold if it is biodegradable.[11] Drug release profile from fibers is dependents on several factors, fiber properties such as diameter and porosity, composition, arrangement, molecular weight of the polymer, crystallinity of polymer, drug properties such as: drug polymorph, solubility, molecular weight, amount of loaded drug, and interaction between drug, polymer and solvent. It is also important to note that mechanical properties of nanofiber such as fiber alignment, fusion, and structural morphology are also effective in drug release and have to be considered. Drug release profile from non-biodegradable polymers is mainly dependent on their diffusion rate, while when biodegradable polymers are used more deciding factors contribute to release profile. Biodegradable electrospun fibers have potential to be used in clinical applications. Some examples of biodegradable polymers are lactic acid polymers (PLA, PLLA, PDLA, PDLLA), PVA, PLGA, and chitosan. By capsulating rifampin in biodegradable electrospun poly L-lactic acid (PLLA) fibers, a constant drug release along with degradation of fibers with enzyme obtained.[13,37,38]

Apart from mixing the drug directly with the polymer as in co-electrospinning or embedding the drug in core–shell structures obtained by coaxial electrospinning and emulsion electrospinning, the drug can be physically or chemically immobilized on the surface of electrospun fibers. Incorporation of drug in electrospun fibers can avoid changes in bioactivity of drug due to organic solvents and applied electric field during electrospinning process. Electrostatic interactions, hydrophobic interactions, hydrogen bonding, and van der Waals interactions are the main interactions between physically immobilized drug and electrospun fibers. In a study done by Bolgen *et al.*[39] an antibiotic (Biteral®) was physically absorbed on the surface of PCL electrospun mats. Although complete release of drug occurred within 18 hours, 80% of drug released in first 3 hours. This initial burst release can be advantageous for prevention of infection after surgery. Chemical immobilization of drug on electrospun fibers can be accomplished by introducing amine, carboxyl, hydroxyl, or thiol functional groups on fibers. Co-electrospinning of poly(ethyleneglycol) (PEG) and poly(ethyleneimine) (PEI) with main polymer have been used to produce fibers with functional groups in different studies. Chemical immobilization of drug reduces the initial burst release and provides a better sustained release profile in comparison to physical immobilization of drug.

Other than drug delivery and formulation, electrospun fibers are also promising structures for tissue engineering. A tissue scaffold should provide structural and mechanical cues for guiding cells. Cells also need specific biochemical molecules for promoting tissue generation or repair. Electrospun fibers can provide both structural and a controlled release of needed biomolecules for cells.[40]

## III. Layer-by-Layer Assembled Films

### A. Introduction to LbL assembled films

Layer-by-layer (LbL) assembly can be used to fabricate multilayer films for drug delivery. Lipid vesicles, polypeptides, proteins, and nucleic acids can incorporate in films while their functionality is preserved.[41] Hydrogen bond, charge transfer between two types of nonionic molecules when one type is electron-donor and the other is electron-acceptor, host–guest interaction, biospecific interaction, covalent bond, and electrostatic interactions are most important interactions that are the driving forces for LBL assembly. Controlled release of a therapeutic agent can be achieved through insertion of barrier layer in LbL films or by using stimuli sensitive materials. Several parameters affect LBL assembly process such as temperature, PH, ionic strength, and solvent. Material properties can exert profound effects on release profile. Hong *et al.*[43] fabricated films using poly ($\beta$-amino ester) and poly (L-lysine) (PLL) as positively charged components and ovalbumin was incorporated in films as a model for negatively charged protein. As a result of lower electrostatic interaction between poly ($\beta$-amino ester) and ovalbumin in comparison to PLL and ovalbumin and also due to hydrolytic degradation of poly ($\beta$-amino ester), a rapid release of ovalbumin was observed from poly ($\beta$-amino ester) films, while slow release was obtained from PLL films. A controlled release profile was achieved when both polymers were used in the multilayer film. In another study, Hong *et al.*[43]

highlighted the effect of barrier layer insertion on release profile of ovalbumin from poly ($\beta$-amino ester). Multilayers of graphene oxide were used as barrier layer and sustained release of ovalbumin was obtained for 30–90 days, depending on the number of graphene oxide layers. By using sensitive materials to specific environmental factors, stimuli-sensitive LbL films can be fabricated. Temperature changes, pH alteration, electric potential variations, light, hydrolysis, and enzymes can all be used as stimuli. For instance, in a study done by Wood *et al.*[44] Heparin was incorporated in poly($\beta$-amino ester) films. Due to pH-dependent degradation of this polymer, sustained release of drug can be achieved.[41,45]

## B. Biomedical use of LbL films

Layer-by-layer films can be used for both local and systemic drug delivery. Special attention has been devoted to their application in skin drug delivery systems and vaccines. In one study, ovalbumin, as a model for antigen, and CpG (cytosine-phosphate diester-guanine-rich) oligonucleotide were incorporated in a layer-by-layer film. They were both released and were uptaken by Langerhans cells in barrier-disrupted skin. Langerhans cells are antigen-presenting immune cells in skin that can initiate immune response.[46] Biosensors are an other interesting area that LbL films can have a special place in. The aforementioned uses are just few examples of LbL film potentials in biomedical applications and drug delivery.[41]

## References

1. Huang, Z.-M.; *et al.*, A review on polymer nanofibers by electrospinning and their applications in nanocomposites. *Compos. Sci. Technol.* **2003**, *63* (15), 2223–2253.
2. Pillay, V.; *et al.*, A review of the effect of processing variables on the fabrication of electrospun nanofibers for drug delivery applications. *J. Nanomater.* **2013**, 22.
3. Deitzel, J. M.; *et al.*, The effect of processing variables on the morphology of electrospun nanofibers and textiles. *Polymer* **2001**, *42* (1), 261–272.

4. Megelski, S.; *et al.*, Micro- and Nanostructured surface morphology on electrospun polymer fibers. *Macromolecules* **2002**, *35* (22), 8456–8466.
5. Zong, X.; *et al.*, Structure and process relationship of electrospun bioabsorbable nanofiber membranes. *Polymer* **2002**, *43* (16), 4403–4412.
6. Matthews, J. A.; *et al.*, Electrospinning of collagen nanofibers. *Biomacromolecules* **2002**, *3* (2), 232–238.
7. Teo, W. E.; Ramakrishna, S. A review on electrospinning design and nanofibre assemblies. *Nanotechnology* **2006**, 17 (14), R89.
8. Lu, T.; Li, Y.; Chen T. Techniques for fabrication and construction of three-dimensional scaffolds for tissue engineering. *Int. J. Nanomedicine.* **2013**, *8*, 337–350.
9. Kenawy, E.-R.; *et al.*, Release of tetracycline hydrochloride from electrospun poly(ethylene-co-vinylacetate), poly(lactic acid), and a blend. *J. Control. Release* **2002**, *81* (1–2), 57–64.
10. Wang, C.; Tong, S. N.; Tse, Y. H.; Wang, M. Conventional electrospinning vs. emulsion electrospinning: a comparative study on the development of nanofibrous drug/biomolecule delivery vehicles. *In Advanced Materials Research* **2012** (Vol. 410, pp. 118–121). Trans Tech Publications.
11. Goonoo, N.; Bhaw-Luximon, A.; Jhurry, D. Drug loading and release from electrospun biodegradable nanofibers. *J. Biomed. Nanotechnol.* **2014**, *10* (9), 2173–2199.
12. Gupta, P.; Wilkes, G. L. Some investigations on the fiber formation by utilizing a side-by-side bicomponent electrospinning approach. *Polymer* **2003**, *44* (20), 6353–6359.
13. Lu, Y.; *et al.*, Coaxial electrospun fibers: applications in drug delivery and tissue engineering. *Wiley Interdisciplinary Reviews: Nanomedicine and Nanobiotechnology*, **2016**, *8* (5), 654–677.
14. Kim, K.; *et al.*, Incorporation and controlled release of a hydrophilic antibiotic using poly(lactide-co-glycolide)-based electrospun nanofibrous scaffolds. *J. Control. Release* **2004**, *98* (1), 47–56.
15. Tang, C.; *et al.*, Effect of pH on protein distribution in electrospun PVA/BSA composite nanofibers. *Biomacromolecules* **2012**, *13* (5), 1269–1278.
16. Xu, X.; *et al.*, BCNU-loaded PEG–PLLA ultrafine fibers and their in vitro antitumor activity against Glioma C6 cells. *J. Control. Release* **2006**, *114* (3), 307–316.

17. Liu, S.; *et al.*, Inhibition of orthotopic secondary hepatic carcinoma in mice by doxorubicin-loaded electrospun polylactide nanofibers. *J. Mater. Chem. B*, **2013**, *1* (1), 101–109.
18. Xie, J.; Wang, C.-H. Electrospun micro- and nanofibers for sustained delivery of paclitaxel to treat C6 glioma in vitro. *Pharm. Res.* **2006**, *23* (8), 1817–1826.
19. Liu, D.; *et al.*, Necrosis of cervical carcinoma by dichloroacetate released from electrospun polylactide mats. *Biomaterials* **2012**, *33* (17), 4362–4369.
20. Xie, J.; Tan, R. S.; Wang, C.-H. Biodegradable microparticles and fiber fabrics for sustained delivery of cisplatin to treat C6 glioma in vitro. *J. Biomed. Mater. Res. Part A*, **2008**, *85A* (4), 897–908.
21. Zhang, Z.; *et al.*, Controlled Antibiotics release system through simple blended electrospun fibers for sustained antibacterial effects. *ACS Appl. Mater. Interfaces* **2015**, *7* (48), 26400–26404.
22. Yarin, A. Coaxial electrospinning and emulsion electrospinning of core–shell fibers. *Polymer Adv. Tech.* **2011**, *22* (3), 310–317.
23. Jiang, H.; Wang, L.; Zhu, K. Coaxial electrospinning for encapsulation and controlled release of fragile water-soluble bioactive agents. *J. Control. Release* **2014**, *193*, 296–303.
24. He, C.-L.; Huang, Z.-M.; Han, X.-J. Fabrication of drug-loaded electrospun aligned fibrous threads for suture applications. *J. Biomed. Mater. Res. Part A* **2009**, *89A (1),* 80–95.
25. Jiang, H.; *et al.*, Modulation of protein release from biodegradable core–shell structured fibers prepared by coaxial electrospinning. *J. Biomed. Mater. Res. Part B: Appl. Biomater.* **2006**, *79B* (1), 50–57.
26. Geun Hyung, K.; *et al.*, Coaxially electrospun micro/nanofibrous poly (ε-caprolactone)/eggshell-protein scaffold. *Bioinspir. Biomim.* **2008**, *3* (1), 016006.
27. Wei, S.; *et al.*, Coaxial PCL/PVA electrospun nanofibers: osseointegration enhancer and controlled drug release device. *Biofabrication* **2013**, *5* (3), 035006.
28. Zhang, Y. Z.; *et al.*, Coaxial Electrospinning of (Fluorescein Isothiocyanate-Conjugated Bovine Serum Albumin)-Encapsulated Poly(ε-caprolactone) Nanofibers for Sustained Release. *Biomacromolecules* **2006,** *7* (4), 1049–1057.

29. Xu, X.; *et al.*, Preparation of core-sheath composite nanofibers by emulsion electrospinning. *Macromol. Rapid Commun.* **2006**, 27 (19), 1637–1642.
30. Tian, L.; *et al.*, Emulsion electrospun vascular endothelial growth factor encapsulated poly(l-lactic acid-co-ε-caprolactone) nanofibers for sustained release in cardiac tissue engineering. *J. Mater. Sci.* **2012**, *47* (7), 3272–3281.
31. Xu, X.; *et al.*, The release behavior of doxorubicin hydrochloride from medicated fibers prepared by emulsion-electrospinning. *Eur. J. Pharm. Biopharm.* **2008**, *70* (1), 165–170.
32. Xu, X.; *et al.*, Ultrafine PEG–PLA fibers loaded with both paclitaxel and doxorubicin hydrochloride and their in vitro cytotoxicity. *Eur. J. Pharm. Biopharm.* **2009**, *72* (1), 18–25.
33. Yang, Y.; *et al.*, Release pattern and structural integrity of lysozyme encapsulated in core–sheath structured poly(dl-lactide) ultrafine fibers prepared by emulsion electrospinning. *Eur J. Pharm. Biopharm.* **2008**, *69* (1), 106–116.
34. Li, X.; *et al.*, Encapsulation of proteins in poly(l-lactide-co-caprolactone) fibers by emulsion electrospinning. *Coll. Surf. B: Biointerfaces.* **2010**, *75* (2), 418–424.
35. Qi, H.; *et al.*, Encapsulation of drug reservoirs in fibers by emulsion electrospinning: morphology characterization and preliminary release assessment. *Biomacromolecules* **2006**, 7 (8), 2327 2330.
36. Camerlo, A.; *et al.*, Fragrance encapsulation in polymeric matrices by emulsion electrospinning. *Eur. Polym. J.* **2013**, *49* (12), 3806–3813.
37. Zeng, J.; *et al.*, Biodegradable electrospun fibers for drug delivery. *J. Control. Release* **2003**, *92* (3), 227–231.
38. Zeng, J.; *et al.*, Influence of the drug compatibility with polymer solution on the release kinetics of electrospun fiber formulation. *J. Control. Release* **2005**, *105* (1–2), 43–51.
39. Bölgen, N.; *et al.*, In vivo performance of antibiotic embedded electrospun PCL membranes for prevention of abdominal adhesions. *J. Biomed Mater. Res. B: Appl. Biomater.* **2007**, *81B* (2), 530–543.
40. Szentivanyi, A.; *et al.*, Electrospun cellular microenvironments: understanding controlled release and scaffold structure. *Adv. Drug Delivery Rev.* **2011**, *63* (4–5), 209–220.

41. Tang, Z.; *et al.*, Biomedical applications of layer-by-layer assembly: from biomimetics to tissue engineering. *Adv. Mater.* **2006**, *18* (24), 3203–3224.
42. Xiao, F.-X.; *et al.*, Layer-by-layer assembly of versatile nanoarchitectures with diverse dimensionality: a new perspective for rational construction of multilayer assemblies. *Chem. Soc. Rev.* **2016**, *45* (11), 3088–3121.
43. Hong, J.; *et al.*, Inherent charge-shifting polyelectrolyte multilayer blends: a facile route for tunable protein release from surfaces. *Biomacromolecules* **2011**, *12* (8), 2975–2981.
44. Wood, K. C.; *et al.*, Tunable drug release from hydrolytically degradable, layer-by-layer thin films. *Langmuir* **2005**, *21* (4), 1603–1609.
45. Borges, J.; *et al.*, Layer-by-layer assembly of light-responsive polymeric multilayer systems. *Adv. Funct. Mater.* **2014**, *24* (36), 5624–5648.
46. Su, X.; *et al.*, Layer-by-Layer-Assembled Multilayer Films for Trans cutaneous Drug and Vaccine Delivery. *ACS Nano* **2009**, *3* (11), 3719–3729.
47. Moghe A. K.; Gupta B. S. *Polymer Reviews* **2008**, *48*, 353–377.

# Chapter 8

# Controlled Released from Hydrogel

Hossein Riahinezhad*, Vida Araban†, and Mohammad Raoufi‡

** Department of Chemical Engineering, Dupuis Hall, Queen's University, Kingston, ON, Canada*
*† Faculty of Pharmacy, Tehran University of Medical Science, Tehran 1417614411, Iran*
*‡ Nanotechnology Research Center, Faculty of Pharmacy, Tehran University of Medical Science, Tehran 1417614411, Iran*

## Abbreviations

| | |
|---|---|
| BMA | butyl methacrylate |
| EPA | environmental protection agency |
| FDA | food and drug administration |
| IPN | interpenetrating polymer network |
| LCST | low critical solution temperature |
| PAAm | poly (acrylamide) |
| PAAc | poly (acrylic acid) |
| PEO | poly (ethylene oxide) |
| PNIPAAm | poly (*N*-isopropylacrylamide) |
| PPO | poly (propylene oxide) |
| TRG | thermoreversible gel |
| UV | ultraviolet |

# I. Introduction to Hydrogels

## A. Definition

Hydrogels are hydrophilic[1] polymeric networks that imbibe large amounts of water or biological fluids and remain insoluble in aqueous solution. They could be physical or chemical regarding to their crosslink type, actually these physical (physical entanglements) or chemical (Covalent or hydrogen bonding, van der Walls interactions) crosslinking of polymer chains restrict the amount of water that they absorb.[1,2] Hydrogels are very interesting in the field of medical and pharmaceutics because of their softness and their resemblance to the natural living tissues and high amount of water absorption that contributes to their biocompatibility.[2,3] They are excellent candidates for contact lenses, materials for artificial skin, membranes for biosensors, and drug delivery devices. Their special place in drug delivery systems arises from their appropriate properties for formulation of macromolecules such as proteins and oligonucleotides. These macromolecules are highly susceptible to enzymatic degradation and should be protected. It has shown that hydrophobic polymers have potential impact on structure and function of proteins, so hydrophilic ones are our polymers of choice. Hydrogels can protect drugs from the specific body environments, e.g. presence of enzymes and low or high pH, on the other hand, they can also release the drug in a controllable pattern or in response to the stimulation of the hostile environments by swelling, degradation of the structure, gel-sol phase transition.[4]

## B. Classification

Hydrogels can be classified differently according to their vast properties and characteristics. They can be classified according to the nature of backbone polymers (natural or synthetic polymers), nature of polymer crosslinks (chemical or physical hydrogels), nature of the side groups (neutral or ionic), mechanical and structural characteristics (affine or phantom networks), method of preparation (homopolymer or copolymer networks), their degree of porosity, physical structure of the networks (amorphous, semi-crystalline, hydrogen-bonded structures, supramolecular structures and hydrocolloidal aggregates), and the way of their response to the physical and chemical stimulations or their size scale (nano, micro, or bulk hydrogels).[1,2,4–6]

Each classification has its own characteristics and applications. Hydrogels that are prepared from natural polymers inherent biodegradability and biocompatibility; meanwhile, they may not show sufficient mechanical properties comparing to the synthetic polymers.[1,7] Nature of the side group, anionic or cationic polymers, causes hydrogel to respond to the pH environment regarding to their acidic or basic pendant group.[4] Smart hydrogels can be synthesized by using several homopolymer to build a copolymer network that respond to more than one stimulus.

## II. Controlled Release Mechanism in Hydrogel Devices

Understanding and modeling the mechanism of drug release is very important and challenging for both designing a hydrogel system and also pharmaceutical formulation.[1,2]

Drug release mechanisms from hydrogel devices can be categorized in three major parts based on the rate limiting step for controlled release of drug.

### A. Diffusion-controlled

Drug can be incorporated to the hydrogel by post-loading and *in situ* loading.[1] In the first method, drug is added to the formed hydrogel network and usually the mechanism of loading and release of drug is determined by diffusion. In the second method, drug is mixed with polymer precursor solution and the network formation and the drug loading/encapsulation are done simultaneously. In the second system, the drug release can be controlled by diffusion, swelling of the hydrogel, or cleavage of the polymer chains due to the chemical reactions.

The detail information about structure, shape, and characteristics of hydrogel lead toward a better mathematical model for drug release. For example, the diffusion coefficient is higher for porous hydrogel than non-porous hydrogel.[1,8,9] When the pore sizes are larger than molecular dimensions of the drug, the diffusion coefficient is higher comparable to the hydrogel without pore or with smaller pore sizes (smaller than drug molecular size). As a result, the average free volume per molecule that is available to the drug decreased, and drug has to diffuse toward a larger path due to the less or smaller pores.

Fick's law of diffusion can be applied to hydrogel to model the release of the drug. The following formula show the first and second Fick's law.

$$\text{First law: } J_A = -\mathrm{D}\frac{\mathrm{d}C_A}{\mathrm{d}x}$$

$$\text{Second law: } \frac{\mathrm{d}\ C_A}{\mathrm{d}t} = \mathrm{D}\frac{d^2 C_A}{\mathrm{d}x^2}$$

D is the drug diffusion coefficient, $C_A$ is the concentration of drug, $J_A$ is the flux of the drug in the $x$ direction.

The first and second Fick's law can be used to simulate reservoir and matrix systems, respectively.[1,10,11] In the reservoir system, a polymeric membrane separates the drug depot from the external environment, while in the matrix system; drug is dispersed in the matrix. In both first formulas the drug diffusion coefficient consider as constant for mathematical simplification. To improve the Fick's law and to predict the drug release more precisely, it is better to assume that drug diffusion coefficient is concentration dependent. The following equation is used when D is concentration dependent.

$$\text{Concentration dependent: } \frac{\partial C_A}{\partial t} = \frac{\partial}{\partial x}\left(D(C_A)\frac{\partial C_A}{\partial x}\right)$$

Many attempts are done in order to formulize and understand the diffusion coefficient, because once you have it, the above formulas can be solved with applying several boundary conditions. According to the Eyring analysis, the diffusion coefficient is not continuous and the diffusion of a solute is described as a series of jumps.[12] In another theory, Fujita mentioned that free volumes in polymers affects exponentially on the drug diffusion coefficient.[13]

In general, drug diffusion coefficient depends on many different parameters such as degree of swelling of hydrogel, crosslinking of the gels, degradation of polymer network, environmental changes, and drug concentration; therefore, it is difficult to formulize it, specifically for the carriers with complex geometry or when non-constant drug diffusivities occur.[1,2,14] In some cases, as a result of polymer degradation the drug diffusion coefficient varies over time and it is not constant. Detail information and study about network structure of drug carrier help towards understanding the mechanism of release and as a result, leading to design of complex drug delivery systems. The most important parameters that explain network structure are polymer volume fraction in swollen state ($\nu_{2,s}$), molecular weight of polymer chain between two neighboring cross linking points ($M_c$), and mesh size ($\xi$). The first parameter determines the amount of liquid absorbs and retained by hydrogel and it is calculated by dividing polymer volume by swollen gel volume. Degree of cross linking of the polymer can be determined by the second parameter that is reported as average value. Mesh size is a valuable parameter to determine physical properties of the hydrogels. Mesh size of hydrogels mostly varies from 5 to 100 nm in their swollen state and macromolecules will trap in these cavities. Diffusion rate could be changed through structure and mesh size of the hydrogel. Several factors can affect mesh size such as the structure of precursor monomers, degree of crosslink between polymer chains. In the environment sensitive hydrogels, mesh size could vary in response to stimuli. Depending on the mesh size and its information about pore size, hydrogels can be divided into non-porous, micro-porous, and macro-porous. For example, when size of pores is larger than the size of drug molecule in a hydrogel carrier, then the release mechanism is controlled by diffusion instead of swelling control.

## B. Swelling-controlled

Hydrogel swelling can control the release of the drug and it happens when hydrogel swelling is smaller than diffusion of drug, while in the diffusion control system, the rate limiting step is drug diffusion.[1,2,15] In the swelling controlled systems, the drug is dispersed in the glassy polymers and when the polymer network is subjected to the body fluid or water, it undergoes a transition from glassy state to rubbery state. One of the important parameters that influence on release of drug is polymer relaxation time in the interface between glassy and rubbery states. One of the challenges in formulizing swelling controlled systems is moving boundary conditions.

## C. Chemically-controlled

Based on the mechanism of drug release, the chemically controlled system is divided into two major categories of pendent chain and erodible systems.[1,2] In the first one, drugs are bonded covalently to the hydrogel and they can be detached from polymer matrix hydrolytically or enzymatically. In this case, the rate of cleavage of drugs from hydrogel determines the rate of drug release. In some cases, drug is bonded to degradable hydrogel, therefore, drug release depend both on bond cleavage and hydrogel degradation. In the second system, erosion and diffusion are rate determinant step. Depending on the rate of matrix erosion or drug diffusion from matrix, the release control mechanism can be determined. For example, the drug release is erosion controlled when the rate of erosion of hydrogel is much faster than diffusion of drug. Depending on the hydrophilic or hydrophobic properties of hydrogel, erosion can be done homogeneous or heterogeneous, respectively. In the hydrophilic polymers, water can absorb in the whole structure, therefore, erosion is occurred in the whole network. On the other hand, for hydrophobic polymers, water molecule cannot diffuse in the bulk and it results in surface erosion.

It should be noticed that many hydrogel systems don't follow just one release mechanism and these classifications are for simplifying. As an example, chemical release could be dominant release mechanism but diffusion of load is probable in parallel.

## III. Stimuli Sensitive Hydrogels in Drug Delivery

Hydrogels are excellent candidates for using them in diverse applications, such as drug delivery carrier, encapsulating growth factors and making artificial muscles.[2,4,5] In biological systems many hormones and bioactive molecules release in a pulsatile manner and their fluctuating concentration attributes to their physiological effect as in many cases biological response differs for continuous or fluctuating concentrations. In order to design drug delivery systems for these type of molecules it should be taken into consideration that their release profile should be mimicked as certain amount of molecule released during a short period of time after a certain lag time. This goal could be achieved using smart drug delivery systems which can change their properties by applying external stimuli. Smart hydrogels are popular because they can sense their environment and they can respond to one or several stimuli factors vary from physical stimuli i.e. pressure, magnetic fields, temperature, electric fields, light and sound to chemical stimuli i.e. pH change and specific enzyme or ion identifications.

Some stimuli have presented in the following scheme 1:

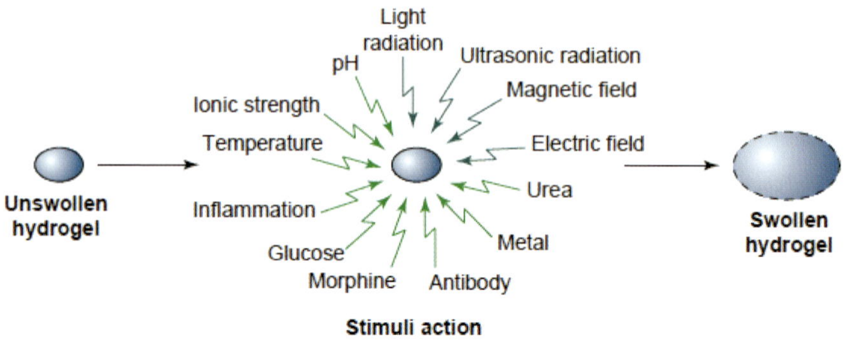


Scheme 1. Stimuli sensitive hydrogels in drug delivery.

## A. Temperature sensitive hydrogels

Temperature sensitive hydrogels show reaction to the environmental temperature by changing their volume or sol–gel transition.[16,17] According to their properties, they can be used in different applications such as drug delivery by using on-off drug release feature,[16,18,19] biosensors and smart cell culture dishes.[20–23] Depending on their response to temperature, they can be classified into three main classes of positive and negative temperature sensitive systems and thermoreversible gels.

Negative thermo-sensitive hydrogels are swelling by decreasing temperature below the low critical solution temperature (LCST), and they are shrinking above this temperature. The reason of this behavior is the presence of hydrophilic and hydrophobic groups together. It is possible to change the LCST by adjusting the ratio of hydrophobic and hydrophilic segment of the polymer. Interaction between hydrophobic parts of polymer becomes robust by increasing temperature above LCST, while interaction between hydrophilic segments and water molecule are strengthened below the LCST, therefore hydrogel experience shrinking and swelling, respectively. Poly (*N*-isopropylacrylamide), (PNIPAAm), is very famous example for negative-sensitive hydrogel and show well defined LCST around

32°C that can be increased to body temperature by adding ionic co-monomers.[2,4,6,24] Many different researches are done by using PNIPAAm and its copolymers or conjugate it with other polymers. In one study by Inoue *et al.* in order to have a hydrogel with two different LCSTs, the synthesized hydrogel grafted with carboxy-terminated oligo NIPAAm, oligo (N-vinylcaprolactam and a random co-oligomer of NIPAAm and AAm.[25] It is also possible to conjugate PNIPAAm with different enzymes.[26,27] Conjugates of oligomer of PNIPAAm and trypsin are synthesized to catalyze enzymatic reactions and they recovered afterwards by thermal precipitation. PNIPAAm hydrogel microparticles can be placed inside a rigid matrix or they can even graft to the surface of rigid membranes. Then, permeability of the membrane can be controlled through shrinking and swelling of the microparticles.[18] Although many studies are done based on NIPAAm and its derivatives, but the NIPAAm is not known to be biocompatible and biodegradable.[4]

Positive thermo-sensitive hydrogels swell above their transition temperature and shrink below it. Katono *et al.* synthesized interpenetrating polymer network (IPN) composed of poly (acrylic acid) (PAAc) and poly (acrylamide(AAm)-co-butyl methacrylate (BMA)).[28] These IPNs behave like positive temperature sensitive hydrogel and they observed that increasing BMA content leads to higher transition temperature, and IPNs with limited BMA content showed reversible swelling changes in response to stepwise temperature changes.

The last category of temperature sensitive hydrogel is thermoreversible gel (TRG). They behave like free flowing liquid at ambient temperature and gels at body temperature.[6] One of the advantages of this type of gels is to load drug by mixing it with thermoreversible polymer, then inject it to the body to form crosslinked gels. Block copolymers of poly (ethylene oxide) (PEO) and poly (propylene oxide) (PPO) are the most common TRG systems which approved by food and drug administration (FDA) and environmental protection agency (EPA) for applications in food additives, pharmaceutical ingredients, and agricultural products.

## B. pH-sensitive hydrogels

The pH-sensitive hydrogels react to the environmental pH by swelling due to repulsion between their pendant acidic group (such as carboxylic acid and sulfuric acid) or basic group (such as ammonium salts) in their ionized state.[29–32] The pH-sensitive hydrogels are anionic or cationic. In the anionic hydrogels once environmental pH is higher than the pKa (acidic dissociation constant) of the pendant group, they ionize to anionic form and hydrogels imbibe more water because of the electrostatic repulsion of charges.[29,30,33,34] On the other side, for cationic hydrogels, ionization occurs when pH is below the pKb of the ionizable groups. The amount of swelling can be adjusted by controlling pKa, number of pendant groups, and degree of polymer crosslinks.

Figure 1 shows ionization of Poly (acrylic acid) and poly (*N,N'*-diethylaminoethyl methacrylate) in the low and high pH.

The pH-sensitive hydrogels can be used in different applications such as controlled drug delivery, and biosensors.[33,35] In order to prepare biosensors, usually the pH-sensitive hydrogel is incorporated with enzymes that can change the local microenvironment pH and cause the hydrogel swell and release the drug.[36] One of the enzymes which used in the insulin delivery is glucose oxidase. By adding this enzyme

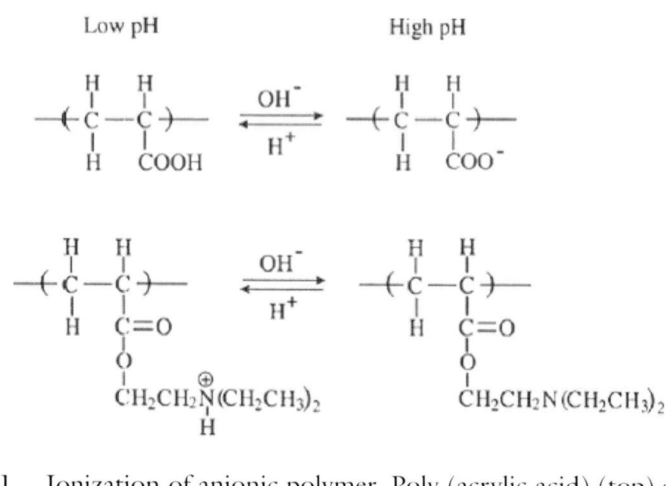


Figure 1. Ionization of anionic polymer, Poly (acrylic acid) (top) and cationic polymer, and poly (*N,N'*-diethylaminoethyl methacrylate) (below) in the low and high pH. Taken from Ref. [4] with permission of Elsevier.

to the ionic hydrogel membrane, it starts to oxidize glucose to gluconic acid in the body and leads to pH change, followed by swelling of the cationic membrane and release of drug, specifically insulin.[37,38]

It is also possible to design hydrogels that respond to both temperature and pH by incorporating ionizable and hydrophobic functional groups. Copolymers of P(NIPAAm-g-AA) prepared by Chen *et al.* responding to both temperature and pH changes.[25] In another study, PNIPAAm and PAA were loaded with the protein drug, calcitonin, and the drug released effectively due to change in temperature and pH.[39]

## C. Light sensitive hydrogels

Light sensitive hydrogels respond to the radiation of light and based on the nature of the light, they can be divided into ultraviolet (UV) sensitive and visible light sensitive hydrogels.[40,41] Compare to the temperature and pH-sensitive hydrogels, it is much more convenient to stimulate light sensitive hydrogels due to irradiation of light with different power to the specific parts of body in order to activate the hydrogel. In one study on visible light sensitive hydrogel, a photo sensitive molecule which produce local heat by light radiation, incorporated into a temperature sensitive polymer.[41] As a result of applying visible light, temperature-sensitive polymer reacts to the local increase in temperature and undergoes volume change. In another study, Zhang *et al.* used $CO_2$ infrared laser to stimulate a straight cylindrical *N*-isopropylacrylamide gel which is a temperature-sensitive polymer.[42] Based on the infrared absorbance of water molecule in the hydrogel, they didn't add any photo sensitive molecule to the polymer and as a result of radiation, they observed not only volume phase transition of the gel, but also bending of the gel towards the beam line.

## D. Electrical sensitive hydrogels

Like pH-sensitive hydrogels, most of the electrically sensitive hydrogels are synthesized from polyelectrolytes.[4] This type of hydrogel is interesting in the field of drug delivery because of its possibility to have "on–off" drug release.[43] Kwon *et al.* loaded negatively charged hydrogel made of poly (2-acrylamide-2-methylpropanesulfonic acid-co-butyl methacrylate) (P(AMPS-co-BMA)) with positively charged drug, edrophonium chloride.[44] They observed an increase in the drug release by applying an electric field due to the electrostatic force, squeezing effect, and the electro-osmosis of the gel; while no drug released in the absence of electric field which prove the on-off drug release system.

Electrically sensitive hydrogels are also interesting to be used as actuators and artificial muscles due to their properties to convert chemical energy into mechanical energy.[4]

## E. Other stimuli sensitive hydrogel

In addition to the discussed stimulation above, there are several stimuli that hydrogels respond to them such as antigen,[45] pressure,[46] specific ions or molecule,[47] magnetic field,[48] and ultrasound.[49]

In a study about antigen responsive hydrogel, the hydrogel consist of antibody-grafted linear poly (acrylamide) (PAAm) and the crosslinked PAAm grafted with the corresponding antigen.[45] As shown in Figure 2, in the absence of the free antigen, the hydrogel shrink due to the intra-chain antigen-antibody binding, while it

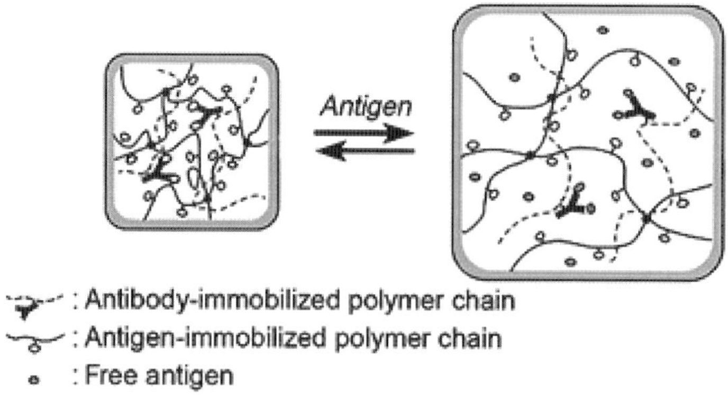


Figure 2. Swelling (right) and shrinking (left) of an antigen-responsive hydrogel. Taken from Ref. [45] with permission of American Chemical Society.

swells by adding free antigen to the system, followed by exchange of the grafted antigen with free antigen and dissociation of the binding.

## References


1. Lin, C.-C.; Metters, A. T. Hydrogels in controlled release formulations: network design and mathematical modeling. *Adv. Drug Deliv. Rev.* **2006**, *58,* 1379–1408.
2. Peppas, N. A.; Bures, P.; Leobandung, W.; Ichikawa, H. Hydrogels in pharmaceutical formulations. *Eur. J. Pharm. Biopharm.* **2000**, *50,* 27–46.
3. Hoffman, A. S.; Ratner, B. D. American chemical society, polymer preprints, division of polymer chemistry. *Synthetic hydrogels for biomedical applications — A review.* **1975**, *16*(2), 272–275.
4. Qiu, Y.; Park, K. Environment-sensitive hydrogels for drug delivery. *Adv. Drug Deliv. Rev.* **2012**, *64,* 49–60.
5. Kikuchi, A.; Okano, T. Pulsatile drug release control using hydrogels. *Adv. Drug Deliv. Rev.* **2002**, *54,* 53–77.
6. Bromberg, L. E.; Ron, E. S. Temperature-responsive gels and thermogelling polymer matrices for protein and peptide delivery. *Adv. Drug Deliv. Rev.* **1998**, *31,* 197–221.
7. Davis, K. A.; Anseth, K. S. Controlled release from crosslinked degradable networks; *Crit. Rev. Ther. Drug Carrier Syst.* **2002**, *19,* 385–424.
8. Masaro, L.; Zhu, X. X. Physical models of diffusion for polymer solutions, gels and solids. *Prog. Polym. Sci.* **1999**, *24,* 731–775.
9. Amsden, B. Solute diffusion within hydrogels: mechanisms and models. *Macromolecules* **1998**, *31,* 8382–8395.
10. Lowman, A. M.; Peppas, N. A. Solute transport analysis in pH-responsive, complexing hydrogels of poly(methacrylic acid-g-ethylene glycol). *J. Biomater. Sci. Polym. Ed.* **1999**, *10,* 999–1009.
11. Crank, J. *The Mathematics of Diffusion* (2nd edn.). Oxford University Press Inc., New York, Clarendon Press, **1975**, 1–27.
12. Eyring, H. Viscosity, Plasticity, and diffusion as examples of absolute reaction rates. *J. Chem. Phys.* **1936**, *4,* 283–291.
13. Fujita, H. *Fortschritte Hochpolym.-Forsch.* Springer Berlin Heidelberg, 1961, 1–47.

14. Peppas, N. A.; Huang, Y.; Torres-Lugo, M.; Ward, J. H.; Zhang, J. Physicochemical foundations and structural design of hydrogels in medicine and biology. *Annu. Rev. Biomed. Eng.* **2000**, *2,* 9–29.
15. Siepmann, J.; Peppas, N. A. Modeling of drug release from delivery systems based on hydroxypropyl methylcellulose (HPMC). *Adv. Drug Deliv. Rev.* **2012**, *64,* (Suppl.), 163–174.
16. Dinarvand, R.; D'Emanuele, A. The use of thermoresponsive hydrogels for on-off release of molecules. *J. Control. Release* **1995**, *36,* 221–227.
17. Jeong, B.; Bae, Y. H.; Kim, S. W. Thermoreversible gelation of PEG–PLGA–PEG triblock copolymer aqueous solutions. *Macromolecules* **1999**, *32,* 7064–7069.
18. Ichikawa, H.; Fukumori, Y. A novel positively thermosensitive controlled-release microcapsule with membrane of nano-sized poly (N-isopropylacrylamide) gel dispersed in ethylcellulose matrix. *J. Control. Release Off. J. Control. Release Soc.* **2000**, *63,* 107–119.
19. Gutowska, A.; Bark, J. S.; Kwon, I. C.; Bae, Y. H.; Cha, Y.; Kim, S. W. Squeezing hydrogels for controlled oral drug delivery. *J. Control. Release* **1997**, *48,* 141–148.
20. von Recum, H. A.; Kim, S. W.; Kikuchi, A.; Okuhara, M.; Sakurai, Y.; Okano, T. Novel thermally reversible hydrogel as detachable cell culture substrate. *J. Biomed. Mater. Res.* **1998**, *40,* 631–639.
21. Sakai, H.; Doi, Y.; Okano, T.; Yamada, N.; Sakurai, Y. *Advanced Biomaterials in Biomedical Engineering and Drug Delivery Systems* (eds. Ogata, N.; Kim, S. W.; Feijen, J.; Okano, T.), Springer Japan, 1996, 229–230.
22. Okano, T.; Yamada, N.; Sakai, H.; Sakurai, Y. A novel recovery system for cultured cells using plasma-treated polystyrene dishes grafted with poly(N-isopropylacrylamide). *J. Biomed. Mater. Res.* **1993**, *27,* 1243–1251.
23. Okano, T.; Yamada, N.; Okuhara, M.; Sakai, H.; Sakurai, Y. Mechanism of cell detachment from temperature-modulated, hydrophilic-hydrophobic polymer surfaces. *Biomaterials* **1995**, *16,* 297–303.
24. Schild, H. G. Poly(N-isopropylacrylamide): Experiment, theory and application. *Prog. Polym. Sci.* **1992**, *17,* 163–249.
25. Inoue, T.; Chen, G.; Nakamae, K.; Hoffman, A. S. Temperature sensitivity of a hydrogel network containing different LCST oligomers grafted to the hydrogel backbone. *Polym. Gels Netw.* **1998**, *5,* 561–575.

26. Ding, Z.; Chen, G.; Hoffman, A. S. Unusual properties of thermally sensitive oligomer–enzyme conjugates of poly(N-isopropylacrylamide)–trypsin. *J. Biomed. Mater. Res.* **1998**, *39*, 498–505.
27. Matsukata, M.; Aoki, T.; Sanui, K.; Ogata, N.; Kikuchi, A.; Sakurai, Y.; Okano, T. Effect of Molecular Architecture of Poly (N-isopropylacrylamide)–Trypsin Conjugates on Their Solution and Enzymatic Properties. *Bioconjug. Chem.* **1996**, *7*, 96–101.
28. Katono, H.; Maruyama, A.; Sanui, K.; Ogata, N.; Okano, T.; Sakurai, Y. Thermo-responsive swelling and drug release switching of interpenetrating polymer networks composed of poly (acrylamide-co-butyl methacrylate) and poly (acrylic acid). *J. Control. Release* **1991**, *16*, 215–227.
29. Katchalsky, A.; Michaeli, I. Polyelectrolyte gels in salt solutions. *J. Polym. Sci.* **1955**, *15*, 69–86.
30. Brannon-Peppas, L.; Peppas, N. A. Equilibrium swelling behavior of pH-sensitive hydrogels. *Chem. Eng. Sci.* **1991**, *46*, 715–722.
31. Katchalsky, A. Rapid swelling and deswelling of reversible gels of polymeric acids by ionization. *Experientia* **1949**, *5*, 319–320.
32. Katchalsky, A.; Lifson, S.; Exsenberg, H. Equation of swelling for polyelectrolyte gels. *J. Polym. Sci.* **1951**, *7*, 571–574.
33. Khare, A. R; Peppas, N. A. Swelling/deswelling of anionic copolymer gels. *Biomaterials* **1995**, *16*, 559–567.
34. Ohmine, I. Salt effects on the phase transition of ionic gels. *J. Chem. Phys.* **1982**, *77*, 5725.
35. Vakkalanka, S. K.; Brazel, C. S; Peppas, N. A. Temperature- and pH-sensitive terpolymers for modulated delivery of streptokinase. *J. Biomater. Sci. Polym. Ed.* **1997**, *8*, 119–129.
36. Peppas, N. A.; Merrill, E. W. Differential scanning calorimetry of crystallized PVA hydrogels. *J. Appl. Polym. Sci.* **1976**, *20*, 1457–1465.
37. Albin, G.; Horbett, T. A.; Ratner, B. D. Glucose sensitive membranes for controlled delivery of insulin: Insulin transport studies. *J. Control. Release* **1985**, *2*, 153–164.
38. Ishihara, K.; Kobayashi, M.; Ishimaru, N.; Shinohara, I. Glucose induced permeation control of insulin through a complex membrane consisting of immobilized glucose oxidase and a poly(amine). *Polym. J.* **1984**, *16*, 625–631.
39. Serres, A.; Baudyš, M.; Kim, S. W. Temperature and pH-sensitive Polymers for Human Calcitonin Delivery. *Pharm. Res.* **1996**, *13*, 196–201.

40. Mamada, A.; Tanaka, T.; Kungwatchakun, D.; Irie, M. Photoinduced phase transition of gels. *Macromolecules* **1990**, *23*, 1517–1519.
41. Suzuki, A; Tanaka, T. Phase transition in polymer gels induced by visible light. *Nature* **1990**, *346*, 345–347.
42. Park, H; Park, K. Biocompatibility Issues of Implantable Drug Delivery Systems. *Pharm. Res.* **1996**, *13*, 1770–1776.
43. Kwon, I. C.; Bae, Y. H.; Okano, T.; Kim, S. W. Drug release from electric current sensitive polymers. *J. Controlled Release* **1991**, *17, 149*, 156–153.
44. Kwon, I. C.; Bae, Y. H; Kim, S W. Electrically erodible polymer gel for controlled release of drugs. *Nature* **1991**, *354*, 291–293.
45. Miyata, T.; Asami, N.; Uragami, T. Preparation of an Antigen-Sensitive Hydrogel Using Antigen–Antibody Bindings. *Macromolecules* **1999**, *32*, 2082–2084.
46. Lee, K. K.; Cussler, E. L.; Marchetti, M.; McHugh, M. A. Pressure-dependent phase transitions in hydrogels. *Chem. Eng. Sci.* **1990**, *45*, 766–767.
47. Zhong, X.; Wang, Y.-X.; Wang, S.-C. Pressure dependence of the volume phase-transition of temperature-sensitive gels. *Chem. Eng. Sci.* **1996**, *51*, 3235–3239.
48. Zrınyi, M., Szabó, D. & Kilian, H.-G. Kinetics of the shape change of magnetic field sensitive polymer gels. *Polym. Gels Netw.* **1998**, *6*, 441–454.
49. Hsu, C. S.; Block, L. H. Anionic gels as vehicles for electrically-modulated drug delivery. I. Solvent and drug transport phenomena. *Pharm. Res.* **1993**, *13*, 1865–1870.

# Chapter 9

# Nano Delivery Systems

Sophie Laurent*[,†], Afsaneh Lahooti[‡],
Saeed Shanehsazzadeh*[,§], and Robert N. Muller*[,†]

** NMR and Molecular Imaging Laboratory, General, Organic, and Biomedical Chemistry Unit, University of Mons, Avenue Maistriau, 19, B-7000 Mons, Belgium*
*[†]Center for Microscopy and Molecular Imaging (CMMI), Rue Adrienne Bolland, 8 B-6041 Gosselies, Belgium*
*[‡]Department of Medical Physics and Biomedical Engineering, Faculty of Medicine, Tehran University of Medical Science, Iran*
*[§]Radiation Application Research School, Nuclear Science and Technology Research Institute (NSTRI), Tehran, Iran*

## Abbreviations

| | |
|---|---|
| Ad | Adenovirus |
| BBB | blood brain barrier |

| | |
|---|---|
| BMV | brome mosaic virus |
| CCMV | cowpea chlorotic mottle virus |
| CPMV | cowpea mosaic virus |
| CNTs | carbon nanotubes |
| Cb | chlorambucil |
| CMC | critical micellar concentration |
| DDS | drug delivery system |
| EPR | enhanced permeability and retention |
| ELPs | elastin-like synthetic peptides |
| FHV | Flock House virus |
| AuNPs | gold nanoparticles |
| HCRSV | Hibiscus chlorotic ringspot virus |
| IOs | iron oxide nanoparticles |
| Ir | Irinotecan |
| LCST | lower critical solution temperature |
| MRI | magnetic resonance imaging |
| MOFs | metal organic frameworks |
| MPS | mononuclear phagocyte system |
| MSNs | mesoporous silica nanoparticles |
| NIR | near infra-red |
| NLNs | nanostructured lipid nanoparticles |
| HPMA | *N*-(2-hydroxypropyl) methacrylamide |
| PAMAM | polyamidoamine dendrimers |
| PVX | Potato virus X |
| PEG | polyethylene glycol |
| PACA | poly (alkylcyanoacrylate) |
| PLA | poly (lactide) |
| PLGA | poly (lactide co-glycolide) |
| PCL | poly (e-caprolactone) |
| QDs | quantum dots |
| RCNMV | red clover necrotic mottle virus |

| | |
|---|---|
| RES | reticuloendothelial system |
| SLN | solid lipid nanoparticles |
| SDDS | smart drug delivery system |
| NLSs | solid lipid nanoparticles |
| TEOS | tetraethoxysilane |
| TMV | Tobacco mosaic virus |
| TYMV | Turnip yellow mosaic virus |
| VNPs | viral nanoparticles |
| UV | ultra violet |

## I. Drug Delivery Systems

Delivery of a pharmaceutical compound through systemic circulation to the site of action to produce the desired therapeutic effect is the goal of drug delivery. In a biological system, numerous mechanisms exist to protect the body from exposure to foreign substances. A drug delivered in different ways will encounter physiological and biochemical barriers before reaching the site of action. Passage through these barriers depends on the physicochemical and biochemical properties of the drug molecule. Among these properties, solubility is very important for oral drug delivery. The lipophilic characteristics of the molecule are also essential for crossing cell membranes by diffusion. The molecule also has to survive biodegradation caused by the gastrointestinal system and the liver.[1] Figure 1 illustrates the importance of these physicochemical properties.[2]

The permeability (related to passive diffusion or a transporter-mediated process) and metabolic stability (related to intrinsic clearance) of a drug molecule are two important factors in drug delivery when the compound is in solution.

The goal of a drug delivery system is to provide enhance efficacy and reduced the toxicity of drug molecules. Long-circulating nanoparticles, such as liposomes, micelles, and polymeric nano-objects, can

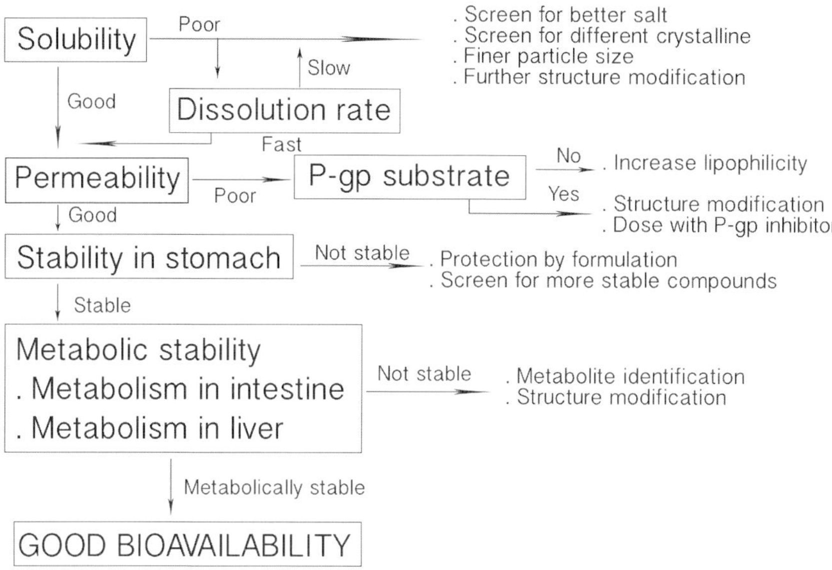


Figure 1. Importance of the physicochemical properties. Taken from Ref. [2] with permission of Wiley and Sons.

exploit the "enhanced permeability and retention" (EPR) effect for preferential extravasation from tumor vessels.[3]

The currently approved stable formulations of liposomal drug delivery systems allow for improved pharmacokinetics and passive targeting to tumor tissues. Recent studies on drug carriers involve the direct molecular targeting of cancer cells via antibodies or other small ligands, such as peptides.

## A. Nanoparticles in circulation: How to increase blood stability

Conventional liposomes are lipid vesicles essentially composed of various phospholipids (e.g. phosphatidylcholine, phosphatidylethanolamine). These systems suffer from low stability and rapid clearance from the blood.[4,5] Generally, they are rapidly taken up by macrophages in the liver. To overcome these problems, polymers can be used for steric stabilization. Protecting the liposome surface with polyethylene glycol (PEG) or other polar ligands, such as carbohydrates, allowed for the development of a stealth system.[6] PEGylation creates a hydrophilic and non-charged surface that can prevent early clearance of nanoparticles and increase their circulation time.[7] These long-circulating liposomes have improved pharmacokinetics compared to traditional systems. Surface coating prevents the nanosystems from agglomeration and from sticking to blood cells or vascular walls. They are invisible to the immune system and have shown promising results in cancer therapy. There is less uptake by the liver, and the liposomes can stay in circulation longer.

## B. Nanoparticles for drug delivery

### 1. *Virus strategy*

To improve biodistribution and efficacy and to reduce the side effects of treatment, a number of nanostructures have been designed. Among them, the virus can be seen as a core/shell system consisting of an assembly of proteins enclosing genetic material. These nanostructures have naturally evolved to infect host cells with high efficiency and deliver their genetic material. Their size (20–300 nm) increases their chances of reaching the target cells, and some naturally differentiate healthy cells from tumor cells by preferentially targeting the latter. It is therefore possible to transport cargo diagnostics and therapeutic agents. For example, the functionalization of a capsid of a P22 bacteriophage with a contrast agent (Gd-DTPA) for magnetic resonance imaging (MRI) has improved the effectiveness by a factor of 6.[8]

Another interesting example is the ability of an alphavirus to transport molecules of interest preferentially in tumor cells. The advantage of viruses as nanocarriers[9] is that unlike other nanosystems, the number and orientation of the functional groups are well defined on the surface. Viral nanoparticles (VNPs) usually come from plants and bacteria. These particles are widely available and have a monodispersed structure. They have the advantage of being biocompatible and biodegradable and are considered to be non-infectious and safe for humans. Their basic structure allows their cavity to be loaded with active molecules, and their surface can be functionalized with specific ligands.[10] It is also possible to generate VNPs by "genetic engineering" through chemical bio-conjugation or the self-assembly of proteins (Figure 2).

Virus, such as nanoparticles (VLPs), are generally produced from a baculovirus derived from insect cells or a mammalian adenovirus. They may also be produced from viral nanoparticles that are disassembled into protein subunits and reassembled once the nucleic acid has been removed. This type of particle is typically used in vaccinations.[10] The geometric shapes and sizes of VNPs and VLPs can be modulated by varying the pH, ionic strength, and physicochemical properties of the medium in which the protein is collected. The VNPs or their protein shell (VLPs) can also be used to encapsulate other types of particles,[11] such as nano-emulsions, NP polymers, enzymes, and inorganic NPs, forming more complex structures (Figure 3). Therefore, these nanoparticles are involved in many applications, such as gene delivery, catalysis, imaging, and the release of therapeutic agents.[10–14]

## II. Chemical Engineering

Theranostic nanoparticles are multifunctional and are made from various building blocks of organic molecules (e.g. lipids, polymers, proteins, and polysaccharides) or inorganic components (e.g. iron, gold, metal oxides, carbon, and silica).

### A. Drug conjugates

Drug conjugates are built from covalent or reversible interactions between two active ingredients or between a chemical cargo and an active substance. For example, the association between two anticancer drugs, such as Irinotecan (Ir) and Chlorambucil (Cb), via an

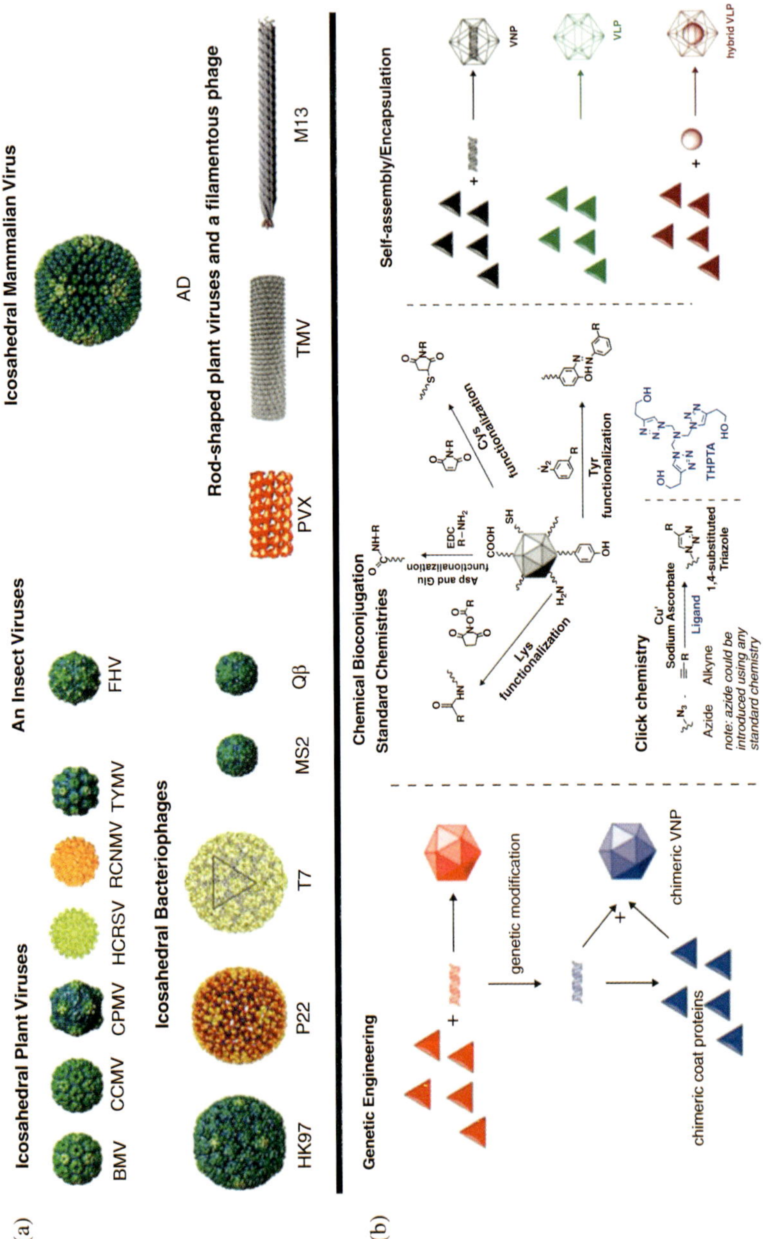
(a)
Icosahedral Plant Viruses
BMV
CCMV
CPMV
HCRSV
RCNMV
TYMV
An Insect Viruses
FHV
Icosahedral Mammalian Virus
AD
Icosahedral Bacteriophages
HK97
P22
T7
MS2
Qβ
Rod-shaped plant viruses and a filamentous phage
PVX
TMV
M13
(b)
Genetic Engineering
genetic modification
chimeric coat proteins
chimeric VNP
Chemical Bioconjugation
Standard Chemistries
Lys functionalization
Asp and Glu functionalization
EDC
R–NH2
Cys functionalization
Tyr functionalization
Click chemistry
Azide
Alkyne
Cu+
Sodium Ascorbate
Ligand
1,4-substituted Triazole
THPTA
note: azide could be introduced using any standard chemistry
Self-assembly/Encapsulation
VNP
VLP
hybrid VLP

ester bond allows the formation of an amphiphilic conjugate. The therapeutic efficacy of this conjugate for breast cancer is increased compared to the efficiency of each constituent taken separately (Figure 4).[15]

For theranostics, two main classes of drug conjugates exploit the interactions between active substances (proteins, peptides or other active substances) and polymers.[16] For example, elastin-like synthetic peptides (ELPs) can be conjugated covalently to doxorubicin hydrazone cleavable linkages in an acid medium to increase their time of vascular remanence. The internalization of these prodrugs and their accumulation in lysozyme cleaves hydrazone bonds and releases active substances.[17] Albumin can also be covalently conjugated with anticancer drugs. For example, a methotrexate-albumin conjugate has been the subject of pre-clinical and clinical studies. The association is established by direct coupling between methotrexate and the lysine residues of albumin.[18] Synthetic polymers have also been developed as an alternative to previous polypeptide polymers. One polymer used

Figure 2. (a). Viral nanoparticles (VNPs) are used in materials science and medicine. Icosahedral plant viruses: *Brome mosaic virus* (BMV), *Cowpea chlorotic mottle virus* (CCMV), *Cowpea mosaic virus* (CPMV), *Hibiscus chlorotic ringspot virus* (HCRSV), *Red clover necrotic mottle virus* (RCNMV), and *Turnip yellow mosaic virus* (TYMV). Icosahedral insect virus: *Flock House virus* (FHV). Icosahedral bacteriophages: HK97, P22, T7, MS2 and Q$\beta$. Note that P22 and T7 are head-tail phages with the tails not shown. Icosahedral mammalian virus: *Adenovirus* (Ad). Rod shaped and filamentous viruses: *Potato virus X* (PVX), *Tobacco mosaic virus* (TMV), and bacteriophage M13. Images of the following VNPs were reproduced from the VIPER Database (http://www.viperdb.scripps.edu/): BMV, CCMV, CPMV, P22, TYMV, FHV, HK97, MS2, Ad, and Q$\beta$. The structures of HCRSV, RCNMV, T7, PVX and TMV were reproduced from refs, respectively. (b). Genetic, chemical, and self-assembly/ encapsulation manipulations of VNPs in biomedical research. Taken from Ref. [10] with permission of Elsevier.

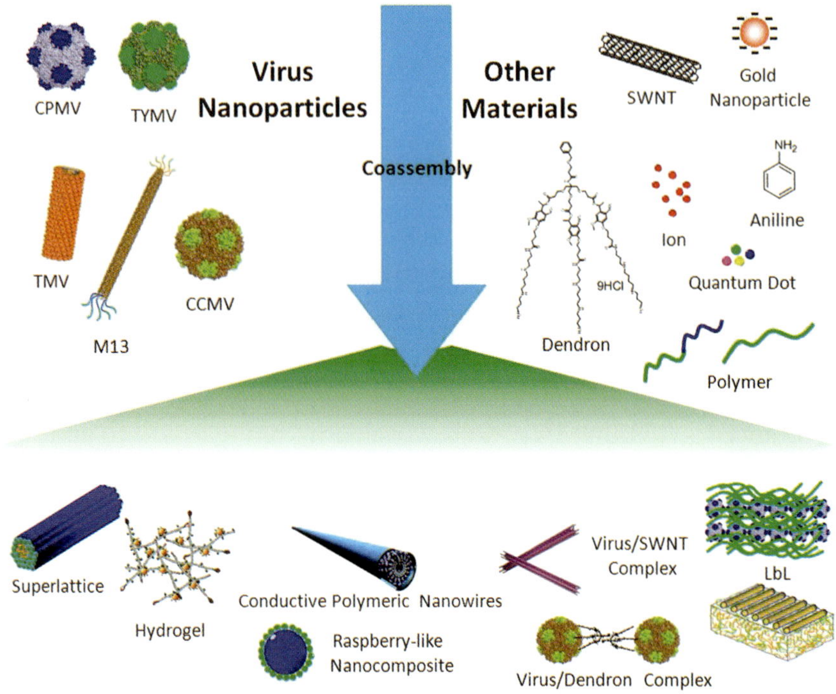


Figure 3. Self-assembly of hybridized VNPs with other materials. Taken from Ref. [11] with permission of Royal Society of Chemistry.

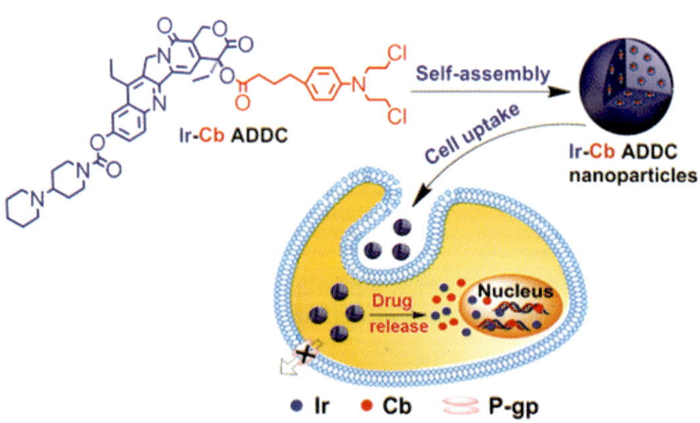


Figure 4. Example of an amphiphilic drug conjugate. Taken from Ref. [15] with permission of American Chemical Society.

is *N*-(2-hydroxypropyl) methacrylamide (HPMA), which is known to be non-toxic, non-immunogenic and very stable. To date, several conjugated, HPMA-active substances are the subject of clinical trials, including HPMA-doxorubicin, HPMA-paclitaxel, and HPMA-platinates.[19, 20] These conjugated HPMA and active substances, such as anticancer and imaging agents, have been obtained by copolymerization[21] or chemical post-conjugation.[22]

## B. Dendrimers

Dendrimers are highly branched synthetic polymers whose dimensions and physicochemical properties are close to those of biomolecules, such as proteins. They are produced by reactions in iterative steps; i.e. a tree.[23, 24] Both divergent and convergent approaches can be used. In the divergent approach,[25–27] the dendrimer is constructed from a heart that repeatedly emanates peripheral subunits. In this approach, the number of reaction sites is very important because it requires the use of highly selective reactions to avoid structural defects. In the convergent approach, the dendrimer is constructed by assembling dendrons.[28–30] In this approach, only a small number of reaction sites are activated at each step, thereby limiting the number of side reactions per step. Therefore, polymerization is better controlled.

Due to their structure and multivalent characteristics, dendrimers can be used as good hosts for guest substrates. The host-guest interactions can occur in the internal cavities of the dendrimer structure landscape (endo-receptor) or on the surface of the dendrimer (exo-receptor). This leads to the encapsulation of guest substrates (Figure 5).[31]

Because of their many surface sites, dendrimers can also be used to graft covalently active substances on their periphery (Figure 6).[32, 33]

Therefore, dendrimers can accommodate molecules of interest to bind with or be conjugated to different active substances for mono- or multi-modal applications. For example, polyamidoamine dendrimers (PAMAM) conjugated to iron oxide nanoparticles have been studied as magneto-guidable transfection vectors of antisense

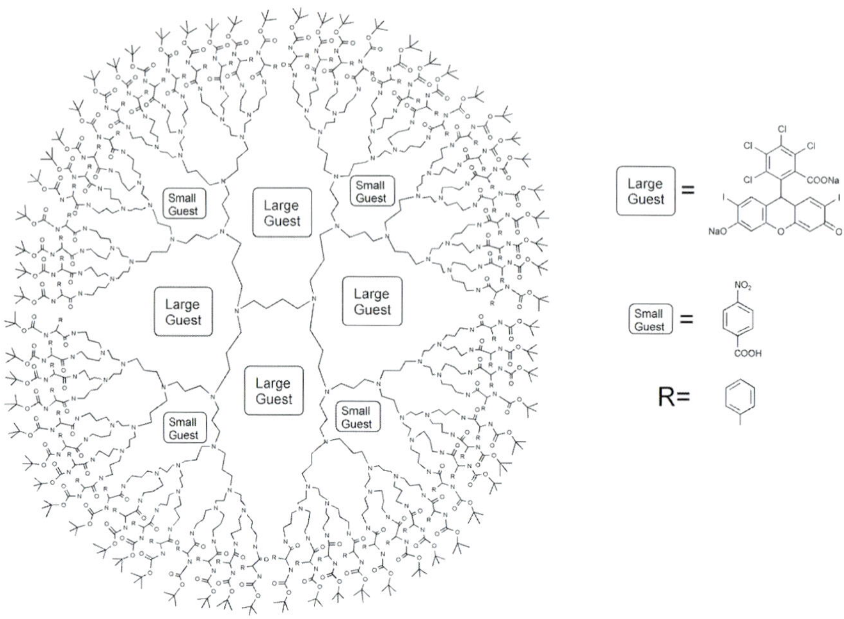


Figure 5. Representation of a dendritic box. Taken from Ref. [31] with permission of American Chemical Society.

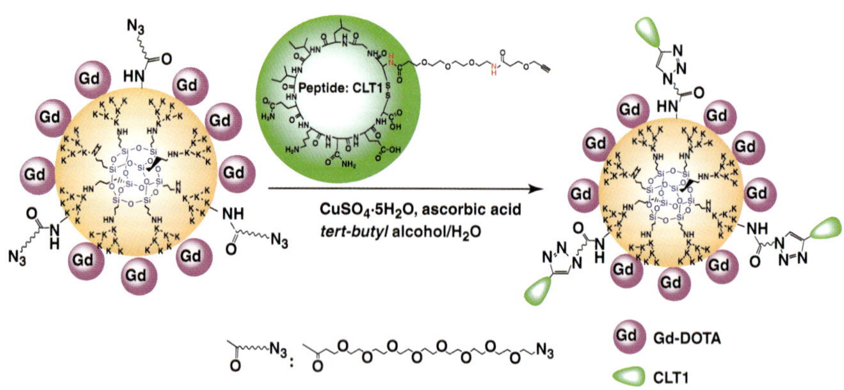


Figure 6. Dendrimer grafted periphery. Taken from Ref. [33] with permission of American Chemical Society.

oligodeoxynucleotides. Another more recent example is the encapsulation of a silicon naphthalocyanine within the hydrophobic cavity of fifth-generation polypropylenimine dendrimers. Depending on the light dose received and the power of the exciting laser, the naphthalocyanine can be used for optical imaging, PDT, PTT or all three simultaneously.[34]

## C. Inorganic nanoparticles

Unlike polymer nanoparticles and dendrimers, which are primarily used as "carriers" of active substances, most inorganic nanoparticles have imaging agent properties and can present a therapeutic effect. For example, iron oxide nanoparticles can be used as contrast agent for MRI or therapeutic agents in the case of hyperthermia therapy.

Several types of inorganic nanoparticles (e.g. carbon nanotubes (CNTs), fullerenes, iron oxide nanoparticles (IOs), quantum dots (QDs), and gold nanoparticles) have been developed with a theranostic goal.[35,36]

## D. Carbon nanotubes and fullerenes

Carbon nanotubes (CNTs) and fullerenes ($C_{60}$) comprise only carbon materials. Carbon nanotubes are made of graphite sheets wound on themselves. They may consist of a single layer of graphite (single-walled carbon nanotubes) or multiple layers (multi-walled carbon nanotubes) to improve their resistance.[37–39] The properties of the CNTs, and their optical properties in particular, depend on their size and the number of sheets that make up the tubes. The synthesis of nanotubes may be particularly oriented to impart good light absorption properties and a broad emission spectrum[39] for use in near infrared and Raman spectroscopy, for example.[40]

CNTs can also be functionalized with contrast agents for MRI and PET imaging applications.[41] For this, the surface of the CNT must be modified to develop graft sites by the introduction of surface defects (i.e. carboxylic acid).[42] For the non-covalent interaction of $\pi$–$\pi$ type and due to a large surface area, active substances can be associated with nanotubes.[43] Therefore, CNTs are credited with significant amounts of doxorubicin. As the $\pi$–$\pi$ interactions are pH dependent, the active compounds can be released by modulating the pH.[43] Fullerenes constitute a family of allotropic atoms in which the carbon atoms of the arrangement are composed of sheets of bonded hexagonal and pentagonal rings that prevent the sheet from being flat. Fullerenes have a hollow spherical structure of nanometric dimensions, which is useful to encapsulate active substances[44, 45] or contrast agents.[46–48] If the carbon nanotubes and fullerenes are interesting from a theranostic point of view, their *in vivo* toxicity is a crucial point to consider before clinical development.

## E. Nanoscale iron oxide

Iron oxide nanoparticles are magnetic nanoparticles consisting of hematite or magnetite. They can be obtained by the co-precipitation of precursors of Fe (II) and Fe (III) in aqueous solution.[49,50] The aggregation of nanoparticles is avoided by covering the magnetic core with hydrophilic polymers, such as dextran or polyvinylpyrrolidone. These polymers also provide functionalizable sites that make it possible to graft on the surface of the active substances or targeting systems.[50] Iron oxide nanoparticles can also be obtained by thermal decomposition of precursors of Fe (II) and Fe (III) in organic solvent.[51] This protocol allows better control of the size of the nanoparticles and their magnetic properties. However, because the nanoparticles that are obtained are hydrophobic, it is necessary to render them hydrophilic by ligand exchange or grafting to the surface.[52] Iron oxide nanoparticles that are 20 nm or smaller in size are characterized by superparamagnetic behavior.[53] They will affect the relaxation mechanism $T_2$ of the water molecules of protons and generate $T_2$-weighted images on MRI.[54, 55]

Iron oxide nanoparticles can also be used for the magnetic guiding of active substances.[56] In this case, the active compounds can be grafted onto the surface of the nanoparticles by amination reaction or co-encapsulated with iron oxides in polymer matrices.[54, 56]

## F. Quantum dots

Quantum dots (QDs) are semiconductor nanocrystals based on CdTe, PbS or $Cd_3P_2$, CdTe/CdSe, InAs/ZnSe, and InAs/InP/ZnSe and are generally covered with ZnS.[57, 58] They behave like a potential well that confines the electrons (and holes) in the three dimensions of space in a region of order size of the wavelength of electrons (wavelength of de Broglie), which is a few tenths of a nanometer. QDs are characterized by a very narrow emission spectrum whose wavelength may be modulated by their composition and size. They are also extremely bright and more photostable than are organic fluorophores, which makes them interesting for biomedical applications.[58] QDs are obtained by heating organometallic precursors in a high boiling solvent in the presence of surfactants (trioctylphosphine or trioctylphosphine oxide) to control particle growth. The organic layer surface may be functionalized to make water-soluble QDs. QDs' affinity for the sulfur can be used for compounds such as mercaptosuccinic acid, glutathione or cysteine.[57,58] Due to the toxicity of the metals entering into the composition of the QD, few *in vivo* tests using QD conjugates with active substances have been reported even if the QD may be functionalized. However, the use of theragnostic QD conjugated to doxorubicin and the RNA aptamer A10 for detecting a specific cell marker for prostate cancer has been reported.[59]

## G. Gold nanoparticles

Gold nanoparticles (AuNPs) are most often obtained by reducing gold salts in the presence of a polar solvent and a surfactant. This acts as an agent to protect metal particles because when it is adsorbed on the particle surface, it prevents particles from agglomerating.[60,61] The morphology and nanoparticle size distribution are controlled by parameters such as reduction kinetics and the nature of the stabilizer. The shape of the AuNPs determines their optical properties, in particular by adjusting the absorption wavelength plasmon nanoparticles (the plasmon band for spherical AuNPs is approximately 500 nm and that for AuNPs formed of rods is between 650 and 900 nm).[62] In terms of the functionalization of AuNPs, the strong interaction of gold to sulfur can be advantageously exploited. The biomolecules can be thiolated and then grafted onto AuNPs.[60,61] Similarly, the AuNPs can be conjugated to antibodies. AuNPs[62] have strengths for use in theragnostics. These nanoparticles are not only stable, biocompatible and easily functionalizable but also have optical exploitable properties, including photoacoustics, photothermal ablation and SERS spectroscopy.[62] An example of AuNPs for theragnostics that is in clinical trials for the photothermal ablation of tumors is AuroLase® (Nanospectra Biosciences).[64] Finally, gold has a high atomic number, a high electron density and a strong attenuation coefficient of X-rays, making it a good contrast agent for X-ray tomography.[60] However AuNPs are non-biodegradable and their possible persistence can cause toxicity in the long term.[65] Even if their biocompatibility is proven, it is possible that the surfactant, reducing and other molecules used for their preparation can lead to inflammatory processes. Currently, the question of *in vivo* and *in vitro* toxicity of AuNPs is quite controversial and hampers their clinical use.[66]

## H. Silica nanoparticles

Silica is a non-toxic, biocompatible material. Silica nanoparticles can be synthesized with good control of their size and shape. Their surface chemistry allows them to functionalize easily with contrast agents, active substances and targeting agents.[67] Silica nanoparticles are obtained by hydrolysis condensation of silicic precursors, such as tetraethoxysilane (TEOS).[68] The synthesis protocol may include co-precursors, such as aminopropyltrimethoxysilane or mercaptopropylmethoxysilane to introduce amine or thiol functions on the surface.[68] Substrates, such as antibodies, contrast agents, and fluorescent probes, can be introduced during synthesis provided that they have previously been coupled to one of the known co-precursors.[68] The nanoparticles may further be rendered mesoporous if a surfactant, such as n-alkyl trialkoxysilane, is introduced at the same time as TEOS in the synthesis.[69–71] Once the particles have formed, the surfactant is removed using a solvent, thereby creating pores with diameters that can be perfectly controlled. The mesoporous silica particles that are obtained have a high surface area that is suitable for functionalization. In addition, their biodegradability can be adjusted, making them of interest for theranostics.[72] However, the capture of these nanoparticles by the reticuloendothelial system can be a potential source of toxicity. Recent studies have suggested that silica nanoparticles are removed rapidly enough or eventually degrade so that their long-term toxicity can be considered low.[73] Finally, the biodistribution of the nanoparticles can be enhanced by PEGylation or by grafting specific biomarkers (Figure 7).[74]

In Figure 7, nanoparticles are functionalized sequentially in three areas: the silica structure with a contrast agent (ATTO647N), the

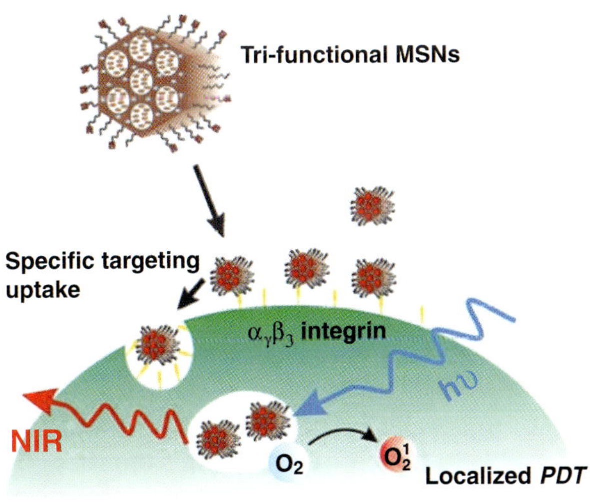


Figure 7. Mesoporous silica nanoparticles for theranostic applications. Taken from Ref. [74] with permission of Royal society of chemistry.

mesopores with a PDT agent (PdTPP), and the surface with ligand targeting tumor cells (cRGD).

## I. Coordination polymers

Coordination polymers or MOFs (metal organic frameworks), represent the last class of ordered porous materials.[75–79] One of their advantages over their organic counterparts, either carbonaceous or inorganic (zeolites, silica), is their ability to adjust their composition tailored by the choice of metal and/or organic constituent brick(s). Organic bricks are extremely varied (polycarboxylates, phosphonates, sulfonates, imidazolates, amines, pyridyl, and phenolates) and functionalizable.[80–82]

Functionalization can also be carried out retroactively on MOFs that are already formed.[80–82] In addition to their great chemical diversity, MOFs are characterized by a great diversity of shapes and pore sizes and there is the possibility of adjusting the pore size reversibly adsorbed to the substrate.[83,84] Therefore, one of their applications is their use as biomedical nanocargos and systems for the controlled delivery of active substances.[83, 84] The use of these porous solids in biomedical applications requires the use of biocompatible components for the development of MOFs. The most suitable metals are those whose toxicity is low, such as Ca, Mg, Zn, Fe, Ti, and Zr. The most common building blocks are synthetic from natural compounds that will be inert with respect to biological cycles *in vivo* (e.g. 2, 5 dihydroxyterephthalate,[85] polycarboxylates,[86] adeninate-4,4–biphenyldicarboxylate).[87] A second option is to use endogenous constituent bricks, such as fumarate, muconate,[88–90] cyclodextrins,[91] or amino acids.[92] Currently, the number of sufficiently porous frameworks and/or MOFs that are suitable for theragnostic applications is small, but the development of new synthetic protocols in the near future should advance this field. Finally, it is possible to obtain nanoparticles from these porous materials. Control of the particle size is a key point because it determines the mechanical properties of these nanoMOFs (NMOFs). Obtaining a stable, homogenous nanoparticle monodisperse population in this area is also crucial. The hydrothermal method,[93,94] which involves the use of microemulsions in the reverse phase,[95–98] the sonochemical path,[99–102] and the path hydro/solvothermal assisted by microwaves,[103,104] allows access. The issue of biodegradability (inherent in the use of inorganic ions) can be addressed through the use of endogenous metals with a control concentration. In general, the choice of components is infinite, making it is possible to select ligands and metals that degrade or eliminate. Therefore, the compounds of iron carboxylate NMOFs are biocompatible and biodegradable and are an example of MOFs compatible with intravenous administration. In addition, their high load capacity and extended release make these NMOFs very attractive nanocarriers.[77]

## J. Micelles and polymersomes

Micelles and polymersomes result from the self-assembly of surfactants or amphiphilic copolymers in solution. In an aqueous solution, the hydrophobic blocks of the amphiphilic molecules orient themselves to minimize adverse interactions with the aqueous environment. This organization takes place only above a certain concentration, called the critical micellar concentration (CMC).[105] The CMC depends on the chemical nature of the amphiphilic molecules. Generally, the higher the molecular weight of the hydrophobic portion, the lower the CMC.[105] As long as the amphiphilic concentration remains above the CMC, the assembly remains thermodynamically stable, but below the CMC, it disassembles with a specific speed that depends on the structure of and interactions between the chains. According to the physicochemical properties of the amphiphile, aggregates formed can be of different sizes and shapes. Therefore, one can find aggregates formed from a single spherical or cylindrical layer (micelles simple type), two flat layers (lamellar), or closed as concentric hollow spheres. In the latter case, this is called vesicle architecture.[106–108] The vesicles, as opposed to micelles, are composed of at least two layers of amphiphilic derivatives. The vesicles may be constituted by a single lamina (uni-lamellar) or two or more sipes (multilamellar).

The micelles may be obtained using either common surfactants, such as Cremophor® and polysorbates (non-polymeric micelles), or amphiphilic copolymers (polymeric micelles).[108] The amphiphilic copolymers compared to low molecular weight surfactants have a lower CMC, and micelles can be more resistant to disassembly under the effect of dilution (particularly in the blood).[109–111] Therefore, polymeric micelles are preferable to non-polymeric micelles. The micelles have a heart-ring architecture in which the heart is composed of the hydrophobic portion, creating a space into which lipophilic molecules of interest may be solubilized. The ring formed by the hydrophilic part of the amphiphile generates a hindered opsonin and therefore makes it possible to increase the half-life in the bloodstream significantly. In addition, this ring may be functionalized by grafting hydrophilic markers or diagnostic agents (Figure 8).[112,113]

Micelles can be obtained in different ways, including direct dissolution, dialysis, oil in water emulsion (O/W), evaporation of solvent, salting-out, or rehydrating a polymer film.[105] These methods can lead to differences in structure and load levels of active ingredients.[114] Their small size (typically <100 nm) reduces their elimination by phagocytes of the liver and spleen and maximizes

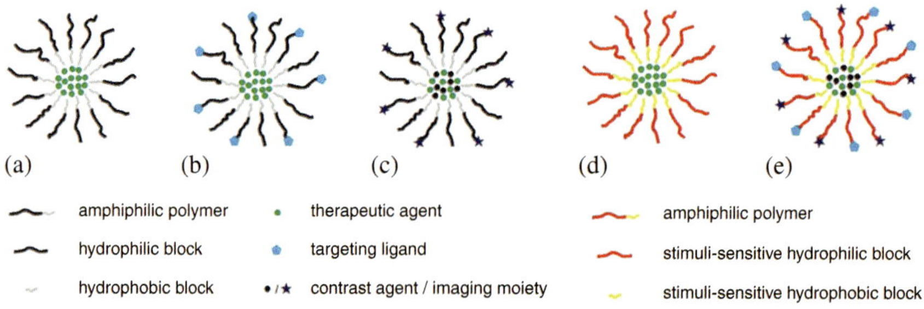


Figure 8. Schematic micelle polymer (a); functionalized micelle with targeting markers (b); theragnostic micelle (c); modified micelle for controlled release (d); and optimized micelle to treat cancer or thernagnostic micelle (e). Taken from Ref. [107] with permission of Royal Society of Chemistry.

the EPR effect. At the same time, the size is sufficient (>10 nm) to prevent renal excretion. Using polymers sensitive to temperature changes, pH, light or magnetic field controls the disassembly of the micelles and therefore the release of the active principle. Grafting recognition systems facilitate the preferential accumulation in the area of interest.[112,115–117] These features make micelles prime candidates to convey anticancer drugs or to serve as theranostic agents.[118–121] For example micelles used in theranostics, such as micelle polymers in which the amphiphilic unit carries a fluorescent group, have been developed and used for the delivery of doxorubicin. Modifying fluorescence micelles, which are controlled by their assembly and disassembly, can be used to track the release of active principle by fluorescence.

Moreover, the encapsulation of doxorubicin in these micelles enhances its antitumor activity *in vitro*.[120] The polymersomes are vesicles synthesized from amphiphilic copolymers of blocks that are organized as a sphere hollow (Figure 9). Two methods are primarily used: solvent change and rehydration of a polymer film.[122,123] In these hollow structures, the aqueous heart is surrounded by a bilayer wall.

Figure 9. Schematic representation of polymersome with a bilayer wall. The hydrophilic crown is shown in blue and the hydrophobic portion is shown in red. Taken from Ref. [122] with permission of Elsevier.

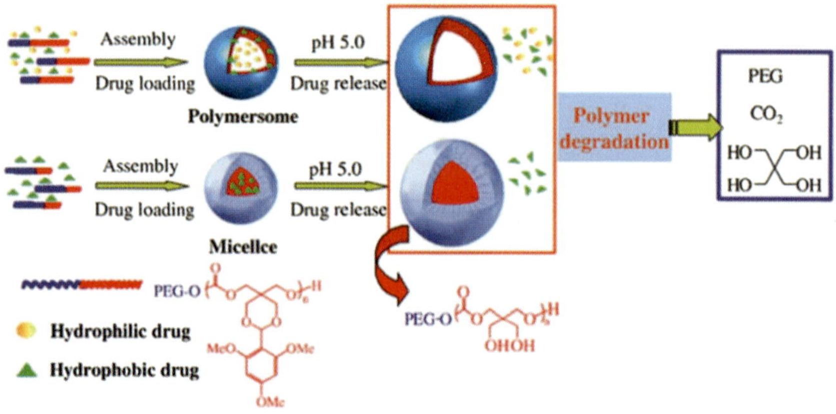


Figure 10. Smart polymersomes and micelles prepared with diblock that are PEG-sensitive to pH and PTMBPEC degradable. Taken from Ref. [123] with permission of Royal society of chemistry.

This membrane displays a hydrophilic interface on both the inner and outer surfaces. Between the two is the hydrophobic portion of the wall.[122] This organization can encapsulate hydrophilic molecules in the cavity and transport hydrophobic molecules and/or amphiphilics in the wall.

With this hydrophilic/hydrophobic duality, the ability to incorporate different molecules, and the robustness of the wall, polymersomes are smart polymers (Figure 10) to activate the release of active substances by polymersomes and their external functionalization may ensure better targeting.[125,126]

## K. Liposomes and lipid nanoparticles

Liposomes are the result of the self-assembly of natural or synthetic lipids (cholesterol, phospholipids and derivatives) in water. They have a structure similar to that of polymersomes. These are spherical vesicles with a hydrophilic cavity protected by one or more membrane bilayers whose polar heads oriented toward the outer surface and the internal volume.[127] Differences between liposomes and polymersomes come from the nature of the constituents. In the case of polymersomes, the polymers have a molar mass greater than that of the phospholipids used for the liposomes, resulting in the formation of a thicker membrane that is stiffer, stronger and less permeable.[111,128] The relative permeability of liposomes can result in premature or uncontrolled release. This weakness can be overcome by incorporating a liposome in other liposomes, similar to a nesting doll. These new structures, called vesosomes, have multiple membranes and allow for a delayed release of the molecule or molecules of interest.[128]

In vesosomes, internal compartments are separated from each other and may have different lipid compositions and encapsulate various substances.[129] Liposomes are classified according to the number of lipid layers (uni- or multilamellar) and size (small, large, and giant).[128,130,131] The size and number of the bilayers affect traffic, the half-life, and the charge rate of encapsulated molecules.[127] Multilamellar liposome sizes are obtained from the hydration of a lipid film. To form smaller unilamellar vesicles, it is necessary to use ultrasound or extrusion. A method involving dispersion solvent and detergent removal may also be employed.[127] Because of their biocompatibility, biodegradability, low toxicity and ability to transport all hydrophilic molecules, both lipophilic and amphiphilic, they are extensively used in various fields, such as cosmetics, pharmaceuticals, and food.[127,130–132]

Liposomes, which were discovered by Bangham in 1965, were cleared by the FDA for the first time for therapy in 1995 as Doxil® (Ben Venue Laboratory).[133] Since then, liposomes have been approved for clinical applications in oncology (DaunoXome® (Novex Pharma) and Myocet® (Cephalon)); virology (Inflexal V® (Crucell)); and ophthalmology (treatment of AMD Visudyne® (Novartis Pharma)).[133–136]

Most of these liposomes are cleared for intravenous and intramuscular administration (in the case of antivirals, such as Epaxal® and Inflexal®). However, the oral route is avoided due to the instability of liposomes in gastric fluids.[133,134]

These systems are also suitable for the encapsulation of diagnostic agents.[137,138] However, few studies have described the combination of therapeutic and diagnostic agents into liposomes.[136] The liposomes and lipid nanoemulsions are composed of excipients that are well tolerated *in vivo*, have a structure and composition similar to those of physiological membranes and oral bioavailability, and can be easily mass produced. However, they suffer from too much permeability and chemical sensitivity to oxidation and degradation compared to other nano-formulations.[131,132] In the mid-1990s, solid lipid nanoparticles (NLSs) were needed as an alternative to existing lipid nanoformulations.[139] NLSs combine the advantages of solid nanoparticles (stability, controlled release) with those of liposomes and nanoparticles liquid lipids (biocompatibility and good tolerance for many routes of administration).[139,140] NLSs are derived from oil-in-water (O/W) emulsion nanoparticles in which the liquid lipid is replaced by a solid lipid at room temperature and up to 37°C. The lipids used are physiological or recognized as biocompatible *in vivo*.[141] NLSs are prepared from lipids, surfactants, organic solvents and an aqueous phase using the one of many possible synthesis methods, including inversion temperature (PIT), high-pressure homogenization, dispersion of a film by ultrasonic emulsion-solvent evaporation, micro-emulsion, and the use of a supercritical fluid. The hydrophobic lipid cores of nanoparticles allow the encapsulation of therapeutic agents and diagnostic lipophilics.[139–142] The encapsulation of hydrophilic molecules, such as certain proteins or anticancer or contrast agents, for MRI is still possible. For example, the encapsulation of Gd-DOTA in an NLS produced by a double water-oil-water (W/O/W) emulsion was achieved.[143] Solid lipid nanoparticles have potential as theranostic agents.[144] An NLS has both advantages and a number of disadvantages, such as problematic reproducibility in the growth of nanoparticles and the possibility of polymorphic transitions that can induce the expulsion of the active substance during storage and lead to a low expense ratio.[139,140,145] These disadvantages can be overcome with second-generation lipid particles of a controlled nanostructure, which are called nanostructured lipid nanoparticles (NLNs). NLNs consist of a mixture of lipid solids and liquids that generate a matrix that remains solid at 37°C, but the melting point decreases compared to an NLS. The solubility of the active ingredient increases.[145] These structures are very stable and have a high capacity for lipophilic agent loading. However, the incorporation of hydrophilic molecules is possible only in limited quantities.[146]

## L. Polymeric nanoparticles

Polymeric nanoparticles are defined as colloidal particles prepared from polymers that range in size from 10 to 1000 nm. There are two types of nanosystems: nanocapsules and nanospheres.[147]

Nanocapsules can be defined as reservoir vesicles consisting of a liquid or semi-liquid heart (water or oil) surrounded by a solid polymer shell. The molecules to be conveyed can be encapsulated into the heart or be adsorbed to the surface of the nanocapsules. The nanospheres are matrix particles that consist of a completely solid polymer entanglement. The molecules of interest can be adsorbed to the surface or be dispersed throughout the matrix network of the nanospheres (Figure 13).[147–151] These nanoparticles are generally spherical. The structure of the nanospheres/nanocapsules and the adsorption mode of incorporation/encapsulation are driven by the choice of synthesis method and the nature of the molecules to be conveyed, respectively.[150, 151]

Fragile molecules, which are desired to preserve integrity *in vivo*, will preferably be encapsulated within the nanoparticle rather than adsorbed on its surface. By contrast, molecules that may lose their integrity at the time of encapsulation will be preferentially adsorbed after the synthesis of nanoparticles.[152–155] Polymeric nanoparticles have applications in many fields, including electronics, medicine, cosmetics, and technology.[147]

## M. Polymeric nanoparticle design

Among all nano-objects of therapeutic and diagnostic interest, polymeric nanoparticles are currently the most studied for their numerous advantages, including biocompatibility, biodegradability, and chemical versatility for the alteration of their physicochemical properties. The development of these nanocarriers has evolved from a simple structure to complex and carefully crafted assemblies.

## III. Different Generations of Nanocarriers

Currently, there are four successive stages in the development of nanocarriers. They are listed in order of increasing responsiveness to the body and cancer pathology.

### A. First-generation passive addressing

The first-generation polymer nanocarriers appeared shortly after the reported use of raw liposomes. These nanoparticles were "naked," with an unmodified surface of nanosphere or nanocapsule types (Figure 13). The active ingredients are dispersed in the molecular state in the matrix (nanospheres) or in an oily or aqueous heart (nanocaspules). A proportion of the active ingredients is adsorbed on the surface of the vectors. The most common polymers used for the synthesis of nanocargos are acrylates, such as poly (alkylcyanoacrylate) (PACA); polyesters, such as poly (lactide) (PLA), poly (lactide co-glycolide) (PLGA), and poly (e-caprolactone) (PCL); and polysaccharides, such as chitosan, gelatin, and hyaluronic acid. All of these polymers share the properties of being biocompatible and biodegradable, which is caused by hydrolysis *in vivo* from acids or sugars naturally present in the body.

The first attempts to intravenously administer polymeric carriers resulted in rather disappointing results because a very rapid clearance (on the order of minutes) and limited biodistribution, mainly in liver and spleen, were observed.[156] Adsorption of plasma proteins on the surface polymer nanocarriers, or opsonins, has been recognized as a factor responsible for the recognition of such particles by the reticuloendothelial system (RES). Hydrophobic interactions that develop these opsonins for polymer surfaces can be reduced by changing the nature of the polymer matrix. Unfortunately, these interactions cannot be prevented completely. Some therapeutic strategies,[157] however, address this situation by targeting cancerous diseases of the liver. Therefore, Verdun *et al.*[158] used nanoparticles (hexylcyanoacrylate) loaded with doxorubicin to treat diseases of the liver and spleen. In addition, the accumulation of nanoparticles in the heart and other organs reduced the cardiotoxicity of doxorubicin, which is a significant advance in their use.

## B. Second-generation stealth

Because the first-generation nanoparticles (lipid or polymer) had an extremely short half-life, solutions based on the work initiated by the team of De Gennes on a "protein interaction surface" in the presence of poly (ethylene oxide) (PEO or PEG) were proposed.[159] These studies reported that the incompressibility of PEG chains grafted onto the surface of a hydrophobic solid repelled proteins and did not allow their adsorption onto the hydrophobic surface. This observation on the role of steric hindrance on the surface of nonionic polymers quickly led to the development of second-generation nanocarriers, which have a hydrophilic PEG crown surface. Because the adsorption of serum proteins on the surface of the nanoparticles is the leading cause of immune recognition and rapid clearance of nanocarriers, the plasma residence time increases significantly in the presence of a PEG surface of nanocarriers.[160] Therefore, these nanoparticles are called "stealth" to distinguish them from those of the first generation. This feature of stealth also alters the biodistribution of second-generation nanocarriers compared to those of the first generation. Because they have a longer circulating in the body, these nanoparticles can accumulate in tissues whose blood supply is looser for pathological reasons. This is the case in inflammatory tumor areas. Under the impulse of tumor growth, neovascularization is established quickly. It is characterized by an abnormal structure, disorganized and a more permeable vascular endothelium with fenestrations (pores) between the endothelial cells ranging from 300 to 500 nm. In this context, nanocarriers that are less than 200 nm may passively reach the tumor tissues. In addition, malfunction of the lymphatic drainage of the tumor areas slows down or even prevents the clearance of these nanocarriers. This is the EPR effect (enhanced permeability and retention, Figure 11).[161] The efficiency of this approach, which was first demonstrated by Matsumura and Maeda,[162] on the accumulation of macromolecules in tumor tissue has been confirmed for various nanocarriers, including polymeric nanoparticles.[163]

The surface PEGylation of the nanoparticles can occur in different ways. The PEG may be covalently grafted to the hydrophobic polymer, or it may be co-precipitated during synthesis or adsorbed onto the surface. From these various steps, surface coatings with densities and various architectures will result. The optimum parameters for better stealth and biodistribution are adequate and not controversial at this time.[165] However, the accumulation of data on the use of PEG has highlighted a number of negatives attributes, including the fact that PEG interferes with cellular internalization and the endosomal escape phenomenon and could cause undue immune reactions.[165] For this reason, other surface-coating polymers have been studied and used successfully, in particular poloxamers, poloxamines and polysosides.[166] Regardless, PEG remains the ultimate stealth material, and numerous ongoing clinical trials are using PEGylated nanoparticles. For example, Abraxane® (Celgene) is an injectable, specialty-based

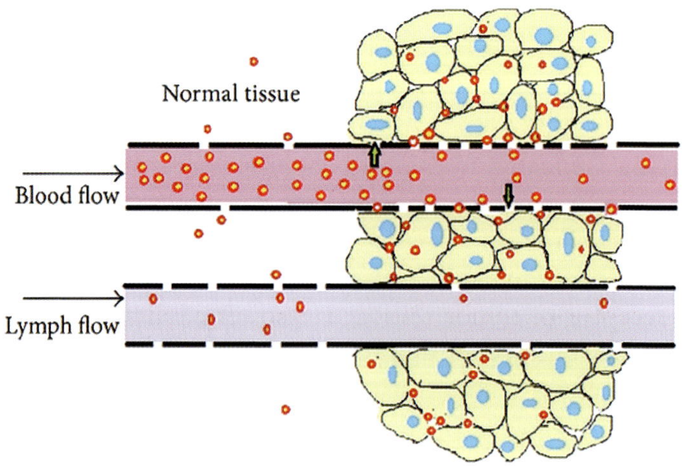


Figure 11. Schematic representation of the EPR effect. Taken from Ref. [164] with permission of Hindawi.

paclitaxel (cleared for marketing since 2005) that is prescribed for the treatment of breast, lung and pancreatic cancer. These albumin nanoparticles carry the antineoplastic agent, and this was the first specialty polymer to obtain clearance to be placed on the market. Although it is devoid of PEG, this formulation may be regarded as a second-generation nanocarrier because albumin naturally attracts the RES, and the half-life of the nanoparticles is extended.[167]

## C. Third generation: active targeting

The concept of active targeting is defined in opposition to that of passive targeting in tumor areas, or the EPR effect. The term "active targeting" refers to second-generation nanocarriers with surface specific ligands that target specific membrane cell receptors.[168]

Two targeting strategies can be employed. One strategy targets the tumor environment (e.g. endothelial cells) to deliver anti-angiogenic molecules or molecules that stimulate antitumor immunity. The other strategy targets cell membrane receptors of tumor cells for the intracellular delivery of antineoplastic molecules.[169] In the case of solid tumors, for active targeting to work properly, the accumulation of nanocarriers in the tumor areas (the EPR effect) is a prerequisite. Molecular recognition between the ligand and target membrane site can then take place, which triggers endocytosis of the nanocarrier and the release of molecules of interest.[170] The targeting entities to the surface of the nanocarrier can be very diverse and include, for example, monoclonal antibodies, lectin, aptamer, folate, and peptides.[171–173] Selecting this entity will, of course, be dictated by the tumor to be treated, and, in particular, the level of affinity between the ligand and its receptor.[174] Therefore, a candidate must have a high affinity for a specific receptor of tumor cells. It is also required that this receptor be present in large amounts on the surface of the target cell as well as on the ligand to the surface of the nanocarrier. This is one reason why the design and development of these vectors is so delicate and is still in the experimental stage. This strategic approach has nevertheless been proven *in vivo*. Farokhzad *et al.*[173] have demonstrated the effectiveness of PLGA-PEG nanoparticles loaded with docetaxel on human tumors implanted in nude mice. The presence of an aptamer that recognizes the extracellular domain of the antigen specific for prostate tumor cells (PSMAs) on the surface of the nanoparticles resulted in the complete regression of tumors, while nanocarriers that had had no aptamers slowed tumor growth.

However, despite these encouraging results, a major problem remains: the low penetration of nanocarriers in tumor areas, which is limited to a few millimeters around the blood vessels.[175] Therefore, the hearts of solid tumors can be reached by antineoplastic molecules, which do not eradicate the tumor.

## D. Fourth-generation "smart" vectors

The term "intelligent vectors" refer to nanoparticles that are capable of responding to exogenous (e.g. magnetic or electric field, temperature rise) or endogenous (e.g. change in pH, change in concentration of an endogenous molecule, presence of an enzyme special) stimuli.[176] In response to the specific stimulus, the polymer matrix undergoes protonation hydrolytic cleavage or a conformational change that activates the release of the active substance.[177] Therefore, these nanocarriers enable the release and/or activation of molecules transported in a controlled manner. It is possible to exercise control over the space where the molecules are released, the release time and the dose released (Figure 12).

The outbreak of desired activity (e.g. the release of an active principle or imaging localization) is made possible by the design of polymers that correspond to the target stimulus. The biophysical properties[179] most promising for biomedical applications include thermal sensitivity. This property is particularly useful for treating tumors because of the beneficial effects of a temperature increase on the destruction of cancer cells. It is therefore easy to combine the two concepts to enable the release of a therapeutic action. This can be achieved by the external application of a heat source because the increase in temperature causes the passage of certain polymer segments of the state of aggregates to the monomer, which releases the polymer network and allows for the distribution of the active

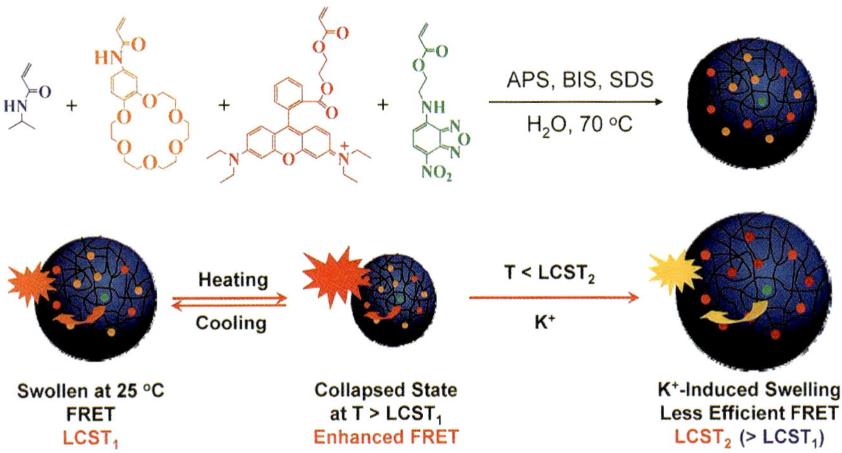


Figure 12. Example of the activation of an intelligent nanovector. The change in temperature causes a reduction in the size of the particle, which enhances the Förster resonance energy transfer (FRET) effect. Taken from Ref. [178] with permission of Royal society of chemistry.

ingredient outside of the nanocarrier.[180] This behavior is possible for polymer units exhibiting a "lower critical solution temperature" (LCST). A significant number of studies have also included the activation of these polymer blends with the release of heat produced by the magnetic induction SPIO.[181–183] SPIO contributes to both the response to the stimulus and the component imaging. This property is often referred to as magneto-sensitivity. The LCST is the critical temperature below which the components of a mixture become miscible in any proportion.

Another important biophysical property is pH sensitivity. The observations that tumor microenvironments have a slightly acidic pH (pH = 6–6.5) and that acidification occurs during the endosomal and lysosomal digestion process (pH = 6–4.5) led to the design of pH-sensitive polymers. The effectiveness of this strategy for the targeted release of active ingredients has already been demonstrated *in vitro* and *in vivo*.[184–186] It is particularly advantageous in the case of gene therapy because the nanocarrier is able to disassemble at a specific pH and therefore avoid lysosomal degradation for better transfection of target cells.[187] The release of the active principle is based on photo-crosslinking of the polymer matrix, which induces tightening the polymer network and the subsequent expulsion of the encapsulated molecules.[188] Of course, other types of responses are being explored. These various strategies are an innovative and promising approach to circumvent the current blockages in the clinical translation of nanotechnology.

## IV. How do Nanoparticles Pass Barriers?

An efficient drug delivery system needs to protect drugs from enzymatic, mechanical, or chemical degradation. It must also have enhanced diffusion through the epithelium, targeted tissue distribution, or increased penetration into the target cells.[189] Therefore, drug delivery nanosystems need to be designed to overcome many physical barriers.

A typical application that involves crossing biological barriers is the development of nanosized carriers for brain cancer treatment. The main limitations are crossing the blood brain barrier (BBB), the transport within the interstitium, the specific targeting of tumor cells and the delivery of the drug in high amounts.

To overcome these problems, controlling the physicochemical properties of the nanocarriers, such as composition, size, and zeta potential, is very important. The nanoparticles have to be smaller than 6–8 $\mu$m (the size of human red blood cells). To be sequestered inside the cellular nucleus, nanoparticles must have a diameter of less than 40 nm.[190,191] Unlike microparticles, nanoparticles delivered intravascularly can bypass the innate immune system by preventing uptake by the mononuclear phagocyte system (MPS).[192] Their size allows good penetration through the smallest pores of the capillaries in the human vasculature (200–1000 nm).[192–196] Spherical nanoparticles promote better surface coverage by hydrophilic polymers and targeting ligands, which is important.[197,198]

An ideal particulate carrier would have the capability to carry therapeutic agents to the target because of conjugated antibodies or other recognition moieties,[199,200] to image diseased tissue, and to avoid biological barriers that can promote clearance from the systemic circulation.[201,202]

Healthy tissues contain blood vessels lined by a smooth layer of endothelial cells with pericytes that maintain the integrity of the vessels.[203] In tumor vasculature, there is a defective endothelial cell barrier with a loose attachment of pericytes that results in a leaky vasculature with fenestrations and irregular vessel diameters.[204] Many tumors lack lymphatic vessels, and tumors with lymphatic vessels have characteristically wider lumens and an increased number of intracellular spaces and sprout endothelial cells, resulting in an increase in interstitial fluid pressure.[203,205,206]

A tumor microvasculature has a characteristic pore cutoff size,[195,207] which is important for the design of a selective colloidal carrier. Transport across the tumor microvasculature has been shown to occur via interendothelial junctions, fenestrations and phagocytosis.[195,204] In normal vessels, interendothelial junctions have an effective size of 6–7 nm, which provides resistance to particulate drug delivery.[195,208] Hobbs *et al.*[195] demonstrated that tumors grown subcutaneously have a tumor-dependent pore cutoff size of 200 nm –1.2 mm. Tumor microvessels have been shown to be hyperpermeable to long-circulating PEG liposomes and polystyrene latex particles of up to 600 nm in diameter.[207] The results obtained by Hobbs, Yuan, and Dreher demonstrated that the size of the nanoparticle should be within a range of 100–780 nm to effectively transport across the microvascular wall into the tumor interstitium.[195,207,209]

Drug delivery can be achieved by passive targeting (following the EPR effect, which promotes the accumulation of drug delivery carriers in the tumor interstitium)[210] or by active targeting because of specific receptors overexpressed on the surface of cancer cells;[211,212] molecules or ligands (e.g. antibodies, lectins, saccharides, hormones, and small molecular weight compounds) can recognize the cellular receptors to trigger internalization via receptor-mediated endocytosis.[211,213]

Drug delivery carriers can be externally guided to cancer cells or tissues with an applied magnetic field through which colloidal carriers bypass physical and cellular barriers. Strategies that incorporate iron oxide nanoparticles and gold nanoparticles within the nanocarrier structure are beneficial for applications in magnetic resonance imaging (MRI), triggered drug release, and localized thermal therapy.[214–217] The concept of a magnetically targeted delivery system emerged 50 years ago,[218] and since then, extensive research has been performed to illustrate the utility of these ferrofluids in cancer treatment.[213] Once the particulate drug delivery has collected in the tumor vasculature, external activation by focused ultrasound, radiofrequency, laser light (photodynamic therapy), or an applied magnetic field can help trigger drug release and cell death (thermal ablation therapy).[219–222] Drug delivery carriers, such as liposomes (MAGfect)[215] and hollow microcapsule-loaded magnetite particles,[223] have been shown through MRI imaging to enable triggered release for anticancer applications.

To deliver high concentrations of drug to the brain, the BBB has to be crossed over by the nanocarriers. This barrier ensures tight regulation of transport between the bloodstream and brain tissue. Several methods have been explored to allow transport of a therapeutic across the BBB.[224]

One method to increase drug delivery is the use of BBB disruption to improve the permeability of the vessels by increasing the local osmotic pressure in brain vasculature using hyperosmotic agents.[225] However, this method does not provide the same results for all patients.[226] All BBB disruption methods cause decreased integrity of the complete BBB and not specifically the vasculature of the tumor, which could adversely affect healthy brain tissue. BBB disruption can also allow leakage of unwanted molecules from the circulation into the brain.[227] To avoid disrupting the entire BBB, various strategies have been employed to penetrate the BBB via nanoparticle drug carriers.

LDL receptors on the endothelial cells of the BBB can facilitate the uptake of nanoparticles coated with ligands for the LDL receptor. Xin *et al.*[228] reported that Angiopep-coated PEG–poly($\varepsilon$-caprolactone) nanoparticles can be accumulated in the tumor due to the passive EPR effect and active targeting through Angiopep. Apolipoprotein-coated particles using apoA-I-coated protamine/oligonucleotide nanoparticles *in vitro*[229] and apoE-coated serum albumin nanoparticles *in vivo*[230] are taken up by binding to the LDL receptor. By grafting an agent that causes absorptive uptake on the nanoparticle surface, this study showed the uptake of drug-containing nanoparticles by the BBB endothelium via adsorption by the cells rather than by increasing gaps between endothelial cells in the entire brain. In another study, Kreuter *et al.*[231] developed poly(butyl cyanoacrylate) nanoparticles as carriers for a peptide that normally could not cross the BBB. By coating the particles with a surfactant that caused apolipoprotein deposition onto the nanoparticles once they were in the plasma, they achieved absorptive uptake of their particles. The use of polysorbate 80[232] and other surfactant coatings, such as poloxamer 188[233,234] and Tween® 80,[235] has also been shown to cause high uptake by BBB endothelial cells in *in vitro* models and to lead to higher accumulation in the brain *in vivo*. Liposomes have also been cited to increase BBB penetration.[236] Other lipid-based nanoparticles have been reported to have BBB-penetrating properties as well as the ability to take advantage of the EPR effect in tumor models.[237]

To exploit the EPR effect or induce internalization by endothelial cells, particles being studied as potential drug carriers to brain tumors are generally less than 150 nm in diameter.

Another advantage of nanoparticle drug carriers in BBB penetration is their chemical versatility. Surfactants or stealth coatings, such as PEG, ligands and other biological or chemical moieties, can be conjugated to the surface to promote active uptake by cells on the luminal side, trafficking of the particle through the endothelial cell, and exocytosis into the brain tissue. For example, PLA nanoparticles were surface modified with PEG for stability and with cationic serum albumin for increased circulation time. The cationized albumin facilitated the interaction of the particles with brain endothelial cells to promote uptake with little or no observed toxicity.[238]

Qin *et al.*[239] used Tat-conjugated cholesterol to formulate liposomes that showed the ability to transcytosis through brain capillary endothelial cells and accumulate in the brain. In another study, poly(amidoamine) dendrimers were conjugated to a peptide derived from the rabies virus glycoprotein, RVG29, through a PEG linker and conjugated with DNA to form nanoparticles. The RVG29-modified accumulated in the brain.[240] Other researchers have taken advantage of toxins that increase vascular permeability, such as the diphtheria toxin.[241] Specificity can also be built into the delivery system by using specific ligands that promote receptor-mediated endocytosis. Insulin was transported across the BBB from the circulation by receptor-mediated transcytosis,[242] and an antibody to the insulin receptor was taken up effectively into the brain *in vivo* in a primate model.[243] Ulbrich *et al.*[244] conjugated drug-loaded human serum albumin nanoparticles to antibodies against insulin receptors and were able to achieve uptake into the brain in a mouse model. Polyester nanoparticles loaded with anticancer drugs, such as taxols, based on materials such as PEG-conjugated PLA or PLGA,[245,266] were able to achieve higher accumulation in brain endothelial cells when conjugated to transferrin. Like transferrin, the folate receptor for the transport of folic acid is also upregulated in rapidly dividing cells, including those in malignant brain cancer.[247]

## V. Controlled Release from Mesoporous Silica Nanoparticles (MSN)

In order to have an ideal cancer therapy with nanoparticles delivery system, there is a need for nano delivery system that keeps anticancer drugs inside the nanoparticles until they reach tumor and the cargo (which is the anticancer drug in this case) is released only when a signal to release is provided. An important feature of nanoparticles which is used in cancer therapy is "zero release until nanoparticles reach tumor."[248] A variety of triggers such as external magnetic field, pH and redox state as well as external cues such as light have been used to accomplish on-command release. Besides carrying out delivery of anticancer drugs, they can transport compounds such as gadolinium complexes and fluorescent dye. Therefore the probe changes to the theranostic agent which can have the imaging and therapy simultaneously.[248,249]

Mesoporous silica nanoparticles are usually synthesized by a sol–gel method to produce homogeneous size nanoparticles.[249]

Mesoporous structures which have a structure with many small pores are the results of addition of surfactant during the synthesis of these nanoparticles. Kuroda and his colleagues are the pioneer and developer of the idea of surfactant template to produce mesoporous materials.[250]

Mesoporous silica nanoparticles (MSNs) have numerous advantages for developing controlled release systems. Firstly, extensive surface area enables various modifications to be carried out. Secondly, they are relatively more stable in comparison with other types of NPs. A variety of chemical modifications have been made on their surface as well as on pore interiors. Thirdly, the pores in the NPs provide storage space for anticancer drugs, therefore these NPs have a great ability in loading.

## A. Mesoporous silica nanoparticle (MSN)-based controlled release systems

There are two main approaches to obtain MSN with control release capabilities. The first approach which was developed by Stoddart and Zink who involve attaching organic molecules (such as rotaxanes and pseudorotaxanes) at the pore opening in order to preventing release of the cargo (here it could be anti-cancer drugs) stored in the pore.[251] This method is a so-called "capping" or "gating." "Nanovalves" can be attached to the pore openings to provide open and close function for the cargo stored in the pores. The anticancer drugs stored in the pores will remain inside nanoparticles by closing the nanovalve. Various other materials have been developed for capping (gating). Polymers have also been used to cover pore openings.

The second approach to prepare MSN with controlled release feature is to attach drugs to the surface of MSNs via stimuli-responsive linkages.[249] Two kinds of stimuli can be employed. Firstly, MSNs that respond to external stimuli such as an external magnetic fields[251–258] or light (such as near infra-red (NIR) or ultra violet (UV)) which leads to increase the temperature of environments[259] have been developed. Use of ultrasound to trigger drug release has also been explored.[260,261] Studies showed that using magnetic field as external stimuli for controlled release has an advantage, as it has much better tissue penetration compare to the light.[249,262] MSNs can be heated up to 42° while maintaining the surrounding temperature at 19°.[252,253] The second stimuli are intracellular or intratumoral conditions which is called autonomous release of anticancer drugs. These triggers could be enzymes, biomolecule[263–266] or low pH (due to hypoxic conditions in the tumor where the pH is low). This feature provides an advantage that the drug release is more limited to the tumor area).[267–271] Redox and biomolecule-activated systems (reducing conditions such as glutathione)[272–277] have also been used.

For more detailed information on chemical features of these and other nanoparticles, please refer to reviews by Mekaru and

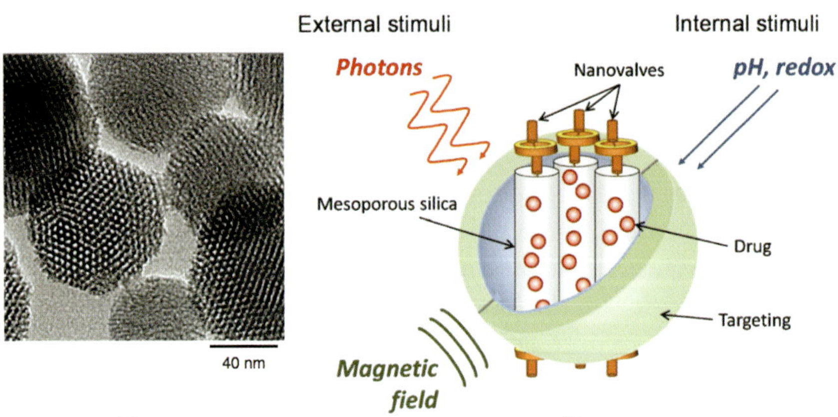


(a) (b)

Figure 13. Mesoporous silica nanoparticles synthesized by the sol–gel method. (a): TEM of MSNs. They are homogeneous with the diameter of approximately 130 nm and contain 1400 pores that can be used to store anticancer drugs. (b): A schematic overview of mechanized nanoparticles based on MSN. Nanovalves will be attached to the opening the pores. Cargo molecules such as anticancer drugs and dyes can be stored in the pore. Surface modifications can be done to target tumor. Nano-machines that respond to internal stimuli such as pH and redox as well as external stimuli such as light and magnetic field will be used to operate nanovalves.[249] Taken from Ref. [250] with permission of Elsevier.

Tamanoi,[249] Song and Yang,[278] Victor Lin *et al.*,[279] Chen *et al.*,[280] or Duguet *et al.*[281]

## VI. Controlled Release from SPIONs

As explained before, since using external magnetic field as a trigger, controlled release had a better penetration depth in comparison with using external light. In order to reach this advantage, it is necessary to use a special type of nanoparticles that contains iron oxide magnetic core.[262] These nanoparticles have usually a core size of 2–10 nm which contains iron oxide ($Fe_3O_4$). A mesoporous silica coating was added to make a particle of about 100 nm.[249] This magnetic core gives numerous properties to the nanoparticle. Firstly, due to the superparamagnetic property of iron oxide core, the nanoparticles have the capabilities to be heated up by the exposure to oscillating magnetic field. Secondly, the magnetic property can be used to recover nanoparticles or to collect them to a desired site. Thirdly, iron oxide has a property to enhance MRI imaging contrasts using $T_2$ weighted imaging protocols.[282,283]

Due to the superparamagnetic property of iron oxide NPs, exposure of iron oxide core of NPs to alternative magnetic field results in heat generation. This temperature increase can be used to allow opening of a nanovalve.[249] To accomplish this, a special type of nanovalve consisting of a stalk and cucurbituril was synthesized. The valve becomes open after the heat generated by the external magnetic field.[255,284,285]

For more detailed information on chemical, and physical features of these nanoparticles, please refer to reviews written by Mekaru and Tamanoi[249] or Kim and Nguyen.[286]

## VII. Controlled Release from Membranes

In medicine, encapsulated drug is release in the body after it is administrated orally or intravenously.[287] Controlled release reduces the loss of an expensive active agent due to leaching at an unnecessarily high rate and eliminates its excessive dispersion to the environment. There are many factors which may affects the rates of drug released by the membrane. Tojo *et al.*[287] investigated the effects of the void fraction, membrane thickness, concentration dependence of diffusivity, and step change in the temperature of the environment on the rate of release. They showed that particles with an excess active agent in the core zone have the profoundly desirable controlled released characteristics. During the past decades, controlled released technology as a new multidisciplinary science has received increasing attention in the pharmaceutical and medical sciences.[288] The unique permeation characteristics of polymeric membranes have begun to be applied for controlled delivery of biologically active agents.[289] It has been predicted that the major controlled release technology will be the polymeric membrane type.[288]

Diffusion controlled membrane tools can be divided into two main categories: the first one is reservoir systems (the active agent is totally encapsulated within a rate controlling membrane) and the second one is monolithic systems (which the active agent is dispersed or dissolved in a rate controlling matrix).[290]

## VIII. Controlled Release from Hydrogels

Hydrogels are polymeric networks with three-dimensional configuration. They can absorb bulky quantities of water or biological fluids while remaining insoluble in aqueous solutions due to chemical or physical crosslinking of individual natural or synthetic polymer chains.[291,292] Their attraction to absorb water is because of the presence of hydrophilic groups such as –OH, –CONH–, $–CONH_2$, and $–SO_3H$ in polymers forming hydrogel structures.[293]

The unique physicochemical characteristics are determined by the water content of a hydrogel. Some of their physical properties such as their high water content, their soft and robbery consistency, and low interfacial tension with water or biological fluids resemble the living tissues, than any other class of synthetic biomaterials.[294]

However, hydrogels show a swelling behavior instead of being dissolved in the aqueous surrounding environment as a consequence of the critical cross-links present in the hydrogel structure.[295] The water content in the hydrogel depends on the distance between polymer chains and the flexibility of those chains together.[296]

Hydrogels can be classified based on a variety of characteristics, including: the nature of side groups (neutral or ionic), mechanical and structural features (affine or phantom), method of preparation (homo- or co-polymer), physical structure (amorphous, semicrystalline, hydrogen bonded, supramolecular, and hydrocollodial), and responsiveness to physiologic environment stimuli (pH, ionic strength, temperature, electromagnetic radiation, etc.).[292,297]

The polymers commonly used in preparation of hydrogels with pharmaceutical and biological applications have natural or synthetic origins.[297] Therefore, there is a classification based on the type of polymeric materials[295]:

1. Chitosan-based hydrogel nanoparticles
2. Alginate-based hydrogel nanoparticles
3. Poly (vinyl alcohol)-based hydrogel nanoparticles
4. Poly (ethylene oxide) and poly (ethyleneimine)-based hydrogel nanoparticles
5. Poly (vinyl pyrrolidone)-based hydrogel nanoparticles
6. Poly-*N*-isopropylacrylamide-based hydrogel nanoparticles
7. Hydrogel nanoparticles of other origins

In general, there are three important parameters in describing the nanostructure of cross-linked hydrogel networks: (1) polymer volume fraction in the swollen state, (2) number average molecular weight between crosslinks, and (3) network mesh size, zeta potential $\xi$.[298]

Drug release mechanisms from hydrogels can be categorized as:

## A. Diffusion-controlled

The most common feasible mechanism for describing drug release from hydrogels is diffusion-controlled. Molecules of different sizes and characteristics can freely diffuse into/out of hydrogel matrix during the loading and storage periods.[295]

The drug diffusion out of a hydrogel matrix depends on the mesh sizes within the matrix of the gel, which is affected by several parameters. Typical mesh sizes reported for biomedical hydrogels range from 5 to 100 nm (in their swollen state), which are much larger than most small-molecule drugs.[295,299] Fick's law of diffusion with either constant or variable diffusion coefficients is commonly used in modeling diffusion-controlled release.[300]

## B. Swelling-controlled

This mechanism is considered as a releasing behavior, when diffusion of a drug is significantly faster than hydrogel distention. The modeling of this mechanism usually involves moving boundary conditions where molecules are released at the interface of rubbery and glassy phases of swollen hydrogels.[299]

## C. Chemically-controlled

Chemically-controlled release is used to describe molecule release determined by reactions occurring within a delivery matrix. The most common reactions that occur within hydrogel delivery systems are cleavage of polymer chains via hydrolytic or enzymatic degradation or reversible or irreversible reactions occurring between the polymer network and releasable drug.[296]

For a successful drug delivery, controlled-release is needed to release the drug of interest at a specific predetermined temporal and/or spatial manner within the body. Therefore, there are several studies about controlled-release systems. The hydrogel-based delivery systems are of two major categories: (i) time-controlled systems and (ii) stimuli-induced release system.[295,296]

Most of controlled-release systems use stimuli-sensitive hydrogels however their response time is considerably slow. For fastening their response, developing thinner and smaller hydrogels could be effective. But this approach causes fragility and loss of mechanical strength in the polymer network.[301]

The stimuli-sensitive hydrogel systems can be further sub-classified into[302]:

(i) Physically-induced release systems: Temperature, electricity, light, pressure, sound, and magnetic field.
(ii) Chemically-induced release systems: pH, solvent composition, ions, and specific molecular recognition events.
(iii) Other stimuli-induced release systems.

Investigations on other controlled release systems from hydrogels and developing ideal system to achieve the most precise drug delivery are continued.

## IX. Controlled Release from Lipid-Protein NPs

Traditional administration routes for drugs are the oral and parenteral administration. In both routes, the drug arrives in the blood stream finally. After that, it will be distributed in the body depending on its physicochemical properties. Therefore, it can be degraded. For the solution of these problems and having targeting, a new approach was suggested: the entrapment of those drugs into a particulate carrier system. One of the carrier systems is lipid nano-emulsions, they are fine oil/water (o/w) dispersions, having droplets covering the size range between 50 and 200 nm for carrying lipophilic drugs.[303] Lipid-like nanoparticles can be protein and gene delivery vehicles and also for RNAi.[304,305] These lipid carriers may be available in different phases such as solid, semi-solid, or liquid state in the form of solid lipid nanoparticles, nanostructured lipid carriers, lipid drug conjugate nanoparticles, liposomes, or nanoemulsions.[305] They can reduce side effects but are thermodynamically unstable.[306] Among the lipid carriers, liposomes are well established and extensively investigated. They are able to control release successfully.[306] Some of the advantages of liposomes are compatibility of the constituent components with the body system thereby presenting low inherent toxicity, facile preparation and easy variation of composition to obtain more efficient preparations.[306]

Solid lipid nanoparticles (SLN) were described by Müller *et al.*,[307] for the first time. They have been introduced as an efficient and non-toxic alternative lipophilic colloidal drug carrier. Two established production techniques are: the high-pressure homogenization described by Müller and Lucks[308] and the microemulsion-based technique by Gasco.[309]

They allow drug protection and administration via parenteral and non-parenteral routes thus emphasizing the versatility of this nanoparticulate carrier because of their colloidal dimensions and the controlled release behavior.[306]

## X. Toxicity of Nanoparticles

Despite our exposure to nanomaterials is growing, there is little understanding of the unique toxicological properties of NPs and their long-term impact on human health. Nanomaterials can be released into the environment via spillages, wear, washing and disposal at a rate proportional to their level of use. The worst thing is that we don't know where they go or what happens to them after they are released. Because of their very small size, they might enter the human body by inhalation, ingestion, skin penetration or injections, and NPs can interact with intracellular structures and macromolecules for long periods of time.[310] Whereas transition from bulk materials to nanomaterials cause changing the properties, the toxicity of a bulk material may not be a good indication of the toxicity of the nanomaterial. Moreover, toxicity is already hard to measure because it depends upon dose and the animals, plants or cells tested. Although there are a lot of studies about the toxicity of variety nanomaterial compounds. Some of them compared two types of toxicity measurements, *in vitro* and *in vivo*. This comparison demonstrated that for identifying characteristics of nanomaterials that can be used as indicators of toxicity and in order to establish a ranking of NP toxicity for mechanistic studies, the *in vitro* systems are principally practical. But for studying aspects that cannot be obtained with *in vitro* systems, such as toxicokinetics in the body, i.e. absorption, distribution, metabolism, and elimination, the *in vivo* tests would be mainly. However, the disadvantages of *in vivo* measurements are time-consuming, expensive, and involve ethical issues.[311]

The another priority of *in vitro* methods in comparison with *in vivo* methods is that they can produce reproducible results rapidly and inexpensively without the use of animals.[312]

Among all *in vitro* toxicity assays, the MTT assay method has been prominent for probing the toxicity of SPION.[313] It allows rapid evaluation of cell viability, cell survival, cell growth and gives good reproducibility.

The physicochemical properties such as size, surface chemistry, shape, protein absorption gradient and surface smoothness or roughness play a critical role in determining the toxicity of nanomaterials. Consequently, a fundamental understanding of the biological interactions of NPs with cells, proteins, and tissues, is vital to the future design of safe nanotechnologies. The main concern is a high degree of biocompatibility of NP-products besides minimum negative effects on blood components, genetic material, and cell viability.[310]

## XI. Protein Corona and its Effects on Targeting Capability of Nanoparticles and their Drug Release Profile

Studies showed that, as the NPs enter in the biological medium, the surface of NPs is covered by various biomolecules (proteins) through a process called "protein corona effect."[314,315] Thus, the primary interaction of organs or cells with NPs is strongly influenced by the "hard corona" (i.e. a long-lived protein layer that strongly adsorbed to the surface of the NP and remains stable for several hours).[316,317] This new biological identity of the NPs can entirely change the biological fate of the NPs. Therefore, it is important to have a deep understanding on the interactions at the nano–bio interfaces to design safe, reliable, and high-yield NPs, for desired biomedical purpose. In this case, extensive studies were dedicated to probe every individual crucial factor, which can be considered at the nano–bio interfaces.[318,319] For instance, it has been discovered that the protein corona can cover the targeting molecules on the surface of NPs and causes loss of specificity in targeting.[320] Mahmoudi *et al.*[321] showed that the corona formation slows down the fibrillation process of the amyloid proteins (e.g. amyloid beta), regardless to the physicochemical properties of the NPs. Another study showed that after getting very promising result in *in vitro* the *in vivo* results were disappointing due to the coverage of targeting site by protein corona and because absorption of opsonin-based proteins at the surface of nanoprobes prohibit the active targeting and the probe was eliminated from the blood immediately after injection by the RES system.[322]

Mahmoudi *et al.*[318] showed that the drug release profile of nanoprobes depends on their interaction with the protein corona (i.e. the amount and types of the associated proteins in the composition of hard corona) in addition to the size/type of the nanoprobe. The formation of a protein buffer layer onto the surface of small nanoparticles *in vivo* acts as a shield and causes a delay in the drug release process and this effect plays a crucial role in the *in vivo* applications. They concluded that, the release profile data that are currently available for various drug-carriers can be modified by incorporating the protein corona effect to have realistic *in vivo* applications.

## XII. Future Perspective on Smart Drug Delivery Systems

A smart system is one that can alter its properties in response to environmental changes, such as pH, temperature, enzymes, and ionic environment. Ideally, a drug delivery system needs to perform multiple functions, including improving the solubility and stability of the drug/payload, reducing the dosage and its frequency, and reducing or eliminating the drug's adverse effects. The carrier should be nontoxic to the biological system, should not trigger adverse immune responses, and, finally, *should be able to deliver the required amount of drug to the desired location over a long period of time.* This last point is the most challenging for a smart drug delivery system (SDDS), which is also called a stimuli-sensitive delivery system.[323]

Two aspects are important. First, the surface modification with a suitable targeting vector on the nanosystem surface delivers the drug at the desired location, and second, the presence of a trigger that can control the amount of drug released at a given instant ensures delivery of required amount of the drug. The signal can be due to an external stimulus (extrinsic triggers such as ultrasound, magnetic or electric field) or to the environment of the system itself (internal triggers such as temperature, proteins, carbohydrates, and pH).

An SDDS has several advantages compared to a conventional drug delivery system (DDS). A DDS release the same amount of the drug independent of the environmental conditions, while an SDDS is based on the release-on-demand strategy, which allows a drug carrier to liberate a therapeutic drug only when it is required in response to a specific stimulation.

Three categories of polymeric systems can be employed exhibit stimuli responsiveness, including linear polymeric chains, cross-linked gels and surface grafted systems.

The solubility of linear polymeric chains can change with a stimulus due to alterations in their hydrophobicity and hydrophilicity. In an aqueous medium, if the stimulus induces an increase in the hydrophilic interactions, the polymer swells, while if the stimulus increases the hydrophobicity, the polymer chains collapse and precipitate out of the solution.

The cross-linked gels exhibit rapid responses to small changes in the stimulus. However, caution should be exercised when tightly regulating the extent of cross-linking because over-crosslinking will lead to a loss in the swelling and deswelling property.

For grafted polymer chain systems, the swelling or deswelling can contribute to a variation in the hydrophilicity or hydrophobicity of the substrate. Such a phenomenon depends on the changes produced at the interface between the polymer chains and the environment and is now referred to as **interfacial engineering**.

A typical example of an SDDS is a self-regulated insulin delivery systems that can respond to changes in the environmental glucose level.[324,325]

Among the numerous SDDSs available, one of the most widely used is polymeric micelles, which can dissolve water-insoluble drugs, such as doxorubicin or paclitaxel, at high concentrations.[326–328] In the body, drug release from micelles depends on the simple diffusion and degradation of the micelles. The kinetics of drug release can be modulated by varying the degradation rate of hydrophobic polymers; however, the degradation rate is usually very slow, and the drug is released by diffusion from micelles. This release by passive diffusion may be undesirable because the polymeric micelles reaching the target site need to release their contents quickly. To avoid this slow release, smart polymeric micelles have been developed to liberate the loaded drug at the targeted site more quickly. For example, Lee *et al.*[329] described poly (ethylene glycol)-b-poly-histidine (PEG-b-PHis) micelles that are stable only over the pKb of the poly-histidine block (pH 6.5–7.0). The pKb can be adjusted by varying the molecular weight of poly-histidine. Because solid tumors have a slightly acidic environment, a small reduction in pH at the tumor site triggers a dissociation of the polymeric micelle to release its contents. The same authors[330] reported that PEG-b- poly-histidine micelles containing doxorubicin killed multi-drug resistant MCF-7 cells at pH 6.8. Similarly, Hruby *et al.*[331] showed that an SDDS can achieve highly localized drug accumulation at target sites.

An SDDS with enhanced targeting properties is highly promising for increasing the efficiency and efficacy of therapy while decreasing side effects.

## References

1. Martinez, M. N.; Amidon, G. L. *J. Clin. Pharmacol.* **2002**, *42*, 620.
2. Drug Delivery: Principles and applications. Ed. By B. Wang, T. Siahaan, R. A. Soltero, Wiley-Interscience, New-Jersey, 2005 (Ch 1 Factors that impact the developability of drug candidates: an overview, C. Han, B. Wang)
3. Fang, J.; Nakamura, H. Maeda, H. *Adv. Drug Del. Rev.* **2011**, *63* (3), 136.
4. Awasthi, V. D.; Garcia, D.; Goins, B. A.; Phillips, W. T. *Int J. Pharm.* **2003**, *253* (1–2), 121.
5. Muthu, M. S.; Kulkarni, S. A.; Xiong, J. Q.; Feng, S. S. *Int. J. Pharm.* **2011**, *421* (2), 332.

6. Hofheinz, R. D.; Gnad-Vogt, S. U.; Beyer, U.; Hochhaus, A. *Anticancer Drugs* **2005**, *16* (7), 691.
7. Owens, D. E.; Peppas, N. A. *Int. J. Pharm.* **2006**, *307* (1), 93.
8. Qazi, S.; Liepold, L. O.; Abedin, M. D. J.; Johnson, B.; Prevelige, P; Frank, J. A.; Douglas, T. *Mol. Pharm.* **2013**, *10* (1), 11.
9. Cheng, F.; Tsvetkova, I. B.; Khuong, Y. L.; Moore, A. W.; Arnold, R. J.; Goicochea, N. L.; Dragnea, B.; Mukhopadhyay, S. *Mol. Pharm.* **2013**, *10*, 51.
10. Yildiz, I.; Shukla, S.; Steinmetz, N. F. *Curr. Opin. Biotechnol.* **2011**, *22*, 901.
11. Liu, Z.; Qiao, J.; Niu, Z.; Wang, Q.; *Chem. Soc. Rev.* **2012**, *41*, 6178.
12. Pokorski, J. K.; Steinmetz, N. F. *Mol. Pharm.* **2011**, *8*, 29.
13. Chen, W.; Cao, Y.; Liu, M.; Zhao, Q.; Huang, J.; Zhang, H.; Deng, Z.; Dai, J.; Williams, D. F.; Zhang, Z. *Biomaterials* **2012**, *33*, 7895.
14. Lee, J. Y.; Choi, D. Y., Cho, M. Y.; Park, K. E.; Lee, S. H.; Cho, S. H.; Hong, K. S.; Lim. Y. T. *Small* **2014**, *10*, 901.
15. Huang, P.; Wang, D.; Su, Y.; Huang, W.; Zhou, Y.; Cui, D.; Zhu, X.; Yan, D. *J. Am. Chem. Soc.* **2014**, *136*, 11748.
16. Janib, M.; Moses, A. S.; MacKay, J. A. *Adv. Drug Deliv. Rev.* **2010**, *62*, 1052.
17. Dreher, M. R.; Raucher, D.; Balu, N.; Colvin, O. M.; Ludeman, S. M., Chilkoti, A. *J. Control. Release* **2003**, *91*, 31.
18. Kratz. F. *J. Control. Release* **2008**, *132*, 171.
19. Lammers, T.; Ulbrich, K. *Adv. Drug Deliv. Rev.* **2010**, *62*, 119.
20. Kopecek, J.; Kopeckova, P. *Adv. Drug Deliv. Rev.* **2010**, *62*, 122.
21. Borgman, M. P.; Coleman, T.; Kolhatkar, R. B.; Geyser-Stoops, S.; Line, B.R.; Ghandehari, H. *J. Control. Release* **2008**, *132*, 193.
22. Lammers, T.; Subr, V.; Ulbrich, K.; Peschke, P.; Huber, P. E.; Hennink, W. E.; Storm, G. *Biomaterials* **2009**, *30*, 3466.
23. Boas, U.; Heegaard, P. M. *Chem. Soc. Rev.* **2004**, *33*, 43.
24. Astruc, D.; Boisselier, E.; Ornelas, C. *Chem. Rev.* 2010, *110*, 1857.
25. Buhleier, E.; Wehner, W.; Vogtle, F. *Synthesis* **1978**, *9*(2), 155–158.
26. Newkome, G. R.; Yao. Z.; Baker, G. R.; Gupta, V. K. *J. Org. Chem.* **1985**, *50*, 2003.

27. Tomalia, D. A.; Baker, H.; Dewald, J.; Hall, M.; Kallos, G.; Martin, S.; Roeck, J.; Ryder, J.; Smith, P. *Polym. J.* **1985**, *17*, 117.
28. Hawker, C. J.; Frechet, J. M. J. *J. Chem. Soc. Chem. Commun.* **1990**, *15*, 1010.
29. Hawker, C. J.; Frechet. J. M. J. *J. Am. Chem. Soc.* **1990**, *112*, 7638.
30. Jayaraman, M.; Frechet, J. M. J. *J. Am. Chem. Soc.* **1998**, *120*, 12996.
31. Jansen, J. F. G. A.; Meijer, E. W.; de Brabander-van den Berg, E. M. M. *J. Am. Chem. Soc.* **1995**, *117*, 4417.
32. Ihre, H. R.; Padilla De Jesus, O. L.; Szoka, F. C.; Fréchet, J. M. J. *Bioconj. Chem.* **2002**, *13*, 443.
33. Tan, M.; Wu, X.; Eun-Kee Jeong, E.-K.; Chen, Q.; Lu, Z.-R. *Biomacromolecules.* **2010**, *11*, 754.
34. Taratula, O.; Schumann, C.; Duong, T.; Taylor, K. L.; Taratula, O. *Nanoscale* **2015**, *7* (9), 3888.
35. Xie, J.; Lee, S.; Chen. X. *Adv. Drug Deliv. Rev.* **2010**, *62*, 1064.
36. Ang, C. Y.; Tan, S.Y.; Zhao. Y. *Org. Biomol. Chem.* **2014**, *12*, 4776.
37. Haddon, R. C. *Acc. Chem. Res.* **2002**, *35*, 997.
38. Prato, M.; Kostarelos, K.; Bianco, A. *Acc. Chem. Res.* **2008**, *41*, 60.
39. Dennany, L.; Sherrell, P.; Chen, J.; Innis, P. C.; Wallace, G. G.; Minnett. A. I. *Phys. Chem. Chem. Phys.* **2010**, *12*, 4135.
40. Kam, N. W.; O'Connell, M.; Wisdom, J. A.; Dai, H. *Proc. Natl. Acad. Sci. U.S.A.* **2005**, *102*, 11600.
41. Liu, Z.; Cai, W.; He, L.; Nakayama, N.; Chen, K.; Sun, X.; Chen, X.; Dai, H. *Nat. Nanotechnol.* **2007**, *2*, 47.
42. Shi Kam, N. W.; Jessop, T. C.; Wender, P. A.; Dai, H. *J. Am. Chem. Soc.* **2004**, *126*, 6850.
43. Liu, Z.; Tabakman, S.; Welsher, K.; Dai, H. *Nano Res.* **2009**, *2*, 85.
44. Sitharaman, B.; Tran, L. A.; Pham, Q. P.; Bolskar, R. D.; Muthupillai, R.; Flamm, S. D.; Mikos, A. G.; Wilson, L. J. *Contrast Media Mol. Imaging* **2007**, *2*, 139.
45. Mikawa, M.; Kato, H.; Okumura, M.; Narazaki, M.; Kanazawa, Y.; Miwa, N.; Shinohara, H. *Bioconj. Chem.* **2001**, *12*, 510.
46. Sitharaman, B.; Kissell, K. R.; Hartman, K. B.; Tran, L. A.; Baikalov, A.; Rusakova, I.; Sun, Y.; Khant, H. A.; Ludtke, S. J.; Chiu, W.; Laus, S.; Toth, E.; Helm, L.; Merbach, A. E.; Wilson, L. J. *Chem. Commun.* **2005**, *31*, 3915.

47. Fatouros, P. P.; Corwin, F. D.; Chen, Z. J.; Broaddus, W. C.; Tatum, J. L.; Kettenmann, B.; Ge, Z.; Gibson, H. W.; Russ, J. L.; Leonard, A. P.; Duchamp, J. C.; Dorn, H. C. *Radiology* **2006**, *240*, 756.
48. Hartman, K. B.; Laus, S.; Bolskar, R. D.; Muthupillai, R.; Helm, L.; Toth, E.; Merbach, A. E.; Wilson, L. J. *Nano Lett.* **2008**, *8*, 415.
49. Duguet, E.; Vasseur, S.; Mornet, S.; Devoisselle, J. M. *Nanomedicine* **2006**, *1*, 157.
50. Erathodiyil, N.; Ying, J. Y. *Acc. Chem. Res.* **2011**, *44*, 925.
51. Park, J.; Lee, E.; Hwang, N. M.; Kang, M. S.; Kim, S. C.; Hwang, Y.; Park, J. G.; Noh, J. H.; Kini, J. Y.; Park, J. H.; Hyeon, T. *Angew. Chem. Int. Ed.* **2005**, *44*, 2872.
52. Xie, J.; Huang, J.; Li, X.; Sun, S.; Chen, X. *Curr. Med. Chem.* **2009**, *16*, 1278.
53. Baumgartner, J.; Bertinetti, L.; Widdrat, M.; Hirt, A. M.; Faivre, D. *Plos One* **2013**, *8*, e57070
54. Jun, Y. W.; Seo, J. W.; Cheon, A. *Acc. Chem. Res.* 2008, *41*, 179.
55. Lee, N.; Hyeon, T. *Chem. Soc. Rev.* **2012**, *41*, 2575.
56. Chouly, C.; Pouliquen, D.; Lucet, I.; Lejeune, J. J.; Jallet, P. J. *Microencapsulation* **1996**, *13*, 245.
57. Kim, S.; Lim, Y. T.; Soltesz, E. G.; De Grand, A. M., Lee, J.; Nakayama, A.; Parker, J. A.; Mihaljevic, T.; Laurence, R. G.; Dor, D. M.; Cohn, L. H.; Bawendi, M. G.; Frangioni, J. V. *Nat. Biotechnol.* **2004**, *22*, 93.
58. Bailey, R. E.; Smith, A. M.; Nie, S. M. *Phys. E: Low-Dimens. Syst. Nanostruct.* **2004**, *25*, 1.
59. Bagalkot, V.; Zhang, L.; Levy-Nissenbaum, E.; Jon, S.; Kantoff, P. W.; Langer, R.; Farokhzad, O. C. *Nano Lett.* **2007**, *7*, 3065.
60. Daniel, M. C.; Astruc, D. *Chem. Rev.* **2004**, *104*, 293.
61. Dreaden, E. C.; Alkilany, A. M.; Huang, X.; Murphy, C. J.; El-Sayed, M. A. *Chem. Soc. Rev.* **2012**, *41*, 2740.
62. Hu, M.; Chen, J. Y.; Li, Z. Y.; Au, L.; Hartland, G. V.; Li, X. D.; Marquez, M.; Xia, Y. N. *Chem. Soc. Rev.* **2006**, *35*, 1084.
63. Huang, X. H.; El-Sayed, I. H.; Qian, W.; El-Sayed, M. A. *Nano Lett.* **2007**, 7, 1591.

64. Schwartz, J. A.; Shetty, A. M.; Price, R. E.; Stafford, R. J.; Wang, J. C.; Uthamanthil, R. K.; Pham, K.; McNichols, R. J.; Coleman, C. L.; Payne, J. D. *Cancer Res.* **2009**, *69*, 1659.
65. Etheridge, M. L..; Campbell, S. A.; Erdman, A. G.; Haynes, C. L.; Wolf, S. M.; McCullough, J.; *Nanomedicine* **2013**, *9*, 1.
66. Jain, S.; Hirst, D. G.; O'Sullivan, J. M. *Brit. J. Radiol.* **2012**, *85*, 101.
67. Jana, N. R.; Earhart, C.; Ying, J. Y. *Chem. Mater.* **2007**, *19*, 5074.
68. Ow, H.; Larson, D. R.; Srivastava, M.; Baird, B. A.; Webb, W. W.; Wiesner, U. *Nano Lett.* **2005**, *5*, 113.
69. Slowing, I. I.; Vivero-Escoto, J. L.; Wu, C. W.; Lin, V. S. *Adv. Drug Deliv. Rev.* **2008**, *60*, 1278.
70. Lee, J. E.; Lee, N.; Kim, T.; Kim, J.; Hyeon, T. *Acc. Chem. Res.* **2011**, *44*, 893.
71. Li, Z.; Barnes, J. C.; Bosoy, A.; Fraser Stoddart, J.; Zink, J. I. *Chem. Soc. Rev.* **2012**, *41*, 2590.
72. Hon, N. K.; Shaposhnik, Z.; Diebold, E. D.; Tamanoi, F.; Jalali, B. *J. Biomed. Mater. Res. Part A.* **2012**, *100A*, 3416.
73. Tang, L.; Cheng, J. *Nano Today* **2013**, *8*, 290.
74. Cheng, S. H.; Lee, C. H.; Chen, M. C.; Souris, J. S.; Tseng, F. G.; Yang, C. S.; Mou, C. Y.; Chen, C. T.; Lo, L. W. *J. Mater. Chem.* **2010**, *20*, 6149.
75. Férey, G.; Mellot-Draznieks, C.; Serre, C.; Millange, F. *Acc. Chem. Res.* **2005**, *38*, 217.
76. Férey, G. *Chem. Soc. Rev.* **2008**, *37*, 191.
77. McKinlay, A. C.; Morris, R. E.; Horcajada, P.; Férey, G.; Gref, R.; Couvreur, P.; Serre, C. *Angew. Chem. Int. Ed.* **2010**, *49*, 6260.
78. Horcajada, P.; Gref, R.; Baati, T.; Allan, P. K.; Maurin, G.; Couvreur, P.; Férey, G.; Morris, R. E.; Serre, C. *Chem. Rev.* **2012**, *112*, 1232.
79. Della Rocca, J.; Liu, D.; Lin, W. *Acc. Chem. Res.* **2011**, *44*. 957.
80. Eddaoudi, M.; Kim, J.; Rosi, N.; Vodak, D.; Wachter, J.; O'Keeffe, M.; Yaghi, O. M. *Science* **2002**, *295*, 469.
81. Li, J. R.; Tao, Y.; Yu, Q.; Bu, H.; Sakamoto, H.; Kitagawa, S. *Chem. Eur. J.* **2008**, *14*, 2771.
82. Devic, T.; Horcajada, P.; Serre, C.; Salles, F.; Maurin, G.; Moulin, B.; Heurtaux, D.; Clet, G.; Vimont, A.; Grenèche, J. M.; Le Ouay, B.;

Moreau, F.; Magnier, E.; Filinchuk, Y.; Marrot, J.; Lavalley, J. C.; Daturi, M.; Férey, G. *J. Am. Chem. Soc.* **2010**, *132*, 1127.

83. Kitagawa, S.; Kitaura, R.; Noro, S. I. *Angew. Chem., Int. Ed.* **2004**, *43*, 2334.
84. Férey, G.; Serre, C. *Chem. Soc. Rev.* **2009**, *38*, 1380.
85. Dietzel, P. D. C.; Blom, R.; Fjellva, H. *Eur. J. Inorg. Chem.* **2008**, 3624.
86. Horcajada, P.; Surblé, S.; Serre, C.; Hong, D. Y.; Seo, Y. K.; Chang, J. S.; Grenèche, J. M.; Margiolaki, I.; Férey, G. *Chem. Commun.* **2007**, 2820.
87. An, J.; Geib, S. J.; Rosi, N. L. *J. Am. Chem. Soc.* **2009**, *131*, 8376.
88. Weber, R.; Bergerhoff, G. *Z. Kristallogr.* **1991**, *195*, 87.
89. Serre, C.; Millange, F.; Surblé, S.; Férey, G. *Angew. Chem. Int. Ed.* **2004**, *43*, 6286.
90. Serre, C.; Surblé, S.; Mellot-Draznieks, C.; Filinchuk, Y.; Férey, G. *Dalton Trans.* **2008**, 5462.
91. Smaldone, R. A.; Forgan, R. S.; Furukawa, H.; Gassensmith, J. J.; Slawin, A. M. Z; Yaghi, O. M.; Stoddart, J. F. *Angew. Chem Int. Ed.* **2010**, *122*, 8812.
92. Rabone, J.; Yue, Y. F.; Chong, S. Y.; Stylianou, K. C.; Bacsa, J.; Bradshaw, D.; Darling, G. R.; Berry, N. G.; Khimyak, Y. Z.; Ganin, A. Y.; Wiper, P.; Claridge, J. B.; Rosseinsky, M. J. *Science* **2010**, *329*, 1053.
93. Horcajada, P.; Chalati, T.; Serre, C.; Gillet, B.; Sebrie, C.; Baati, T.; Eubank, J. F.; Heurtaux, D.; Clayette, P.; Kreuz, C.; Chang, J. S.; Hwang, Y. K.; Marsaud, V.; Bories, P. N.; Cynober, L.; Gil, S.; Férey, G.; Couvreur, P.; Gref, R. *Nat. Mater.* **2010**, *9*, 172.
94. Chalati, T.; Horcajada, P.; Gref, R.; Couvreur, P.; Serre, C.; *J. Mater. Chem.* **2011**, *21*, 2220.
95. Rieter, W. J.; Taylor, K. M. L.; An, H.; Lin, W. *J. Am. Chem. Soc.* **2006**, *128*, 9024.
96. Taylor, K. M. L.; Jin, A.; Lin, W. *Angew. Chem. Int. Ed.* **2008**, *47*, 7722.
97. Rieter, W. J.; Pott, K. M.; Taylor, K. M. L.; Lin, W. *J. Am. Chem. Soc.* **2008**, *130*, 11584.

98. Taylor, K .M. L.; Rieter, W. J.; Lin, W. *J. Am. Chem. Soc.* **2008**, *130*, 14358.
99. Haque, E.; Khan, N. A.; Park, J. H.; Jhung, S. H. *Chem. Eur. J.* **2009**, *16*, 1046.
100. Schlesinger, M.; Schulze, S.; Hietschold, M.; Mehring, M. *Microporous Mesoporous Mater.* **2010**, *132*, 121.
101. Li, Z. Q.; Qiu, L. G.; Su, T.; Wu, Y.; Wang, W.; Wu, Z. Y.; Jiang, X. *Mater. Lett.* **2009**, *63*, 78.
102. Qiu, L. G.; Li, Z. Q.; Wu, Y.; Wang, W.; Xu, T.; Jiang, S. *Chem. Commun.* **2008**, 3642.
103. Park, S. E.; Hwang, Y. K.; Kim, D. S.; Jhung, S. H.; Hwang, J. S.; Chang, J. S. *Catal. Surv. Asia* **2004**, *8*, 91.
104. Tompsett, G. A.; Conner, W. C.; Yngvesson, K. S. *Chem- PhysChem.* **2006**, *7*, 296.
105. Tyrrell, Z. L.; Shen, Y.; Radosz, M. *Prog. Polym. Sci.* **2010**, *35*, 1128.
106. Guida, V. *Adv. Colloid Interface Sci.* **2010**, *161*, 77.
107. Holder, S. J.; Sommerdijk, N. A. J. M. *Polym. Chem.* **2011**, *2*, 1018.
108. Sagar, G. H.; Arunagirinathan, M. A.; Bellare, J. R. *Indian. J. Exp. Biol.* **2007**, *45*, 133.
109. Oerlemans, C.; Bult, W.; Bos, M.; Storm, G.; Nijsen, J. F. W.; Hennink, W. E. *Pharm. Res.* **2010**, *27*, 2569.
110. Torchilin, V. P. *J. Control. Release* **2001**, *73*, 137.
111. Cho, H. K.; Cheong, I. W.; Lee, J. M.; Kim, J. H. *Korean J. Chem. Eng.* **2010**, *27*, 731.
112. Kim, D.; Gao, Z. G.; Lee, E. S.; Bae, Y. H. *Mol. Pharm.* **2009**, *6*, 1353.
113. Grogna, M.; Cloots, R.; Luxen, A.; Jérôme, C.; Passirani, C.; Lautram, N.; Desreux, J. F.; Detrembleur, C. *Polym. Chem.* **2010**, *1*, 1485.
114. Yokoyama, M.; Opanasopit, P.; Okano, T.; Kawano, K.; Maitani, Y. *J. Drug Target* **2004**, *12*, 373.
115. Kim, K. S.; Park, W.; Hu, J.; Bae, Y. H.; Na, K. *Biomaterials* **2014**, *35*, 337.
116. Husseini, G. A.; Pitt, W. G. *J. Pharm. Sci.* **2009**, *98*, 795.
117. Prabaharan, M.; Grailer, J. J.; Steeber, D.A.; Gong, S. *Macromol. Biosci.* **2009**, *9*, 744.
118. Cohen, E. M.; Ding, H.; Kessinger, C. W.; Khemtong, C.; Gao, J.; Sumer, B. D. *Otolaryngol. Head Neck Surg.* **2010**, *143*, 109.

119. Yang, H.; Mao, H.; Wan, Z.; Zhu, A.; Guo, M.; Li, Y.; Li, X.; Wan, J.; Yang, X.; Shuai, X.; Chen. H. *Biomaterials* **2013**, *34*, 9124.
120. Zhang, C.; Jin, S.; Li, S.; Xue, X.; Liu, J.; Huang, Y.; Jiang, Y.; Chen, W. Q.; Zou, G.; Liang, X. J. *ACS Appl. Mater. Interfaces* **2014**, *6*, 5212.
121. Cabral, H.; Kataoka. K. *J. Control. Release* **2014**, *190*, 465.
122. Lee, J. S.; Feijen, J. *J. Control. Release* **2012**, *161*, 473.
123. Liu, G. Y.; Chen, C. J.; Ji, J. *Soft Matter* **2012**, *8*, 8811.
124. C. Sanson, C.; Diou, O.; Thévenot, J.; Ibarboure, E.; Soum, A.; Brûlet, A.; Miraux, S.; Thiaudière, E.; Tan, S.; Brisson, A.; Dupuis, V.; Sandre, O.; Lecommandoux, S. *ACS Nano* **2011**, *5*, 1122.
125. Oliveira, H.; Pérez-Andrés, E.; Thevenot, J.; Sandre, O.; Berra, E.; Lecommandoux, S. *J. Control. Release* **2013**, *169*, 165.
126. Liu, G.; Wang, X.; Hu, J.; Zhang, G.; Liu, S. *J. Am. Chem. Soc.* **2014**, *136*, 7492.
127. Akbarzadeh, A.; Rezaei-Sadabady, R.; Davaran, S.; Joo, S. W.; Zarghami, N.; Hanifehpour, Y.; Samiei, M.; Kouhi, M.; Nejati-Koshki, K. *Nanoscale Res. Lett.* **2013**, *8*, 102.
128. Marguet, M.; Edembe, L.; Lecommandoux, S. *Angew. Chem. Int. Ed.* **2012**, *51*, 1173.
129. Kisak, E. T.; Coldren, B.; Evans, C. A.; Boyer, C.; Zasadzinski, J. A. *Curr. Med. Chem.* **2004**, *11*, 1241.
130. Mishra, G. P., Bagui, M.; Tamboli, V.; Mitra, A. K.; *J. Drug Del.* **2011**, *2011*, 863734.
131. Gibis, M.; Zeeb, B.; Weiss, J. *Food Hydrocolloids* **2014**, *38*, 28.
132. Lasic, D. D. *Trends Biotechnol.* **1998**, *16*, 307.
133. Fan, Y.; Zhang, Q. *Asian. J. Pharmacol. Sci.* **2013**, *8*, 81.
134. Chang, H. I.; Yeh, M. K. *Int. J. Nanomedicine* **2012**, *7*, 49.
135. Pillai G. *SOJ. Pharm. Pharm. Sci.* **2014**, *1*, 1.
136. Al-Jamal, W. T.; Kostarelos, K. *Acc. Chem. Res.* **2011**, *44*, 1094.
137. Torchilin, V. P. *Mol. Med. Today* **1996**, *2*, 242.
138. Béalle, G.; Di Corato, R.; Kolosnjaj-Tabi, J.; Dupuis, V.; Clément, O.; Gazeau, F.; Wilhelm, C.; Ménager, C. *Langmuir.* **2012**, *28*, 11834.
139. Patidar, A.; Thakur, D. S.; Kumar, P.; Verma, J. *Int. J. Pharm and Pharm. Sci.* **2010**, *2*, 30.
140. Hajare, A. A.; Mali, S. S.; Ahir, A.A.; Thorat, J. D.; Salunkhe, S. S.; Nadaf, S. J.; Bhatia, N. M. *J. Adv. Drug Del.* **2014**, *1* (1), 30.

141. Ekambaram, P.; Sathali, A. A. H.; Priyanka, K. *Sci. Revs. Chem. Commun.* **2012**, *2*, 80.
142. Carbone, C.; Tomasello, B.; Ruozi, B.; Renis, M.; Puglisi, G. *Eur. J. Med. Chem.* **2012**, *49*, 110.
143. Andreozzi, E.; Wang, P.; Valenzuela, A.; Tu, C.; Gorin, F.; Dhenain, M.; Louie, A. *Bioconjugate. Chem.* **2013**, *24*, 1455.
144. Shuhendler, A. J.; Prasad, P.; Leung, M.; Rauth, A. M.; DaCosta, R. S.; Wu, X. Y. *Adv. Health Mater.* **2012**, *1*, 600.
145. Valetti, S.; Mura, S.; Stella, B.; Couvreur, P. *J. Nanobiotechnol.* **2013**, *11*, S6.
146. Radtke, M.; Souto, E. B.; Müller, R. H. *Pharm. Technol. Eur.* **2005**, *17*, 45.
147. Rao, J. P.; Geckeler, K. E. *Prog. Polym. Sci.* **2011**, *36*, 887.
148. Christoforidis, J. B.; Chang, S.; Jiang, A.; Wang, J.; Cebulla, C. M. *Mediat. Inflamm.* **2012**, ID 936721.
149. Soppimath, K. S.; Aminabhavi, T. M.; Kulkarni, A. R.; Rudzinski, W. E. *J. Control. Release* **2001**, *70*, 1.
150. Natarajan, J.; Meyyanathan, S. N. *Int. J. Health Allied Sci.* **2012**, *1*, 217.
151. Vauthier, C.; Bouchemal, K. *Pharm. Res.* **2009**, *26*, 1025.
152. Lambert, G.; Fattal, E.; Pinto-Alphandary, H.; Gulik, A.; Couvreur, P. *Pharm. Res.* **2000**, *17*, 707.
153. Toub, N.; Malvy, C.; Fattal, E.; Couvreur, P. *Biomedicine & Pharmacotherapy* **2006**, *60*, 607.
154. Chavany, C.; Le Doan, T.; Couvreur, P.; Puisieux, F.; Hélène, C. *Pharm. Res.* **1992**, *9*, 441.
155. Zobel, H. P.; Junghans, M.; Maienschein, V.; Werner, D.; Gilbert, M.; Zimmermann, H.; Noe, C.; Kreuter, J.; Zimmer, A. *Eur. J. Pharm. Biopharm.* **2000**, *49*, 203.
156. Lenaerts, V; Nagelkerke, J. F.; van Berkel, T. J. C.; Couvreur, P.; Grislain, L.; Roland, M.; Speiser, P. *J. Pharm. Sci.* **1984**, *73*, 980.
157. Fernandez-Urrusuno, R.; Fattal, E.; Rodrigues, J. M.; Féger, J.; Bedossa, P.; Couvreur, P. *J. Biomed. Mater. Res. Part A* **1996**, *31*, 401.
158. Verdun, C.; Brasseur, F.; Vranckx, H.; Couvreur, P.; Roland, M. *Cancer Chemother. Pharmacol.* **1990**, *26*, 13.

159. Jeon, S. I.; Lee, J. H.; Andrade, J. D.; De Gennes, P. G. *J. Colloid. Interface Sci.* **1991**, *142*, 149.
160. Li, S. D.; Huang, L. *J. Control. Release* **2010**, *145*, 178.
161. Kobayashi, H.; Watanabe, R.; Choyke, P. L. *Theranostics.* **2014**, *4*, 81.
162. Matsumura, Y.; Maeda, H. *Cancer Res.* **1986**, *46*, 6387.
163. Moghimi, S. M.; Hunter, A. C.; Murray, J. C. *Pharmacol. Rev.* **2001**, *53*, 283.
164. Babu, A.; Templeton, A. K.; Munshi, A.; Ramesh, R. *J. Nanomater.* **2013**, ID 863951.
165. Amoozgar, Z.; Yeo, Y. *Rev. Nanomed. Nanobiotechnol.* **2012**, *4*, 219.
166. Owens, D. E.; Peppas, N. A. *Int. J. Pharm.* **2006**, *307*, 93.
167. Miele, E.; Spinelli, G. P.; Miele, E.; Tomao, F.; Tomao, S. *Int. J. Nanomedicine* **2009**, *4*, 99.
168. Bourzac, K. *Nature* **2012**, *491*, S58.
169. Alexis, F.; Pridgen, E. M.; Langer, R.; Farokhzad, O. C. *Handb. Exp. Pharmacol.* **2010**, *197*, 55.
170. Wang, M.; Thanou, M. *Pharmacol. Res.* **2010**, *62*, 90.
171. Fay, F.; Scott, C. J. *Immunotherapy* **2011**, *3*, 381.
172. Gu, F. X.; Karnik, R.; Wang, A. Z.; Alexis, F.; Levy-Nissenbaum, E.; Hong, S.; Langer, R. S.; Farokhzad, O. C. *Nanotoday* **2007**, *2*, 14.
173. Farokhzad, O. C.; Cheng, J.; Teply, B. A.; Sherifi, I.; Jon, S.; Kantoff, P. W.; Richie, J. P.; Langer, R. *PNAS* **2006**, *103*, 6315.
174. Xiao, Z.; Levy-Nissenbaum, E.; Alexis, F.; Luptak, A.; Teply, B. A.; Chan, J. M.; Shi, J.; Digga, E.; Cheng, J.; Langer, R.; Farokhzad, O.C. *ACS Nano* **2012**, *6*, 696.
175. Mikhail, A. S.; Eetezadi, S.; Ekdawi, S. N.; Stewart, J.; Allen, C. *Int. J. Pharm.* **2014**, *464*, 168.
176. Bamrungsap, S.; Zhao, Z.; Chen, T.; Wang, L.; Li, C.; Fu, T.; Tan, W. *Nanomedicine* **2012**, *7*, 1253.
177. Mura, S.; Nicolas, J.; Couvreur, P. *Nat. Mater.* **2013**, *12*, 991.
178. Pasparakis, G.; Vamvakaki, M. *Polym. Chem.* **2011**, *2*, 1234.
179. Crucho, C. I. *Chem. Med. Chem.* **2015**, *10* (1), 24–38.
180. Li, F.; Pei, D.; Huang, Q.; Shi, T.; Zhang, G. *Carbohydr. Polym.* **2014**, *99*, 728.
181. Chipper, M.; Aubert, K. H.; Augé, A.; Fouquenet, J. F.; Saucé, M.; Chourpa, I. *Nanotechnology* **2013**, *24* (39), 395605.

182. Chou, F. Y.; Lai, J. Y.; Shih, C. M.; Tsai, M. C.; Lue, S. J. *Colloids. Surf. B. Biointerfaces* **2013**, *104*, 66.
183. Pernia, M. L.; Torti, A.; Riedinger, A.; le Fleur, R.; Petti, D.; Cingolani, R.; Bertacco, R.; Pellegrino, T. *ACS Nano* **2012**, *6*, 10535.
184. MacKay, J. A.; Chen, M.; McDaniel, J. R.; Liu, W.; Simnick, A. J.; Chilkoti, A. *Nat. Mater.* **2009**, *8*, 993.
185. Li, S.; Hu, K.; Cao, W.; Sun, Y.; Sheng, W.; Feng, L.; Wu, Y.; Liang, X. J. *Nanoscale* **2014**, *6*, 13701.
186. Zan, M.; Li, J.; Luo, S.; Ge, Z. *Chem. Commun.* **2014**, *50*, 7824.
187. Du, Z.; Xiang, S.; Zang, Y.; Zhou, Y.; Wang, C.; Tang, H.; Jin, T.; Zhang, X. *Mol. Pharm.* **2014**, *11*, 3300.
188. Shi, D.; Matsusaki, M.; Akashi, M. *J. Control. Release* **2011**, 149, 182.
189. Couvreur, P. *Pharm. Res.* **2006**, *23* (7), 1417.
190. Pante, N.; Kann, M. *Mol. Biol. Cell* **2002**, *13*, 425
191. Moghimi, S. M.; Hunter, A. C.; Murray, J. C. *FASEB J.* **2005**, *19*, 311.
192. Sahoo, S. K.; Labhasetwar, V. *Drug Discovery Today* **2003**, *8*, 1112.
193. Duncan, R. *Nature Reviews* **2003**, *2*, 247.
194. Hansen, C. B.; Kao, G. Y.; Moase, E. H.; Zalipsky, S.; Allen, T. M. (1995). *Biochim. Biophys. Acta* **1995**, *1239*, 133.
195. Hobbs, S. K.; Monsky, W. L.; Yuan, F.; Roberts,W. G.; Griffith, L.; Torchilin,V. P.; Jain, R. K. (1998). *Proceeding of the National Academy of Sciences of the United States of America* **1998**, *95*, 4607.
196. Kingsley, J. D.; Dou, H.; Morehead, J.; Rabinow, B.; Gendelman, H. E.; Destache, C. J. *J. Neuroimmunol. Pharmacol.* **2006**, *1*, 340.
197. Stolnik, S. I.; Davis, S. S. *Adv. Drug Del. Rev.* **1995**, *16* (2–3), 195.
198. Storm, G.; Belliot, S. H.; Daemen, T.; Lasic, D. D. *Adv. Drug Del. Rev.* **1995**, *17*, 31.
199. Allen, T. M. *Nature* **2002**, *2*, 750.
200. Neri, D., Bicknell, R. *Nature Reviews* **2005**, *5*, 436.
201. Ferrari, M. *Nature Reviews*, **2005**, *5*, 161.
202. Stolnik, S.; Illum, L.; Davis, S. S. *Adv. Drug Del. Rev.* **1995**, *16* (2–3), 195.
203. Heldin, C.-H.; Rubin, K.; Pietras, K.; Ostman, A. *Nat. Rev.: Cancer* **2004**, *4*, 806.

204. di Tomaso, E.; Capen, D.; Haskell, A.; Hart, J.; Logie, J. J.; Jain, R. K.; McDonald, D. M.; Jones, R.; Munn, L. L. *Cancer Research* **2005**, *65* (13), 5740.
205. Dua, R. S.; Gui, G. P. H.; Isacke, C. M. *J. Cancer Surgery* **2005**, *31*, 824.
206. Minchinton, A. I.; Tannock, I. F. *Nature Reviews* **2006**, *6*, 583.
207. Yuan, F.; Dellian, M.; Fukumura, D.; Leunig, M.; Berk, D. A.; Torchilin, V. P.; Jain, R. K. *Cancer Research* **1995**, *55* (17), 3752.
208. Saltzman, W. M. Controlled Drug Delivery Systems (Vol. 1). Oxford, Oxford University Press, Inc., 2001.
209. Dreher, M. R.; Liu, W.; Michelich, C. R.; Dewhirst, M. W.; Yuan, F.; Chilkoti, A. *J. Nati. Cancer Inst.* **2006**, *98* (5), 335.
210. Maeda, H.; Wu, J.; Sawa, T.; Matsumura, Y.; Hori, K. *J. Control Release* **2000**, *65*, 271.
211. Allen, T. M. *Nature* **2002**, *2*, 750.
212. Neri, D.; Bicknell, R. *Nature Reviews* **2005**, *5*, 436.
213. Kingsley, J. D.; Dou, H.; Morehead, J.; Rabinow, B.; Gendelman, H. E.; Destache, C. J. *J. Neuroimmunol. Pharmacol.* **2006**, *1*, 340.
214. Oliver, M.; Ahmad, A.; Kamaly, N.; Perouzel, E.; Caussin, A.; Keller, M.; Herlihy, A.; Bell, J.; Miller, A. D.; Jorgensen, M. R. *Org. Biomol. Chem.* **2006**, *4* (18), 3489.
215. Perez, J. M.; Josephson, L.; Weissleder, R. *Chem. Bio. Chem.* **2004**, *5*, 261.
216. Sukhishvili, S. A. *Curr. Op. Colloid Interface Sci.* **2005**, *10*, 37.
217. Sunderland, C. J.; Steiert, M.; Talmadge, J. E.; Derfus, A. M.; Barry, S. E. *Drug Development Res.* **2006**, *67* (1), 70.
218. de Jaeghere, F.; Doelker, E.; Gurny, R. Nanoparticles. In E. Mathiowitz (Ed.), *Encyclopedia of Controlled Drug Delivery.* New York, John Wiley & Sons, Inc., **1999**, 641–664.
219. Hirsch, L. R.; Halas, N. J., West, J. L. *Proceeding of the National Academy of Sciences of the United States of America* **2003**, *100*, 13549.
220. May, D. J.; Allen, J. S.; Ferrara, K. W. *IEEE Trans.* **2002**, *49*, 1400.
221. Roy, L.; Ohulchanskyy, T. Y.; Pudavar, H. E.; Bergey, E. J.; Oseroff, A. R.; Morgan, J.; Dougherty, T. J.; Parasad, P. N. *J. Amer. Chem. Soc.* **2003**, *125*, 7860.

222. Unger, E. C.; Porter, T.; Culp, W.; Labell, R.; Matsunaga, T.; Zutshi, R. *Adv. Drug Del. Rev.* **2004**, *56*, 1291.
223. Lu, Z.; Prouty, M. D.; Guo, Z.; Golub, V. O.; Kumar, C. S. S. R.; Lvov, Y. M. *Langmuir* **2005**, *21*, 2042.
224. Gaillard, P. J.; Visser, C. C.; Appeldoorn, C. C. M.; Rip J. *Drug Discov. Today Technol.* **2012**, *9* (2), e155.
225. Blanchette, M.; Fortin, D. *Methods Mol. Biol.* **2011**, *686*, 447.
226. Joshi, S.; Ergin, A.; Wang, M.; Reif, R.; Zhang, J.; Bruce, J. N.; Bigio I. J. *J. Neurooncol.* **2011**, *104* (1), 11.
227. Muldoon, L. L.; Soussain, C.; Jahnke, K.; Johanson, C.; Siegal, T., Smith, Q. R.; Hall, W. A.; Hynynen, K.; Senter, P. D.; Peereboom, D. M.; Neuwelt, E. A. *J. Clin. Oncol.* **2007**, *25* (16), 2295.
228. Xin, H. L.; Jiang, X. Y.; Gu, J. J.; Sha, X.; Chen, L.; Law, K.; Chen, Y.; Wang, X.; Jiang, Y.; Fang, X. *Biomaterials* **2011**, *32* (18), 4293.
229. Kratzer, I.; Wernig, K.; Panzenboeck, U.; Bernhart, E.; Reicher, H.; Wronski, R.; Windisch, M.; Hammer, A.; Malle, E.; Zimmer, A.; Sattler, W. *J. Control. Release* **2007**, *117* (3), 301.
230. Zensi, A.; Begley, D.; Pontikis, C.; Legros, C.; Mihoreanu, L.; Wagner, S.; Büchel, C.; von Briesen, H.; Kreuter, J. *J. Control. Release* **2009**, *137* (1), 78.
231. Kreuter, J.; Alyautdin, R. N.; Kharkevich, D. A.; Ivanov, A. A. *Brain Res.* **1995**, *674* (1), 171.
232. Wang, C. X.; Huang, L. S.; Hou, L. B.; Jiang, L.; Yan, Z. T.; Wang, Y. L.; Chen, Z. L. *Brain Res.* **2009**, *1261*, 91.
233. Gelperina, S.; Maksimenko, O.; Khalansky, A.; Vanchugova, L.; Shipulo, E.; Abbasova, K.; Berdiev, R.; Wohlfart, S.; Chepurnova, N.; Kreuter, J. *Eur. J. Pharm. Biopharm.* **2010**, *74* (2), 157.
234. Chen, Y. C.; Hsieh, W. Y.; Lee, W. F.; Zeng, D. T. *J. Biomater. Appl.* **2011**, *27* (7), 909.
235. De Mendoza, A. E. H.; Preart, V.; Mollinedo, F.; Blanco-Prieto, M. J. *J. Control. Release* **2011**, *156* (3), 421.
236. Xie, Y.; Ye, L. Y.; Zhang, X. B.; Cui, W.; Lou, J.; Nagai, T.; Hou, X. *J. Control. Release* **2005**, *105* (1–2), 106.
237. Lasa-Saracibar, B.; De Mendoza, A. E. H.; Guada, M.; Dios-Vieitez, C.; Blanco-Prieto, M. J. *Expert Opin. Drug Del.* **2012**, *9* (10), 1245.

238. Lu, W.; Tan, Y. Z.; Hu, K. L.; Jiang, X. G. *Int. J. Pharm.* **2005**, *295* (1–2), 247.
239. Qin, Y.; Chen, H. L.; Yuan, W. M.; Kuai, R.; Zhang, Q.; Xie, F.; Zhang, L.; Zhang, Z.; Liu, J.; He, Q. *Int. J. Pharm.* **2011**, *419* (1–2), 85.
240. Liu, Y.; Huang, R. Q.; Han, L.; Ke, W.; Shao, K.; Ye, L.; Lou, J.; Jiang, C. *Biomaterials* **2009**, *30* (25), 4195.
241. Van Rooy, I.; Mastrobattista, E.; Storm, G.; Hennink, W. E.; Schiffelers, R. M. *J. Control. Release.* **2011**, *150* (1), 30.
242. Laron, Z. *Arch. Physiol. Biochem.* **2001**, *15* (2), 112.
243. Boado, R. J.; Zhang, Y. F.; Zhang, Y.; Pardridge, W. M. *Biotechnol. Bioeng.* **2007**, *96* (2), 381.
244. Ulbrich, K.; Knobloch, T.; Kreuter, J. *J. Drug Target* **2011**, 19 (2), 125.
245. Gan, C. W.; Feng, S. S. *Biomaterials* **2010**, *31* (30), 7748.
246. Sahoo, S. K.; Labhasetwar, V. *Mol. Pharm.* **2005**, *2* (5), 373.
247. Lu, Y. J.; Sega, E.; Leamon, C. P.; Low, P. S. *Adv. Drug Del. Rev.* **2004**, *56* (8), 1161.
248. Ambrogio, M. W.; Thomas, C. R.; Zhao, Y.-L.; Zink, J. I.; Stoddart, J. F. *Accounts Chem. Res.* **2011**, *44*, 903.
249. Mekaru, H.; Lu, J.; Tamanoi, F. *Adv. Drug Deliv. Rev.* **2015**, *95*, 40.
250. Inagaki, S.; Fukushima, Y.; Kuroda, K. *J. Chem. Soc. Chem. Commun.* **1993**, 8, *680.*
251. Hernandez, R.; Tseng, H. R.; Wong, J. W.; Stoddart, J. F.; Zink, J. I. *J. Am. Chem. Soc.* **2004**, *126*, 3370.
252. Saint-Cricq, P.; Deshayes, S.; Zink, J. I.; Kasko, A. M. *Nanoscale* **2015**, *7*, 13168.
253. Chen, P. J.; Hu, S. H.; Hsiao, C. S.; Chen, Y. Y.; Liu, D. M.; Chen, S. Y. *J. Mater. Chem.* **2011**, *21*, 2535.
254. Lee, J. E.; Lee, N.; Kim, H.; Kim, J.; Choi, S. H.; Kim, J. H.; Kim, T.; Song, I. C.; Park, S. P.; Moon, W. K.; Hyeon, T. *J. Am. Chem. Soc.* **2010**, *132*, 552.
255. Thomas, C. R.; Ferris, D. P.; Lee, J. H.; Choi, E.; Cho, M. H.; Kim, E. S.; Stoddart, J. F.; Shin, J. S.; Cheon, J.; Zink, J. I. *J. Am. Chem. Soc.* **2010**, *132*, 10623.
256. Baeza, A.; Guisasola, E.; Ruiz-Hernández, E.; Vallet-Regí, M. *Chem. Mater.* **2012**, *24*, 517.

257. Hsia, Y.; Sivasubramanian, M.; Chen, N.-T.; Lo, L.-W. In *Nanomaterials for Tumor Targeting Theranostics: A Proactive Clinical Perspective*; World Scientific, 2016, 143.
258. Moorthy, M. S.; Kim, H.-B.; Bae, J.-H.; Kim, S.-H.; Ha, C.-S. *RSC Adv.* **2016**, *6*, 29106.
259. Santha Moorthy, M.; Kim, H.-B.; Bae, J.-H.; Kim, S.-H.; Ha, C.-S. *RSC Adv.* **2016**, *6*, 29106.
260. Lee, S. F.; Zhu, X. M.; Wang, Y. X. J.; Xuan, S. H.; You, Q.; Chan, W. H.; Wong, C. H.; Wang, F.; Yu, J. C.; Cheng, C. H. K.; Leung, K. C. F. *ACS Appl. Mater. Interfaces* **2013**, *5*, 1566.
261. Zheng, Y.; Shi, D.; Wu, X.; Gu, M.; Ai, Z.; Tang, K.; Ye, L.; Wang, X. *Evid. Based Complement. Alternat. Med.* **2014**, *2014.*
262. Liong, M.; Lu, J.; Kovochich, M.; Xia, T.; Ruehm, S. G.; Nel, A. E.; Tamanoi, F.; Zink, J. I. *ACS Nano* **2008**, *2*, 889.
263. Mondragón, L.; Mas, N.; Ferragud, V.; De La Torre, C.; Agostini, A.; Martínez-Máñez, R.; Sancenón, F.; Amorós, P.; Pérez-Payá, E.; Orzáez, M. *Chem-Eur. J.* **2014**, *20*, 5271.
264. Bernardos, A.; Mondragón, L.; Aznar, E.; Marcos, M. D.; Martínez-Máñez, R.; Sancenón, F.; Soto, J.; Barat, J. M.; Pérez-Payá, E.; Guillem, C.; Amorós, P. *ACS Nano* **2010**, *4*, 6353.
265. Liu, J.; Zhang, B.; Luo, Z.; Ding, X.; Li, J.; Dai, L.; Zhou, J.; Zhao, X.; Ye, J.; Cai, K. *Nanoscale* **2015**, *7*, 3614.
266. Climent, E.; Bernardos, A.; Martínez-Máñez, R.; Maquieira, A.; Marcos, M. D.; Pastor-Navarro, N.; Puchades, R.; Sancenón, F.; Soto, J.; Amorós, P. *J. Am. Chem. Soc.* **2009**, *131*, 14075.
267. Xu, Z.; Liu, S.; Kang, Y.; Wang, M. *Nanoscale* **2015**, *7*, 5859.
268. Lee, C. H.; Cheng, S. H.; Huang, I. P.; Souris, J. S.; Yang, C. S.; Mou, C. Y.; Lo, L. W. *Angew. Chem. Int. Edit.* **2010**, *49*, 8214.
269. Gan, Q.; Lu, X.; Yuan, Y.; Qian, J.; Zhou, H.; Lu, X.; Shi, J.; Liu, C. *Biomaterials* **2011**, *32*, 1932.
270. Popat, A.; Liu, J.; Lu, G. Q.; Qiao, S. Z. *J. Mater. Chem.* **2012**, *22*, 11173.
271. Chen, F.; Zhu, Y. *Microporous Mesoporous Mater.* **2012**, *150*, 83.
272. Lai, C. Y.; Trewyn, B. G.; Jeftinija, D. M.; Jeftinija, K.; Xu, S.; Jeftinija, S.; Lin, V. S. Y. *J. Am. Chem. Soc.* **2003**, *125*, 4451.

273. Cui, Y.; Dong, H.; Cai, X.; Wang, D.; Li, Y. *ACS Appl. Mater. Interfaces* **2012**, *4*, 3177.
274. Zhang, Q.; Liu, F.; Nguyen, K. T.; Ma, X.; Wang, X.; Xing, B.; Zhao, Y. *Adv. Funct. Mater.* **2012**, *22*, 5144.
275. Zhu, C. L.; Song, X. Y.; Zhou, W. H.; Yang, H. H.; Wen, Y. H.; Wang, X. R. *J. Mater. Chem.* **2009**, *19*, 7765.
276. Luo, Z.; Cai, K.; Hu, Y.; Zhao, L.; Liu, P.; Duan, L.; Yang, W. *Angew. Chem. Int. Edit.* **2011**, *50*, 640.
277. Ambrogio, M. W.; Pecorelli, T. A.; Patel, K.; Khashab, N. M.; Trabolsi, A.; Khatib, H. A.; Botros, Y. Y.; Zink, J. I.; Stoddart, J. F. *Org. Lett.* **2010**, *12*, 3304.
278. Song, N.; Yang, Y.-W. *Chem. Soc. Rev.* **2015**, *44*, 3474.
279. Slowing, I. I.; Vivero-Escoto, J. L.; Wu, C.-W.; Lin, V. S.-Y. *Adv. Drug. Deliv. Rev.* **2008**, *60*, 1278.
280. Chen, G.; Roy, I.; Yang, C.; Prasad, P. N. *Chem. Rev.* **2016**, *116*, 2826.
281. Liu, J.; Detrembleur, C.; Mornet, S.; Jérôme, C.; Duguet, E. *J. Mater. Chem. B* **2015**, *3*, 6117.
282. Najafian, N.; Shanehsazzadeh, S.; Hajesmaeelzadeh, F.; Lahooti, A.; Gruettner, C.; Oghabian, M. A. *Appl. Magn. Reson.* **2015**, *46*, 685.
283. Somayeh, M.; Saeed, S.; Fatemeh, A.; Rassoul, D. *Mater. Sci. Eng. C* **2016**, *59*, 390.
284. Hwang, A. A.; Lu, J.; Tamanoi, F.; Zink, J. I. *Small* **2015**, *11*, 319.
285. Bringas, E.; Koysuren, O.; Quach, D. V.; Mahmoudi, M.; Aznar, E.; Roehling, J. D.; Marcos, M. D.; Martinez-Manez, R.; Stroeve, P. *Chem. Commun.* **2012**, *48*, 5647.
286. Nguyen, D. T.; Kim, K.-S. *Kona Powder Part. J.* **2016**, *advpub.*
287. Tojo, K.; Fan, L. T. *Math. Biosci.* **1981**, *57*, 279.
288. Xu, P.; Wang, S.; Greenberg, D. B. *Angew. Makromol. Chem.* **1989**, *169*, 211.
289. El-Aassar, M. R.; El Fawal, G. F.; Kamoun, E. A.; Fouda, M. M. G. *Int. J. Bio. Macromol.* **2015**, *77*, 322.
290. Baker, R. W. *Controlled release of biologically active agents.* J. Wiley and sons, New York, **1987**.
291. Davis, K. A.; Anseth, K. S. *Crit. Rev. Ther. Drug Carrier Syst.* **2002**, *19*.

292. Peppas, L.; Elsevier, Amsterdam: 1990, p. 45.
293. Peppas, N. A.; Khare, A. R. *Adv. Drug Deliv. Rev.* **1993**, *11*, 1.
294. Blanco, M. D.; García, O.; Trigo, R. M.; Teijón, J.; Katime, I. *Biomaterials* **1996**, *17*, 1061.
295. Hamidi, M.; Azadi, A.; Rafiei, P. *Adv. Drug Deliv. Rev.* **2008**, *60*, 1638.
296. Lin, C.-C.; Metters, A. T. *Adv. Drug Deliv. Rev.* **2006**, *58*, 1379.
297. Coviello, T.; Matricardi, P.; Marianecci, C.; Alhaique, F. *J. Control. Release* **2007**, *119*, 5.
298. Peppas, N.; Huang, Y.; Torres-Lugo, M.; Ward, J.; Zhang, J. *Annu. Rev. Biomed.* **2000**, *2*, 9.
299. Siepmann, J.; Peppas, N. *Adv. Drug Deliv. Rev.* **2012**, *64*, 163.
300. Amsden, B. *Macromolecules* **1998**, *31*, 8382.
301. Qiu, Y.; Park, K. *Adv. Drug Deliv. Rev.* **2012**, *64*, 49.
302. Miyata, T.; Uragami, T.; Nakamae, K. *Adv. Drug Deliv. Rev.* **2002**, *54*, 79.
303. Martins, S. *Int. J. Nanomedicine* **2007**, *2*, 595.
304. Wang, M.; Zuris, J. A.; Meng, F.; Rees, H.; Sun, S.; Deng, P.; Han, Y.; Gao, X.; Pouli, D.; Wu, Q. *Proc. Natl. Acad. Sci.* **2016**, *113*, 2868.
305. Torrecilla, J.; Rodríguez-Gascón, A.; Solinís, M. Á.; del Pozo-Rodríguez, A. *BioMed. Res. Int.* **2014**, *2014*.
306. Almeida, A. J.; Souto, E. *Adv. Drug Deliv. Rev.* **2007**, *59*, 478.
307. Müller, R.; Mehnert, W.; Lucks, J.-S.; Schwarz, C.; Zur Mühlen, A.; Meyhers, H.; Freitas, C.; Rühl, D. *Eur. J. Pharm. Biopharm.* **1995**, *41*, 62.
308. Mueller, R.; Maeder, K.; Lippacher, A.; Jenning, V. Patent DE19945203A1, **1999**.
309. Gasco M. R. Patent US5250236A, **1981**.
310. Sharifi, S.; Behzadi, S.; Laurent, S.; Forrest, M. L.; Stroeve, P.; Mahmoudi, M. *Chem. Soc. Rev.* **2012**, *41*, 2323.
311. Sayes, C. M.; Reed, K. L.; Warheit, D. B. *Toxicol. Sci.* **2007**, *97*, 163.
312. Sabbioni, M. In *Proceedings of the Workshop: Research Needs on Nanoparticles, Brussels*, **2005**, 22.
313. Mahmoudi, M.; Simchi, A.; Milani, A.; Stroeve, P. *J. Colloid Interface Sci.* **2009**, *336*, 510.

314. Lynch, I.; Salvati, A.; Dawson, K. A. *Nat. Nanotechnol.* **2009**, *4*, 546.
315. Mahmoudi, M.; Lynch, I.; Ejtehadi, M. R.; Monopoli, M. P.; Bombelli, F. B.; Laurent, S. *Chem. Rev.* **2011**, *111*, 5610.
316. Walczyk, D.; Bombelli, F. B.; Monopoli, M. P.; Lynch, I.; Dawson, K. A. *J. Am. Chem. Soc.* **2010**, *132*, 5761.
317. Mahmoudi, M.; Lohse, S. E.; Murphy, C. J.; Fathizadeh, A.; Montazeri, A.; Suslick, K. S. *Nano Lett.* **2014**, *14*, 6.
318. Behzadi, S.; Serpooshan, V.; Sakhtianchi, R.; Müller, B.; Landfester, K.; Crespy, D.; Mahmoudi, M. *Colloids Surf. B* **2014**, *123*, 143.
319. Mahmoudi, M.; Laurent, S.; Shokrgozar, M. A.; Hosseinkhani, M. *ACS Nano* **2011**, *5*, 7263.
320. Mirshafiee, V.; Mahmoudi, M.; Lou, K.; Cheng, J.; Kraft, M. L. *Chem. Commun.* **2013**, *49*, 2557.
321. Mahmoudi, M.; Monopoli, M. P.; Rezaei, M.; Lynch, I.; Bertoli, F.; McManus, J. J.; Dawson, K. A. *ChemBioChem* **2013**, *14*, 568.
322. Shanehsazzadeh, S.; Gruettner, C.; Lahooti, A.; Mahmoudi, M.; Allen, B. J.; Ghavami, M.; Daha, F. J.; Oghabian, M. A. *Contrast Media Mol. I* **2015**, *10*, 225.
323. Skirtach, A. G.; Kreft, O. Stimuli-sensitive nanotechnology for drug delivery, in Nanotechnology in drug delivery. (Eds: de Villiers, M. M.; Aramwit, P.; Kwon, G. S. *American Association of Parmaceutical Scientists.* Springer, 2009.
324. Chu, L. Y.; Liang, Y. J.; Chen, W. M.; Ju, X. J.; Wang, H. D. *Colloids Surf. B: Biointerfaces* **2004**, *37*, 9.
325. Kim, J. J.; Park, K. *J. Control. Release* **2001**, 77, 39.
326. Bawa, P.; Pillay, V.; Choonara, Y. E.; du Toit, L. C. *Biomed. Mater.* **2009**, *4*.
327. Roy, D.; Cambre, J. N.; Sumerlin, B. S. *Prog. Polym. Sci.* **2010**, *35*, 278.
328. Liu, F.; Urban, M. W. *Prog. Polym. Sci.* **2010**, *35*, 3.
329. Lee, E. S.; Shin, H. J.; Na, K.; Bae, Y. H. *J. Control. Release* **2003**, *90*, 363.
330. Lee, E. S.; Na, K.; Bae, Y. H. *J. Control. Release* **2005**, *103*, 405.
331. Hruby, M.; Konak, C.; Ulbrich, K. *J. Control. Release* **2005**, *103*, 137.

# Chapter 10

# Legal Framework for Protection of Pharmaceutical Trade Marks in Europe and USA

Mohammad Hossein Erfanmanesh* and
Shirin Sharifzadeh†

* *Medical Ethics and Law Research Center, Shahid Beheshti University of Medical Sciences and Health Services, Tehran, Iran*
† *Faculty of Law and Political Science, Allameh Tabataba'i University, Tehran, Iran*

## Abbreviations

| | |
|---|---|
| CDER | Center for Drug Evaluation and Research |
| Directive | Directive 2001/83/EC on the Community code relating to medicinal products for human use |
| EMA | European Medicine Agency |
| EMA Guideline | 'Guideline on the acceptability of names for human medicinal products processed through the centralized procedure' |
| EUIPO | European Union Intellectual Property Office |
| FDA | United States Food and Drug Administration |
| NRG | Name Review Group |
| Regulation | Council Regulation (EC) No. 207/2009 on the Community trade mark |
| USPTO | United States Patent and Trademark Office |

## I. Introduction

A trademark is 'any word, name, symbol, or device, or any combination thereof ... used by a person ... to identify and distinguish his or her goods, including a unique product, from those manufactured or sold by others and to indicate the source of the goods, even if that source is unknown.'[1] Trade mark enables supermarket customers to choose 'COCA-COLA' to the exclusion of other cola's brands, knowing that the famous red-label showcasing white letters refer to a particular and familiar brand of cola.[2] In order to protect the economic value of a trade mark, its owner usually opts to register the mark.

No European trade mark legislation specifically addresses the particularities of the pharmaceutical market. However, both the European Medicine Agency (EMA) and the national authorities have established unique rules governing the nomenclature, non-proprietary naming, and trade mark choice of pharmaceuticals. There is also abundant case law dealing with trade mark issues for pharmaceutical products which has established principles that apply specifically to these products.

The registration of trade marks for pharmaceutical products to be commercialized in Europe and the USA is more challenging than other names. Not only must applicants follow routine procedures for registering a trade mark in trade mark offices, but also placing medicinal products on the market and the use of their brand names are subject to a marketing authorization that requires a long process of the examination. That is because healthcare practitioners as well as patients rely on a product's name as a critical identifier of the appropriate therapy in a market where thousands of products are on the sale.[3] In these circumstances, product names that look or sound alike can lead to medication errors and, potentially, to patient harm as it may increase the risk of a healthcare practitioner's misprescribing or misinterpreting the correct product name, and, as a result, administering a wrong product.[3]

In Europe, the trade mark or intellectual property offices of a EU Member State at national level and the new European Union Intellectual Property Office (EUIPO) at the EU level handle the registration of trade marks.[4] The work at the EUIPO is regulated by Council Regulation (EC) No. 207/2009 on the Community trade mark (Regulation). This task in the US is carried out by the United States Patent and Trademark Office (USPTO) under the Lanham (Trademark) Act (2000). With regard to health and safety check, the centralized marketing authorization procedure in the EU is administered by the EMA, and the United States Food and Drug Administration (FDA) in the US.

The aim of this note is to study and review the particular standards for pharmaceutical trade marks in Europe and the US. For this purpose, the procedures involved in obtaining marketing authorization in Europe will be first studied, and then the next part will examine the registration process of pharmaceutical trade marks. The last part of this chapter will provide an overview of authorization and registration of pharmaceutical trade marks in the US.

## II. Marketing Authorization through EMA

According to Article 6 of the Directive 2001/83/EC on the Community code relating to medicinal products for human use (the Directive),

> 'No medicinal product may be placed on the market of a Member State unless a marketing authorization has been issued by the competent

authorities of that Member State in accordance with this Directive or unless an authorization has been granted in accordance with Regulation 726/2004 . . . [that is through the EMA].'

As it is clear from this provision, within the European Union, there are two paths to authorize pharmaceuticals. The first is by using a centralized authorization procedure, and the second is through national competent authorities. The centralized procedure is carried out by the Committee for Medicinal Products for Human Use of the EMA. This procedure lets pharmaceutical companies submit a single application and, if they meet certain criteria, obtain a marketing authorization that is valid in the EU States as well as Iceland, Liechtenstein and Norway. It is also possible to apply for a marketing authorization using the decentralized procedure (i.e. through national competent authorities) that is valid in countries where the authorization has been applied for.

It is important to note that the centralized procedure is compulsory for human medicines for the treatment of HIV/AIDS, cancer, diabetes, neurodegenerative diseases, auto-immune and other immune dysfunctions, and viral diseases, and for veterinary medicines for use for growth or yield enhancers, and for all other medicines derived from biotechnology processes.[5] Alongside the authorization procedure for pharmaceuticals, the name of the medicinal product is an integral part of the authorization.[6]

Article 1(20) of the Directive provides that the trade mark for which the authorization is sought can be a name, 'which may be either an invented name not liable to confusion with the common name, or a common or scientific name accompanied by a trade mark or the name of the marketing authorization holder.' Common names are the international non-proprietary name recommended by the World Health Organization, or, if one does not exist, the usual common name.[7] Proprietary names facilitate the identification of pharmaceutical substances or active pharmaceutical ingredients.[8] Each proprietary name is a unique name that is globally recognized and is public property.[8] A non-proprietary name is also known as a generic name.[8] In practice, many pharmaceutical companies applying for the EU marketing authorization wish to use invented names for their medicinal products.[9]

The criteria that the EMA applies when reviewing proposed names are reflected in the 'Guideline on the acceptability of names for human medicinal products processed through the centralized procedure' (EMA Guideline). These criteria include:

— 'The (invented) name of a medicinal product should not be liable to cause confusion in print, handwriting or speech with the (invented) name of another medicinal product.'
— 'The (invented) name of a medicinal product should not convey misleading therapeutic and/or pharmaceutical connotations.'
— 'The (invented) name of a medicinal product should not be misleading with respect to the composition of the product.'
— 'Consideration should be given to the phonetics and the potential difficulties a proposed (invented) name may create in terms of pronunciation in the different EU official languages.'
— 'The (invented) name should not convey a promotional message with respect to the therapeutic and/or pharmaceutical characteristics and/or the composition of the medicinal product.'
— 'The invented name should not comprise wholly of initial letters (acronyms) or code numbers nor include punctuation marks.'
— 'The importance of other elements such as labeling and pack design should be taken into consideration as contributing factors for the safe use of a medicinal product.'[10]

The EMA has set up a review process that is performed by the Name Review Group (NRG) to ensure that the relevant provisions are met. Representatives from the European Commission and the World Health Organization as well as relevant experts may participate in the group's activities and may be consulted on naming issues.[11] The proposed names with all the background information are sent to national competent authorities of the EU Member States where they are asked to inform the EMA of any objections or comments to the proposed names on grounds of safety concerns.[11] After assessment of all relevant factors, the EMA will decide if the proposed name of a medicinal product is accepted.

## III. Registration of Trade Marks Before EUIPO

The purpose of the checking of the trade marks under the EMA is for the safety of medicinal products within the authorization procedure since the proposed names could create a public health concern or potential safety risk.[12] Such an assessment is performed based on evidence and research.[12] The EMA, however, will not take into consideration aspects of intellectual property rights or trade mark registration within its review.[12] This is carried out by the EUIPO.

A EU trade mark is obtained by registration before the EUIPO.[13] Any natural or legal person may apply for, and be the proprietor of a EU trade mark. According to the Article 4 of the Regulation, a trade mark may consist 'of any signs capable of being represented graphically, particularly words, including personal names, designs, letters, numerals, the shape of goods or of their packaging, provided that such signs are capable of distinguishing the goods or services of one undertaking from those of other undertakings.'

EU trade marks have a unitary character and have equal effect throughout the Union.[14] Trade mark protection confers exclusive rights to the proprietor such that s/he will be entitled to prevent all third parties not having her/his consent from using (1) any sign which is identical with the trade mark in relation to goods which are identical with those for which the trade mark is registered; (2) any sign where, because of its identity with, or similarity to, the trade mark and the identity or similarity of the goods covered by the mark and the sign, there exists a likelihood of confusion on the part of the public; (3) any sign which is identical with, or similar to, the trade mark in relation to goods which are not similar to those for which the trade mark is registered, where the latter has a reputation in the European Union and where use of that sign without due cause takes unfair advantage of, or is detrimental to, the distinctive character or the reputation of the trade mark.[15]

When an application is submitted, the EUIPO assesses it on the basis of absolute grounds for refusal. According to those grounds, trade marks are not to be registered if they are devoid of distinctive character, or if they consist exclusively of signs or indications which may serve to 'designate the kind, quality, quantity, intended purpose, value, geographical origin, time of production of the goods or of rendering of the service, or other characteristics of the goods.'[16] For example, the word 'ACNESTICK' has been refused registration on absolute grounds for the lack of distinctiveness. That was because the trade mark was composed of the words 'ACNE' and 'STICK,' and it was proposed to cover for pharmaceutical preparations for the diagnosis and treatment of acne.[17]

In contrast, the court *in Celltech v Office for Harmonization in the Internal Market (OHIM)* allowed the registration of the word CELLTECH as a trade mark covering, *inter alia*, pharmaceutical preparations and surgical and veterinary apparatus. In this case, the Board of Appeal within the OHIM rejected the application where it argued that CELLTECH, which is the combination of the English word 'cell' and 'tech,' each of which lacks any distinctive character individually, was nothing more than the sum of those two component parts as the order of the words corresponds to a correct syntactical use of the two words, i.e. "cell technology."[18]

But, the court disagreed by noting that a trade mark comprising words may be assessed, in part, in relation to each of its terms or elements; however, the assessment must depend on an appraisal of the whole which they comprise.[19] The mere fact that each of those elements, considered separately, is devoid of distinctive character does not establish that their combination cannot present a distinctive character.[19] To the court, it was true that the goods covered by pharmaceutical goods and, on that account, have a relation with bodies which are composed of cells. However, the court held that the Board of Appeal did not show that the relevant public would immediately and without further reflection make a definite and direct association between the pharmaceutical products at issue and the meaning of the name CELLTECH.[20]

Relative grounds for refusal are the other challenges that trade mark applicants may face. The rationale behind the relative grounds for refusal is to reduce a likelihood of confusion between the proposed mark and the earlier trade mark. The evaluation of these grounds ensures that the proposed mark does not take unfair advantage of, or is not detrimental to, the distinctive character or the repute of a registered trade mark, and that the public are protected against believing that the goods or services in question come from the same undertaking or from economically-linked undertakings.[21]

It should be noted that, like a conflict with an earlier trade mark, relative grounds can only be inspected by the EUIPO if an opposition is submitted by the owner(s) of the earlier mark(s) on the basis of Article 8 of the Regulation.[22] According to this Article,

> 'Upon opposition by the proprietor of an earlier trade mark, the trade mark applied for shall not be registered: ... (b) if because of its identity with, or similarity to, the earlier trade mark and the identity or similarity of the goods or services covered by the trade marks there exists a likelihood of confusion on the part of the public in the territory in which the earlier trade mark is protected; the likelihood of confusion includes the likelihood of association with the earlier trade mark.'[23]

As it is clear, the three key factors in the assessment of relative grounds for refusal are the relevant public, similarity of the goods and similarity of the trade marks. Since a substantial number of legal battles arise from the challenges between the proposed mark and the earlier trade mark, these factors will be studied in more detail.

## A. The relevant public

According to a settled case-law, the perception of the trade marks in the mind of the average consumers of the category of goods at issue plays a decisive role in the assessment of the likelihood of confusion between the marks.[24] In fact, the relevant public is composed of the average consumers, 'reasonably well-informed' and reasonably observant and circumspect.[25] With regard to pharmaceuticals, the two groups of people can be distinguished from the rest of the public that deal with such products; professionals in the health-care and medical sector (specialist doctors, general practitioners, pharmacists and nurses) and end consumers (patients). Professionals, by reason of their knowledge, are more attentive to the differences between marks than end consumers.[26] Therefore, the level of the risk of confusion that the goods in question come from the same undertaking or from economically-linked undertakings can be much lower than patients as medical professionals display a high degree of attention when prescribing or, generally, dealing with pharmaceuticals.[27] Accordingly, it is important to determine the group or groups making up the relevant public in order to decide whether there is a likelihood of confusion.

In this regard, there are two different approaches. According to the first approach, when defining the relevant public, differentiation is made between pharmaceutical preparations that require medical prescription and products without it (over-the-counter medicines). An example of this approach can be seen in *Alcon v OHIM* where the court stated that with respect to the sterile solutions used in ophthalmologic operations, the relevant public could consist of professionals because such solutions were administered by highly qualified persons, like surgeons or specialist doctors of ophthalmology.[28] But, the court expressed that this is not true with respect to products such as eye drops or solutions used to clean contact lenses since those products are used by average consumers.[28] Thus, according to this approach, end-users and patients would be only included within the relevant public if pharmaceutical products could be purchased without medical prescription.

The second approach which has been taken recently is based on defining the relevant public as both the professionals in the medical sector and patients. As a result, even if medicinal products require a doctor's prescription prior to their sale to end-users in pharmacies, the relevant public is comprised of both end-users and health professionals (doctors who prescribe the medicinal preparations and pharmacists who sell the prescribed pharmaceutical products).[29] When adopting this approach, the court in *Ineos Healthcare v OHIM* reasoned that:

> Even though the choice of those products is influenced or determined by intermediaries, a likelihood of confusion can also exist for consumers since they are likely to be faced with those products, even if that takes place during separate purchasing transactions for each of those individual products, at various times.[29]

As is clear from the case-law cited above, even in the case of pharmaceutical preparations available only on prescription, it cannot be excluded that the end-users and patients form part of the relevant public. Accordingly, the relevant public is comprised of both medical and healthcare professionals and average consumers.[30] It seems that the concern to avoid the 'particularly disagreeable consequences' for health which could result from confusion between two medicinal products justifies the adoption of a rather restrictive approach (inclusion of both professionals and end-users in the definition of the relevant public) to the level of attention of the average consumer.[31] The only conceivable exclusion of end-users and patients from the relevant public could relate to pharmaceuticals which are not directly marketed to them, and are chosen and administered exclusively by medical and healthcare professionals, for instance anesthetics which are only used in a hospital environment.[32]

## B. Similarity of trade marks

A likelihood of confusion largely arises from the similarity of trade marks at issue. According to the case law, the likelihood of confusion between trade marks in the mind of the public is assessed on the global scale taking into account all factors relevant including the visual, aural or conceptual similarity of the marks in question, and must be based on the overall impression given by the marks.[33] In fact, the perception of marks in the mind of the average consumer of the type of goods or services in question plays a decisive role in the global appreciation of the likelihood of confusion. An average consumer normally perceives a mark as a whole and does not analyses its various details.[34]

On this basis, how the public understand the marks, and whether they are able to perceive the distinctions is a key issue in decision about the similarity of the marks. For example, regarding the conflict between the later trade mark RESPICUR and the former RESPICORT both of which covered therapeutic preparations for respiratory illnesses, the Court of First Instance noted that the conceptual perception of the two marks were different for the two groups making up the relevant public. The professional public, because of their knowledge and experience, were generally able to understand the conceptual meaning of the different components of the opposing marks, namely 'respiratory' for 'respi,' 'cure' or 'heal' for 'cur', and 'corticoids' for 'cort.'[35] Therefore, by breaking down the two marks into their respective components, they would interpret the mark applied for as 'cure for respiratory problems' and the earlier mark as designating 'corticoids intended for respiratory illnesses.'[35]

To the Court, those two interpretations indicated a degree of conceptual divergence, the earlier mark carrying a more specific meaning than the mark applied for, but they nevertheless shared the general idea of being linked to respiratory diseases.[35] As for the end consumers, it was noted that their level of attention and knowledge is higher than average because of the serious nature of the illnesses.[36] They would, therefore, be able to distinguish the component 'respi' in the two marks in question and to understand its conceptual content, which referred generally to the nature of their health problems. Their limited knowledge of medical terminology, however, would prevent them from being able to discern the conceptual references of the components 'cur' and 'cort.'[36]

The opposing marks would be conceptually similar for them because of the identical component 'respi.'[36] Taking into account the analysis mentioned above, the Court found that the opposing marks were similar for the professional public and highly similar for the end consumers. For the latter, the marks were visually, phonetically, and conceptually similar.[37] As a result, it was held that there was a likelihood of confusion between the mark applied for and the earlier mark in the minds of end consumers.

## C. Similarity of goods

According to the well-established principles of trade mark law, another factor that plays a pivotal role in determining a likelihood of confusion is the similarity of goods at issue. In assessing the similarity of the goods concerned, all the relevant factors relating to those goods should be taken into account. Those factors include, *inter alia*, their nature, their end users, the intended purpose, their method of use and whether they are in competition with each other or are complementary.[38] An example of the way that courts apply these factors can be seen in *Novartis AG v. OHIM.*

In this case, the mark for which registration was sought was TOLPOSAN and the earlier trade mark was TONOPAN. The goods covered under TOLPOSAN were muscle relaxant medicines and muscle relaxant veterinary preparations containing tolperisone, while the other mark covered 'analgesics which soothe pain, in particular in treating headaches and migraines.'[39] The court noted that goods such as medicines have 'the same nature (pharmaceutical products), purpose (treatment of human health problems), consumers (medical professionals and patients) and distribution channels (typically pharmacies).' In this sense, all medicines including the goods at issue belong to the same general category. However, this category is very wide which includes goods that may be different.

To the court, the goods covered by both marks could be distinguished according to their primary therapeutic indication. Therefore, the court took the view that the products concerned came within different sub-categories of medicines.[40] Accordingly, it concluded that since the goods were not in competition with each other (i.e. interchangeable) as a result of the different therapeutic indications, the slight degree of similarity on the phonetic level between the marks could not lead to the conclusion of the existence of a likelihood of confusion within the meaning of Article 8(1)(b) of the Regulation.[41]

Although the medicine's therapeutic indication in *Novartis AG v. OHIM* was of decisive importance as a factor for differentiation between the two marks, the General Court's decisions were not consistent with this view. In *Armour Pharmaceutical Co. v OHIM*, the Court was of the opinion that the similarities between the goods outweigh the difference where such a difference lies only in the therapeutic indication. That is because the court in this case saw the therapeutic indication in a larger context which is the intended purposes of products (treatment of human health problems).

Specifically, the earlier trade mark was CALSYN and the mark sought to be registered was GALZIN. The court found that the goods covered by both marks were of the same nature (pharmaceutical preparations), had the same function or intended purpose (treatment of human health problems), were directed at the same consumers (professionals in the health sector and patients), used the same distribution channels (typically pharmacies) and could be complementary. Their difference, however, laid in their different therapeutic indications.[42] To the court's opinion, given the similarity of the goods concerned and overall impression of the two marks, the difference was not sufficient to eliminate the existence of a likelihood of confusion on the part of the relevant public, and, therefore, GALZIN was refused registration because the conflicting goods were intended for use in the context of therapeutic treatment.[43]

In a similar vein, the General Court in *Kureha v OHIM — Sanofi-Aventis* was to decide about the conflict between KREMEZIN with an earlier trade mark KRENOSIN. KREMEZIN covered pharmaceutical preparations for use in the treatment of different kind of diseases including renal problems. KRENOSIN corresponded to a product consisting of a sterile solution of adenosine for the use in the treatment of a specific heart condition. Although, the therapeutic indications of the two marks differed greatly, and there was a low level of phonetic similarity between the marks, it was held that the goods in question were similar and that the marks were similar.[44]

In spite of the latter cases giving weight to the treatment of human health problems rather than a specific therapeutic indication, it should be noted that the treatment of human health problems always lies in the nature of pharmaceutical products. While the level of attention of the average consumer of pharmaceutical preparations must be determined on a case-by-case basis,[45] it would, therefore, be more appropriate if only a related therapeutic indication, instead of the treatment of human health problems, would be accounted towards a similarity of goods.[46]

## IV. Genuine Use of Pharmaceutical Trade Marks

A trade mark must be used. If the owner fails to put the mark to genuine use, it can be revoked. Article 15(1)(a) of the Community trade mark Regulation provides that:

> If, within a period of five years following registration, the proprietor has not put the Community trade mark to genuine use in the Community in connection with the goods or services in respect of which it is registered, or if such use has been suspended during an uninterrupted period of five years, the Community trade mark shall be subject to the sanctions provided for in this Regulation, unless there are proper reasons for non-use.

This requirement is of particular importance for pharmaceutical trade marks since the authorization procedure for a medicine may take over 5 years.[47] Such a long procedure can threat the validity of a pharmaceutical trade marks registered under the trade mark Regulation. That is because, in this situation, obvious activities that can establish genuine use of a trade mark such as putting the product on the market and other activities described in Article 15(1)(a) and (b) of the trade mark Regulation cannot take place.[47]

In such a circumstance, the question that arises is whether preparatory activities, and, more specifically, the authorization procedure for a medicine could be regarded as 'genuine use.' According to case law, when assessing whether the use of a trade mark is genuine, regards must be paid to all the facts and circumstances relevant to establishing whether the 'commercial exploitation' of the mark is real, particularly whether such a use is viewed as warranted in the economic sector concerned to maintain or create a share in the market for the goods protected by the mark, the nature of the goods at issue and the characteristics of the market.[48] In other words, 'there is genuine use of a trade mark where the mark is used in accordance with its essential function, which is to guarantee the identity of the origin of the goods or services for which it is registered, in order to create or preserve an outlet for those goods.'[48]

Since the product authorization is a prerequisite for 'commercial exploitation' of pharmaceutical products and their nature and characteristics of their market require such a procedure, it would not be too problematic to argue against revocation of a trade mark if the reason for non-use refers to the normal time that is needed for a medicinal product to be authorized.[49]

A provision that could uphold this argument is Article 19(1) of the Agreement on Trade-Related Aspects of Intellectual Property Rights that provides 'Circumstances arising independently of the will of the owner of the trade mark which constitute an obstacle to the use of the trade mark, such as import restrictions on or other government requirements for goods or services protected by the trade mark.'[49] Here, national rules imposing restrictions on the marketing of the goods covered by the trade mark may be regarded as a proper reason for non-use of the mark.[50]

## V. US Experience with Pharmaceutical Trade Marks

In the United States, the Lanham (Trademark) Act (15 U.S.C. § 1051) (2000) provides the legal framework for trade mark protection. The Act grants the USPTO the authority to assess and register proposed trade marks that meet the requirements specified in this Act. Applications for trade marks of pharmaceutical products are evaluated in the same way as other trade mark applications.[51] According to Sections 3, 42, and 43 of the Act, the criteria for registration (absolute grounds for refusal) and the assessment of the likelihood of confusion (relative grounds for refusal) are reviewed in much the same way as they are in the Europe.

Specifically, in examining whether a proposed trade mark is likely to result in confusion among the relevant public, the USPTO takes into account a number of factors known as DuPont factors derived from the famous case *In re E. I. DuPont de Nemours & Co.* The thirteen DuPont factors are:

(1) The similarity or dissimilarity of the marks in their entireties as to appearance, sound, connotation and commercial impression.
(2) The similarity or dissimilarity and nature of the goods or services as described in an application or registration or in connection with which a prior mark is in use.
(3) The similarity or dissimilarity of established, likely-to-continue trade channels.
(4) The conditions under which and buyers to whom sales are made, i.e. "impulse" vs. careful, sophisticated purchasing.
(5) The fame of the prior mark (sales, advertising, length of use).
(6) The number and nature of similar marks in use on similar goods.
(7) The nature and extent of any actual confusion.
(8) The length of time during and conditions under which there has been concurrent use without evidence of actual confusion.
(9) The variety of goods on which a mark is or is not used (house mark, "family" mark, product mark).
(10) The market interface between applicant and the owner of a prior mark:
    (a) A mere "consent" to register or use.
    (b) Agreement provisions designed to preclude confusion, i.e. limitations on continued use of the marks by each party.
    (c) Assignment of mark, application, registration and good will of the related business.
    (d) Laches and estoppel attributable to owner of prior mark and indicative of lack of confusion.
(11) The extent to which applicant has a right to exclude others from use of its mark on its goods.
(12) The extent of potential confusion, i.e. whether de minimis or substantial.

(13) Any other established fact probative of the effect of use.[52]

As a result, the law prohibits the registration of a mark that consists of or comprises a mark which resembles a registered mark in the US when used on or in relation with the goods of the applicant, to cause confusion, or to cause mistake, or to deceive the purchasing public.[53]

Similar to the EMA, the FDA is a scientific, regulatory, and public health agency that supervises 'human and animal drugs, therapeutic agents of biological origin, medical devices, radiation-emitting products for consumer, medical, and occupational use, cosmetics, and animal feed.'[54] The FDA's review of the medicine and its name starts after the four-phase clinical research of the medicine. The purposes of four-phase clinical research include safety and dosage, efficacy, side effects and monitoring of adverse reactions.[55] The FDA's role after a successful clinical research is to allow marketing the drug. Such authority stems from the FDA's task to review false or misleading drug labels.[56]

The Center for Drug Evaluation and Research (CDER) of the FDA has developed internal procedures for evaluating, prior to the marketing of a drug, proposed trade marks (the proprietary names) of drugs and therapeutic biologic products which could cause or contribute to medication errors.[57] CDER's review of trade marks includes inspection of both safety and promotional aspects of a name. CDER's Division of Medication Error Prevention and Analysis conducts the safety review.[57] As for the promotional review, the Division of Drug Marketing, Advertising, and Communications of the CDER evaluates proposed trade marks to decide if they are overly fanciful such that they misleadingly imply unique effectiveness or composition.[57]

CDER inspects the likelihood of confusion between a proposed trade mark and established names and trade marks of medicinal products existing in the marketplace.[57] This evaluation involves (1) expert panel review where experts panel meets regularly to exchange opinions on the safety of new trade marks; (2) handwriting and verbal analyses to decide the degree of confusion in visual appearance or pronunciation between the proposed trade marks and other established names; (3) computer-assisted analysis which involves utilization of existing FDA databases to identify likely sound-alike and/or look-alike names; (4) labeling and packaging analysis to identify areas of possible improvement; (5) overall risk evaluation to weigh the results of each phase of the review.[58]

## References


1. Lanham (Trademark) Act (15 U.S.C. § 1051) (2000).
2. Dana M. Herberholz, 'Curing Confusion: An Overview of the Regulatory Complexities of Obtaining Pharmaceutical Trademarks and a Prescription for Reform,' (2007) 8 Minnesota Journal of Law, Science & Technology, 97, 99.
3. FDA, How FDA Reviews Proposed Drug Names Available at: http://www.fda.gov/downloads/Drugs/DrugSafety/MedicationErrors/ucm080867.pdf. Accessed March 21, 2016.
4. In March 2016, the Office for Harmonization in the Internal Market (OHIM) replaced with the new European Union Intellectual Property Office (EUIPO).
5. Richard Chin and Menghis Bairu, *Global Clinical Trials: Effective Implementation and Management*, (Academic Press, 2011), p. 49.
6. EMA, 'Guideline on the acceptability of names for human medicinal products processed through the centralized procedure,' EMA/CHMP/287710/2014 — Rev. 6 (1 January 2015) p3.
7. Article 1(21) of the Directive.
8. WHO, 'Essential medicines and health products: International Nonproprietary Names,' Available at: http://www.who.int/medicines/services/inn/en/. Accessed January 24, 2016.
9. Helen Mosback, 'Protection of pharmaceutical trade marks in Europe: an overview,' (2013) *8* (2) *Journal of Intellectual Property Law & Practice*, 69, 70.
10. EMA Guideline, pp. 7–9.
11. *Ibid.*, p. 11.
12. *Ibid.*, p. 4.
13. Article 6 of the Regulation.
14. Article 1(2) of the Regulation.
15. Article 9 of the Regulation.

16. Article 7 of the Regulation.
17. Mosback, (n. 9) p. 72 citing CTMA 4528147, refusal letter, 5 October 2006.
18. Case T-260/03 — *Celltech v OHIM* (2005), para. 25.
19. *Ibid.*, para. 43.
20. *Ibid.*, para. 38.
21. Case T-256/04 — *Mundipharma v OHIM — Altana Pharma* (2007), para. 68.
22. Joanna Jowita Sitko, 'Special Criteria of Trade Mark Protection with Regard to Pharmaceutical Products in the European Union Legal System' (2014) 45 IIC — International Review of Intellectual Property and Competition Law 658, 659.
23. Article 8(1)(b) of the Regulation.
24. Case C-412/05 P — *Alcon v OHIM* (2003), para. 59.
25. Joined Cases C-473/01 P and C-474/01 P. *Procter & Gamble v OHIM* (2004), para. 52.
26. Case T-154/03 — *Biofarma v OHIM — Bausch & Lomb Pharmaceuticals* (2005), para. 31.
27. Case T-95/07 — *Aventis Pharma v OHIM — Nycomed* (2008), para. 29.
28. Case T-106/07 — *Alcon v OHIM* (2008), para. 21.
29. Case T-222/09 — *Ineos Healthcare v OHIM — Teva Pharmaceutical Industries* (2011), para. 43.
30. *Ibid.*, para. 44. See also Case T-154/03, (n. 26), para. 46.
31. Case T-201/08 — *Market Watch v OHIM — Ares Trading* (2010), para. 23.
32. Stephan Hanne, 'Case law on pharmaceutical trade marks from the Court of Justice of the European Union,' (2011) 6 (1) *Journal of Intellectual Property Law & Practice* 43, 45.
33. Case T-292/01 — *Phillips-Van Heusen Corp. v OHIM — Pash Textilvertrieb und Einzelhandel* (2003), paras. 45–47.
34. Case C-251/95 *SABEL BV v Puma AG* (2003), para. 23.
35. Case T-256/04, (n. 21), para. 58.
36. *Ibid.*, para. 59.
37. *Ibid.*, para. 62.
38. Case C-39/97 — *Canon Kabushiki Kaisha v Metro-Goldwyn-Mayer* (1998), para. 23.

39. Case T-331/09 — *Novartis v OHIM — Sanochemia Pharmazeutika* (2010), para. 32.
40. *Ibid.*, para. 38.
41. *Ibid.*, para. 62.
42. Case T-483/04 — *Armour Pharmaceutical v OHIM — Teva Pharmaceutical Industries* (2006), para. 70.
43. *Ibid.*, paras. 71–81.
44. Case T-487/08 — *Kureha v OHIM — Sanofi-Aventis* (2010), para. 90.
45. Case T-483/04 (n. 42), para. 79.
46. Hanne, (n. 32), 47.
47. Sitko, (n. 22), 665.
48. Case C-40/01 — *Ansul* (2003), para. 43.
49. Sitko, (n. 29), 668.
50. Case T-194/03 — *Il Ponte Finanziaria v OHIM — Marine Enterprise Projects* (2006), para. 46.
51. Herberholz, (n. 2), 101.
52. *In re E. I. DuPont de Nemours & Co.* 476 F.2d 1357, (C.C.P.A. 1973) para. 1361.
53. Lanham (Trademark) Act (15 U.S.C. § 1051) s. 2(d).
54. John P. Swann, FDA's Origin, Available at: http://www.fda.gov/AboutFDA/WhatWeDo/History/Origin/ucm124403.htm. Accessed March 22, 2016.
55. FDA, Step 3: Clinical Research, Available at: http://www.fda.gov/ForPatients/Approvals/Drugs/ucm405622.htm. Accessed March 22, 2016.
56. 21 U.S. Code § 352 — Misbranded drugs and devices.
57. FDA, How FDA Reviews Proposed Drug Names Available at: http://www.fda.gov/downloads/Drugs/DrugSafety/MedicationErrors/ucm080867.pdf. Accessed March 21, 2016.
58. Carol Holquist & Jerry Phillips, 'How FDA reviews drug names,' (2001) Drug Topics, April, p. 36. Available at: <http://www.fda.gov/downloads/Drugs/DrugSafety/MedicationErrors/ucm115783.pdf> Accessed January 12, 2016.

# Chapter 11

# Future Perspective on the Smart Delivery of Biomolecules

Erick Yu[*,†], Félix Sancenón[‡,§,¶], Elena Aznar[‡,¶], Ramón Martínez-Máñez[‡,§,¶], María Dolores Marcos[‡,§,¶], Mohammad J. Hajipour[||], Morteza Mahmoudi[||,**], and Pieter Stroeve[*,†]

[*] *Department of Chemical Engineering, University of California Davis, Davis, California 95616, USA*

[†] *Department of Materials Science and Engineering, University of California Davis, Davis, California 95616, USA*

[‡] *Instituto Interuniversitario de Investigación de Reconocimiento Molecular y Desarrollo Tecnológico (IDM), Universitat Politècnica de València — Universitat de València, Valencia, Spain*

[§] *Departamento de Química, Universitat Politècnica de València, Camino de Vera s/n, 46022 València, Spain*

[¶] *CIBER de Bioingeniería, Biomateriales y Nanomedicina (CIBER-BBN)*

[||] *Department of Nanotechnology, Faculty of Pharmacy, Tehran University of Medical Sciences, Tehran, Iran*

[**] *Department of Anesthesiology, Brigham and Women's Hospital, Harvard Medical School, Boston, Massachusetts 02115, USA*

## Abbreviations

| | |
|---|---|
| MSN | mesoporous silica nanoparticles |
| PEG | poly(ethylene glycol) |
| PLGA | poly(lactic-co-glycolic acid) |
| FDA | Food and Drug Administration |
| IR | infrared |
| UV | ultraviolet |

Drug delivery via nanoparticles,[1–3] membranes,[4,5] hydrogels,[6–8] and other supports[9–11] has been deeply investigated in the past decades. However, it was not until in the past decade or two that "smart" targeting and release systems have been studied.[12–14] As such, this field of research has been in its infancy and continues to face growth and with it, further challenges and difficulties along the way toward maturity. It can be expected that greater leaps and advances will be made from both the material and engineering design standpoints. Additionally, with the alarming increase in cancer diagnoses and genetic illnesses, traditional drug agents and their delivery media need to be re-evaluated to address a quickly evolving field. With newer smart materials for the controlled release of macromolecules, peptides, genetic material, etc. further complications arise, such as material performance, synthesis, functionalization and targeting, and biocompatibility. Biotechnology and the development of biomaterials is a highly interdisciplinary area of research, and as a result, require a simultaneous and collaborative effort to address both the drug and its delivery in future studies.

As described in earlier chapters, many of the nanoparticles, membrane, gel, or nanostructured supports are produced from several classes of materials, with the majority being highly porous to maximize the surface area and volume. Alternatively, using polymeric, hollow shells or capsules to encapsulate delivery agents have been considerably studied. In some cases, combinations of both an organic and inorganic foundation to form core–shell nanoparticles have been researched as viable drug delivery options. From the foundations set in the previous two decades of research, mesoporous silica nanoparticles (MSNs) have been exhaustively studied. It can be expected that MSNs will continue to be popular as a support for drug delivery due to the greater stability and protection it provides for the drug species, and also as a result, extending the delivery duration to typically several hours to days. While this is a boon toward therapeutic drug delivery, it unfortunately also presents a problem with any residual chemicals and compounds from the synthesis. Longer circulation times in biological environments can lead to accumulation, agglomeration, or undesirable premature interactions in the living organisms. In many instances, compounds used in the synthesis of these nanostructured materials are toxic, such as any unreacted surfactants or silanes. Cellular viability tests indicate that MSNs are biocompatible in appropriate dosages, with, in many instances, mortality rates below 10%. However, minimizing this value to lower thresholds is crucial to ensure the success of MSNs as an effective drug delivery agent in the future.

Alternatively, polymeric and other similarly organic structures, such as dextrans[15] and complex carbohydrates/starches[16] have been studied parallel to their inorganic counterparts as supports for drug delivery. These materials are considerably more biocompatible, with polyethylene glycol (PEG),[17] poly(lactic-co-glycolic acid) (PLGA),[18,19] chitin,[20–22] and other polysaccharides[23] used widely in the literature. In the recent years, hydrogels, membranes, and other scaffolds out of highly biocompatible materials have been developed for topical delivery. Intravenous delivery of these materials also is successful. In recent years, transitioning from micro- to nano-scale particles has enabled these materials to be strong candidates for drug delivery. As many of these classes of materials are naturally derived, they show significant biodegradability and are effectively eliminated through biological processes. Unfortunately, this also hinders the circulation duration and stability in some applications. In instances where easy delivery and non-specific targeting is desirable, nanoparticles comprised of PEG and PLGA are efficient and already approved by the Food and Drug Administration (FDA). Future work and research using these materials need to be centered on expanding their stability, circulation duration, and incorporating targeting mechanisms.

From a design standpoint, it is expected that ensuring homogeneity and purity are significantly important considerations that need to be made when producing nanoparticles or membranes for both research and upscaled manufacturing. Producing highly uniform products that can be made in small, well-controlled synthesis methodologies potentially faces difficulties when up scaling this or across different experimental setups. As such, the synthesis methodology is critical. Template-driven techniques have been to demonstrate incredible versatility and homogeneity, such as the formation of organic or inorganic nanoshells using latex (polystyrene) nanospheres. Microfluidic techniques are utilizing emulsion and droplet-forming methods also are promising for high yield and excellent control over the nanoparticle dimensions. Advancing beyond traditional chemical reactors for the synthesis of biomaterials has significant advantages in automation and the future of the industry.

The agent to be delivered has in recent years expanded toward peptides, proteins, and genetic material such as short strand RNA or DNA as a result of gene therapy. Debilitating and other life-altering illnesses such as cancer, Alzheimer's, and muscular dystrophy require a genetic approach for drug delivery and as such, the design of the delivery agent must adapt to this change. In order to effectively encapsulate these genetic species, which are typically 1–10 nm in size, larger porous or hollow nanostructures are needed. This differs from many porous drug delivery agents where pore sizes can range between 3 and 6 nm. Large, porous nanostructures specifically tailored for this has been extensively studied[24–26] but producing nanoscale, regular porous structures that can effectively contain genetic material has been shadowed by the simple approach of producing hollow nanoparticles.[27,28] When noting previous considerations, these hollow nanoparticles are significantly easier to produce homogeneously and adapted to more biocompatible organic materials.

Research on stimuli-responsive gating mechanisms has been widely successful from many approaches in the recent decade.[1,12] These gated materials make use of varied interactions and different external triggers, such as light, changes in pH, temperature, ultrasound, alternating magnetic fields, enzymes or molecules/biomolecules for the uncapping and controlled mass delivery. The first examples reported centered on the design of the different gating entities whereas currently, gated nanochemistry focuses directly on their use in practical applications, such as drug delivery carriers at cell level and *in vivo* models. Despite the important development of these gated materials, there are drawbacks that still need to be overcome. While many of the external stimuli used are effective for the on-command release of drugs, several are non-specific, such as temperature or pH. However, temperature and or pH-triggered materials would be applied in biological conditions in which these parameters are expected to change (e.g. the acidic pH environment in certain cancer cells). Other gated materials respond to external stimuli, such as magnetic fields or light, and could have potential real applications but more work is needed in order to test the side effects of the application of high intensity fields and irradiation with IR or UV light. Moreover, penetration depth and efficiency issues must be also studied. Additionally, bioaccumulation of the nanoparticles, when used *in vivo*, is a clear problem that must be addressed.

As a whole, research has been steadily gravitating toward new targeting mechanisms or improving the targeting capabilities of drug delivery media. In particular, studies have been accelerating toward the use of aptamers to target only a corresponding genomic sequence. Whereas other stimuli-responsive mechanisms rely on the immediate environment, which often is associated with a particular area to target, aptamers and ligand-receptor pairings are specific and more accurate in pinpointing specific compounds, proteins, or genomic sequences. As a result, advances in this particular field are promising to avoid premature interactions within biological environments. Additionally, these groups can decorate existing nanoparticle surfaces and in some cases, function as the trigger mechanism to initiate the release of the delivery compounds simultaneously. Developing these is a two-fold difficulty in that both the appropriate aptamer must be developed but also attaching it to the nanoparticle surface.

The interaction between the nanoparticle carriers and the biological environment can be a complex, dynamic process, independent of any surface functionalization as well. The biological identity of nanoparticles and how proteins within the biological environment adsorb onto a foreign, high energy surface introduces a bio-interface, or corona, which forms at the nanoparticle surface both due to and regardless of physical considerations such as size and surface properties.[29] Protein adsorption for example will affect the incorporation and performance of the nanoparticle,[30] with possible implications on unintended protein-ligand associations, transporting proteins across membranes into undesirable environments, and deformation of proteins.[31] In order to fully evaluate the impact of introducing nanoparticles as drug delivery carriers, the presence of a protein-corona formation suggests that further studies are needed to ensure that unintended interactions with the local environment do not detract or complicate intended target mechanisms, controlled release kinetics, or potential toxicity of newly introduced protein deformation. In addition, the newly defined concept of "disease-specific protein corona" will add more complications to the filed, as the nanocarriers should be customized for each type of disease to achieve their high therapeutic efficacies.[32–35]

As with other biomaterials, biocompatibility is one of the most significant factors in the success, and ultimately, the applicability of drug delivery media. In many stimuli-responsive nanoparticle systems, the introduction of novel polymeric and organic groups onto the surfaces presents concerns about their biodegradability or compatibility. Similarly, with nanoparticle hybrid systems incorporating inorganic materials have uncertain long-term effects in the body. Materials such as gold and iron oxide are frequently used for light and magnetically driven release mechanisms, but despite prevalent views regarding its biocompatibility, it still faces scrutiny in many studies. A similar argument can be made with popularly used pH and temperature responsive polymers such as poly(acrylic acid) and poly(*N*-isopropylacrylamide). Continual, cell assays and other animal viability tests are needed in future research before extending this to humans and in greater dosages, as this presents many unknown variables that require years of clinical study.

## References

1. Slowing, I. I.; *et al.*, Mesoporous silica nanoparticles as controlled release drug delivery and gene transfection carriers. Advanced drug delivery reviews **2008,** *60* (11), 1278–1288.

2. Kumari, A.; Yadav, S. K.; Yadav, S. C. Biodegradable polymeric nanoparticles based drug delivery systems. Colloids and Surfaces B: Biointerfaces **2010**, *75* (1), 1–18.
3. Allen, T. M.; Cullis, P. R. Drug delivery systems: entering the mainstream. *Science* **2004**, *303* (5665), 1818–1822.
4. Jackson, E. A.; Hillmyer, M. A. Nanoporous membranes derived from block copolymers: From drug delivery to water filtration. *ACS nano* **2010**, *4* (7), 3548–3553.
5. Gultepe, E.; *et al.*, Nanoporous inorganic membranes or coatings for sustained drug delivery in implantable devices. *Advanced drug delivery reviews* **2010**, *62* (3), 305–315.
6. Qiu, Y.; Park, K. Environment-sensitive hydrogels for drug delivery. *Adv. Drug Delivery Rev.* **2001**, *53* (3), 321–339.
7. Hoare, T. R.; Kohane, D. S. Hydrogels in drug delivery: progress and challenges. *Polymer* **2008**, *49* (8), 1993–2007.
8. Jiang, Y.; *et al.*, Click hydrogels, microgels and nanogels: emerging platforms for drug delivery and tissue engineering. *Biomaterials* **2014**, *35* (18), 4969–4985.
9. Merino, S.; *et al.*, Nanocomposite hydrogels: 3D polymer–nanoparticle synergies for on-demand drug delivery. *ACS Nano* **2015**, *9* (5), 4686–4697.
10. Voskerician, G.; *et al.*, Biocompatibility and biofouling of MEMS drug delivery devices. *Biomaterials* **2003**, *24* (11), 1959–1967.
11. Nisar, A.; *et al.*, MEMS-based micropumps in drug delivery and biomedical applications. *Sens. Actuators B: Chem.* **2008**, *130* (2), 917–942.
12. Mura, S.; Nicolas, J.; Couvreur, P. Stimuli-responsive nanocarriers for drug delivery. *Nat. Mater.* **2013**, *12* (11), 991–1003.
13. Liong, M.; *et al.*, Multifunctional inorganic nanoparticles for imaging, targeting, and drug delivery. *ACS Nano* **2008**, *2* (5), 889–896.
14. Sun, T.; *et al.*, Engineered nanoparticles for drug delivery in cancer therapy. *Angewandte Chemie International Edition* **2014**, *53* (46), 12320–12364.
15. Li, Y. L.; *et al.*, Reversibly stabilized multifunctional dextran nanoparticles efficiently deliver doxorubicin into the nuclei of cancer cells. *Angewandte Chemie International Edition* **2009**, *48* (52), 9914–9918.
16. Lenaerts, V.; *et al.*, Cross-linked high amylose starch for controlled release of drugs: recent advances. *J. Control. Release* **1998**, *53* (1), 225–234.

17. Veronese, F. M.; Pasut, G. PEGylation, successful approach to drug delivery. *Drug Discovery Today* **2005**, *10* (21), 1451–1458.
18. Cheng, J.; *et al.*, Formulation of functionalized PLGA–PEG nanoparticles for in vivo targeted drug delivery. *Biomaterials* **2007**, *28* (5), 869–876.
19. Fredenberg, S.; *et al.*, The mechanisms of drug release in poly (lactic-co-glycolic acid)-based drug delivery systems — a review. *Int. J. Pharm.* **2011**, *415* (1), 34–52.
20. Agnihotri, S. A.; Mallikarjuna, N. N.; Aminabhavi, T. M. Recent advances on chitosan-based micro-and nanoparticles in drug delivery. *J. control. Release* **2004**, *100* (1), 5–28.
21. Jayakumar, R.; *et al.*, Biomedical applications of chitin and chitosan based nanomaterials — A short review. *Carbohydr. Polym.* **2010**, *82* (2), 227–232.
22. Prabaharan, M. Review paper: chitosan derivatives as promising materials for controlled drug delivery. *J. Biomater. Appl.* **2008**, *23* (1), 5–36.
23. Liu, Z.; *et al.*, Polysaccharides-based nanoparticles as drug delivery systems. *Adv. Drug Delivery Rev.* **2008**, *60* (15), 1650–1662.
24. Zhao, D.; *et al.*, Triblock copolymer syntheses of mesoporous silica with periodic 50 to 300 angstrom pores. *Science* **1998**, *279* (5350), 548–552.
25. Zhang, J.; *et al.*, Synthesis and characterization of pore size-tunable magnetic mesoporous silica nanoparticles. *J. Colloid Interface Sci.* 2011, *361* (1), 16–24.
26. Kim, M.-H.; *et al.*, Facile synthesis of monodispersed mesoporous silica nanoparticles with ultralarge pores and their application in gene delivery. *ACS Nano* **2011**, *5* (5), 3568–3576.
27. Chen, J.-F.; *et al.*, Preparation and characterization of porous hollow silica nanoparticles for drug delivery application. *Biomaterials* **2004**, *25* (4), 723–727.
28. Yang, J.; *et al.*, Hollow silica nanocontainers as drug delivery vehicles. *Langmuir* **2008**, *24* (7), 3417–3421.
29. Monopoli, M. P.; *et al.*, Biomolecular coronas provide the biological identity of nanosized materials. *Nat. Nano* **2012**, *7* (12), 779–786.
30. Behzadi, S.; *et al.*, Protein corona change the drug release profile of nanocarriers: The "overlooked" factor at the nanobio interface. *Colloids Surf. B: Biointerfaces* **2014**, *123*, 143–149.

31. Mahmoudi, M.; *et al.*, Protein–Nanoparticle Interactions: Opportunities and Challenges. *Chem. Rev.* **2011**, *111* (9), 5610–5637.
32. Hajipour, M. J.; *et al.*, Personalized protein coronas: a "key" factor at the nanobiointerface. *Biomater. Sci.* **2014**, *2* (9), 210–1221.
33. Hajipour, M. J.; *et al.*, Personalized disease-specific protein corona influences the therapeutic impact of graphene oxide. *Nanoscale* **2015**, *7* (19), 8978–8994.
34. Caracciolo, G., Farokhzad, O. C.; Mahmoudi, M. *Biological Identity of Nanoparticles In Vivo: Clinical Implications of the Protein Corona.* Trends in Biotechnology, 2016.
35. Rahman, M.; Mahmoudi, M. *Disease specific protein corona.* in *SPIE BiOS.* 2015. International Society for Optics and Photonics.

# Index